TOTAL SYNTHESIS OF BIOACTIVE NATURAL PRODUCTS

TOTAL SYNTHESIS OF BIOACTIVE NATURAL PRODUCTS

GOUTAM BRAHMACHARI
Department of Chemistry, Visva-Bharati
(A Central University), Santiniketan, India

ELSEVIER

Elsevier
Radarweg 29, PO Box 211, 1000 AE Amsterdam, Netherlands
The Boulevard, Langford Lane, Kidlington, Oxford OX5 1GB, United Kingdom
50 Hampshire Street, 5th Floor, Cambridge, MA 02139, United States

Library of Congress Cataloging-in-Publication Data
A catalog record for this book is available from the Library of Congress

British Library Cataloguing-in-Publication Data
A catalogue record for this book is available from the British Library

ISBN: 978-0-08-102822-3

Publisher: Susan Dennis
Acquisition Editor: Emily Mcclosket
Editorial Project Manager: Carly Demetre
Production Project Manager: Vignesh Tamil
Cover Designer: Greg Harris

Typeset by SPi Global, India

Dedication

In Memory of Santosh K. Brahmachari—My Beloved Father

Contents

Foreword

S. Chandrasekaran
Department of Organic Chemistry, Indian Institute of Science, Bangalore, India

Natural products can be considered as a "gift" of nature to chemists, and the diversity of structures and complexities that one encounters in them is truly amazing. It is a well-known fact that a vast majority of drugs approved for clinical use are either derived from natural products or inspired by research on natural products. In general, organic chemists are enamored by the challenges that the total synthesis of bioactive natural products offers. Since many of the natural products are isolated only in very small quantities from natural sources, there is an additional impetus and necessity for synthetic chemists to design and execute an elegant, efficient, and economically viable protocol for procuring these molecules on a reasonable scale in the laboratory. The strategies and tactics in achieving the goal of synthesis of bioactive natural products, if presented in an appropriate and appealing manner, would be very useful to young students and active practitioners of organic synthesis.

It is in this context that I find *Total Synthesis of Bioactive Natural Products*, authored by Professor Goutam Brahmachari, to be extremely valuable and useful to the scientific community. The author has carefully chosen 60 bioactive natural products and summarized their total syntheses. It is to be appreciated that excellent descriptions of synthetic schemes for each molecule are available in one place. The chapters in this book are arranged logically based on the names of the bioactive molecules in alphabetical order.

The wealth of important information that is available in this book is truly exceptional and I would like to extend my hearty congratulations to the author for his dedicated effort and contribution.

Preface

Mother Nature is an inexhaustible source of complex molecules displaying an incredible range of structural diversity with intriguing biological properties. In fact, the majority of today's approved drugs are either natural product derived or have been developed based on natural product lead structures. On the other side of the coin, synthetic chemistry has also contributed a great deal. From our basic needs for food, clothing, shelter, and medicine to energy, defense, and even art—and most things in between—the advances in synthetic strategy to make molecules has enabled us not only to replicate, often in quantity, what is found in nature but importantly to go beyond nature by producing new chemical entities that fulfill our further needs. The huge ranges of these molecules—both natural and nonnatural—have enormous impacts on our modern civilization as well as on our existence. That is why chemical synthesis is one of the most vibrant areas of modern scientific research, and will continue to supply the drugs, materials, and commodities as per the demand of the day.

Inspiration from natural products has indeed brought new perspectives to organic synthesis. The last 60 years or so have witnessed revolutionary progress in natural product isolation, structural elucidation, total syntheses, and biological studies. Bioactive natural products are often found only in miniscule quantities, and to have a particular bioactive natural product in ample amount is thus one among many objectives of natural products syntheses. Total synthesis, a thorough and didactic presentation of the chemical synthesis of a complex molecule, often a natural product, from simple, commercially available precursors, is regarded as a state of the chemical art at its highest degree of elegance, perception, and aesthetic innovation!

The birth of total synthesis took place in the 19th century when Wöhler accomplished the synthesis of urea *de novo* outside the living system in 1828 [1], simply upon heating an inorganic substance, ammonium cyanate. The synthesis of acetic acid from elemental carbon by Kolbe in 1845 [2] is the second major achievement in the history of total synthesis. The total syntheses of alizarin by Graebe and Liebermann (1869) [3], indigo by Baeyer (1878) [4], and D-(+)-glucose by Fischer (1890) [5] represent landmark accomplishments in the history of total synthesis of the 19th century. Since then, chemists synthesized innumerable bioactive natural products of significance and practical utility. The total synthesis of complex natural products still remains among the most exciting and dynamic areas of research—total synthesis of such complex molecules coupled with advancement in synthetic strategies play a crucial role in modern civilization.

From a fundamental perspective, total synthesis is an authentic platform for proving new methodologies and new strategies or ways of thinking. Total synthesis also cultivates an understanding of the basic principles of chemistry: how and why reactions occur, the relationships between molecular shape and function, stereochemical aspects, and related issues. An ability to synthesize molecules remains an essential training for the next generation of chemists. Woodward once argued, "The unique challenge which chemical synthesis provides for the creative imagination and the skilled hands ensures that it will endure as long as men write books, paint pictures, and fashion things which are beautiful, or practical, or both" [6].

In this context, an outstanding review article, "The art and science of total synthesis at the dawn of the twenty-first century" by Prof. K.C. Nicalaou and his group, may be recommended where the authors demonstrated the elegancy and beauty of this remarkable chemical science coupled with its necessity and accomplishments [7]. It is also felt pertinent to quote from the introductory words of Professor A. Fredga, a member of the Nobel Prize Committee for Chemistry of the Royal Swedish Academy of Sciences, used to introduce Prof. R.B. Woodward at the Nobel ceremonies in 1965, the year in which Woodward received the prize for the art of organic synthesis, "In our days, the chemistry of natural products attracts a very lively interest.... In the course of the investigation of a complicated substance, the investigator is sooner or later confronted by the problem of synthesis, of the preparation of the substance by chemical methods. He can have various motives. Perhaps he wants to check the correctness of the structure he has found. Perhaps he wants to improve our knowledge of the reactions and the chemical properties of the molecule. If the substance is of practical importance, he may hope that the synthetic compound will be less expensive or more easily accessible than the natural product. It can also be desirable to modify some details in the molecular structure. An antibiotic substance of medical importance is often first isolated from a microorganism, perhaps a mould or a germ. There ought to exist a number of related compounds with similar effects; they may be more or less potent, some may perhaps have undesirable secondary effects. It is by no means, or even probable, that the compound produced by the microorganism – most likely as a weapon in the struggle for existence – is the very best from the medicinal point of view. If it is possible to synthesize the compound, it will also be possible to modify the details of the structure and to find the most effective remedies. The synthesis of a complicated molecule is, however, a very difficult task; every group, every atom must be placed in its proper position and this should be taken in its most literal sense. It is sometimes said that organic synthesis is at the same time an exact science and a fine art…." [8].

Synthetic organic chemistry is now associated with biology, medicinal chemistry, and material sciences, and being a composite field of such subdisciplines, organic chemistry would thus be required to attain an appropriate combination of these subdisciplines to have a pronounced overall success of any synthetic endeavor. This field of chemical sciences is facing uninterrupted challenge by new chemical scaffolds isolated from nature's seemingly unlimited library of molecular architectures. At the same time, the practice of total synthesis is also being enriched progressively by newly developed synthetic strategies, reagents, catalysts, as well as advancement in analytical instrumentations for the rapid purification and characterization of compounds.

Thus the original goal of total synthesis to confirm the structure of an isolated natural product has been replaced gradually with prime objectives related more to the exploration and discovery of new chemistry along the pathway to the target molecule. Presently, biological issues are deeply associated with the total synthesis programs. The evolution of drug discovery and development process is closely related to total synthesis—academic research focusing on organic and natural product synthesis offers highly relevant basic knowledge and training to young researchers in this domain. The pharmaceutical industry applies the knowledge gained to discovering and manufacturing new drugs for the benefits of society. That medicinal and combinatorial chemists have so many tools at their disposal today in their quests for huge numbers of novel and diverse small molecules (natural products and/or natural product-like structures) is primarily the result of the contributions of total synthesis and of organic synthesis in combination. The community should consider total synthesis as a vibrant and worthwhile pursuit as long as it enables access to functionally useful molecules and/or powerful new methods and strategies are being invented.

Under the purview of this enormously potent field of total synthesis, *Total Synthesis of Bioactive Natural Products* has been designed to cover the recent cutting-edge development and state of the art of total synthesis of potentially important and useful natural small molecules. The book offers total synthetic approaches for 60 well-judged bioactive natural products under distinct entries with vivid step-by-step descriptions with the help of graphics and schemes so that readers can follow the chemistry involved in each step of the total process to make them familiar with the tactics for their thorough understanding and follow-up. The selections herein are only the tip of the iceberg and are representative of the exciting developments that continue to be reported. This timely volume would not only be helpful to researchers, professionals, and experts involved with this remarkable field but also to advanced students and young chemists in motivating them to the dynamic field of the total synthesis of useful organic molecules.

I would like to express my sincere thanks and deep sense of gratitude to Professor Srinivasan Chandrasekaran, Organic Chemistry Division, Indian Institute of Science, Bangalore, for his keen interest in the manuscript and for writing the foreword to the book.

I would also like to express my deep sense of appreciation to all the editorial and publishing staff members associated with Elsevier for their keen interest in publishing this work as well as their all-round help to ensure that the highest standards of publication have been maintained in bringing out this book. My effort will be successful only when it is found helpful to readers at large. Every step has been taken to make the manuscript error free; in spite of that, some errors might have crept in. Any remaining errors are, of course, my own. Constructive comments and approach of the book from readers will be highly appreciated.

Finally, I would to thank my wife (Piyasi) and my son (Asanjan) for their understanding and allowing me enough time throughout the entire period of writing; without their support, this work would not have been successful.

GOUTAM BRAHMACHARI
Chemistry Department, Visva-Bharati University, Santiniketan, India

References

[1] F. Wöhler, *Ann. Phys. Chem.* 12 (1828) 253.
[2] H. Kolbe, *Ann. Chem. Pharm.* 54 (1845) 145.
[3] a) C. Graebe, C. Liebermann, *Ber. Dtsch. Chem. Ges.* 2 (1869) 332; b) first commercial synthesis: C. Graebe, C. Liebermann, H. Caro, *Ber. Dtsch. Chem. Ges.* 3 (1870) 359.
[4] A. Baeyer, *Ber. Dtsch. Chem. Ges.* 11 (1878) 1296–1297; (b) first commercial production: K. Heumann, *Ber. Dtsch. Chem. Ges.* 23 (1890) 3431.
[5] E. Fischer, *Ber. Dtsch. Chem. Ges.* 23 (1890) 799–805.
[6] P. Ball, *Nature* 528 (2015) 327–329.
[7] K.C. Nicalaou, D. Vourloumis, N. Winssinger, P.S. Baran, *Angew. Chem. Int. Ed.* 39 (2000) 44–122.
[8] Nobel Lectures: Chemistry 1963–1970, Elsevier, New York, 1972, pp. 96–123.

CHAPTER ONE

Aeruginosins 298-A and B

Abbreviations

Alloc allyloxycarbonyl
Bn benzyl
Boc *tert*-butoxycarbonyl
BOP (benzotriazol-1-yloxy)tris(dimethylamino)phosphonium hexafluorophosphate
Bz benzoyl
Cbz benzyloxycarbonyl
CH_2Cl_2 dichloromethane
DEPBT 3-(diethoxyphosphoryloxy)-1,2,3-benzotriazin-4(3*H*)-one
DEPC diethyl cyanophosphinate
(−)-DIP-Cl (−)-B-chlorodiisopinocampheylborane
Et_3N triethylamine
IBCF isobutyl chloroformate
LS-Selectride lithium trisiamylborohydride solution
NMM *N*-methylmorpholine
TBS *tert*-butyldimethylsilyl
TFA trifluoroacetic acid
THF tetrahydrofuran
TIPS triisopropylsilyl

Systematic names: (2*S*,3a*S*,6*R*,7a*S*)-*N*-((*S*)-5-Guanidino-1-hydroxypentan-2-yl)-6-hydroxy-1-((*R*)-2-((*R*)-2-hydroxy-3-(4-hydroxyphenyl)propanamido)-4-methylpentanoyl)octahydro-1*H*-indole-2-carboxamide (**1**; aeruginosin 298-A); (2*S*,3a*S*,6*R*,7a*S*)-6-hydroxy-1-((*R*)-2-((*R*)-2-hydroxy-3-(4-hydroxyphenyl)propanamido)-4-methylpentanoyl)octahydro-1*H*-indole-2-carboxamide (**2**; aeruginosin 298-B)

Compound class: Peptides

Total Synthesis of Bioactive Natural Products
https://doi.org/10.1016/B978-0-08-102822-3.00001-8

Structure:

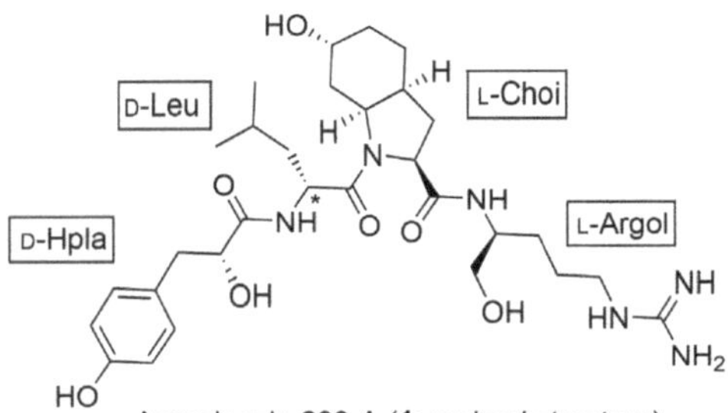

Aeruginosin 298-A (**1**; revised structure)
Aeruginosin 298-A (**1′**; L-Leu epimer at C*; original structure)

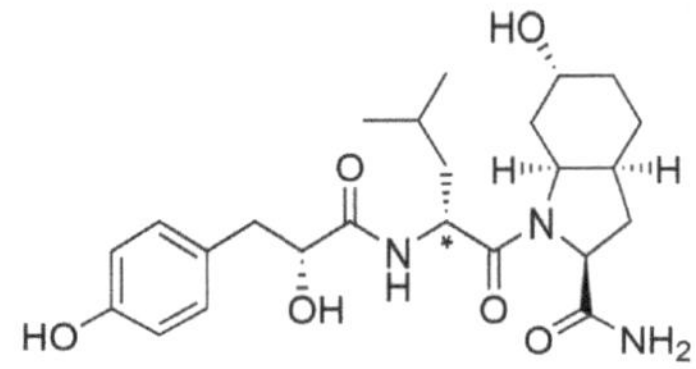

Aeruginosin 298-B (**2**; revised structure)
Aeruginosin 298-B (**2′**; L-Leu epimer at C*; original structure)

Natural source: *Microcystis aeruginosa* NIES-298 (cyanobacterium) [1]
Pharmaceutical potential: Thrombin and trypsin inhibitors [1]
Synthetic routes:

The first total synthesis of the aquatic peptide aeruginosin 298-A (**1**) was reported by Wipf and Methot (Schemes 1–4) [2]. The peptide molecule is composed of four units such as D-leucine (D-Leu; amino acid), 2-carboxy-6-hydroxyoctahydroindole (L-Choi; an unusual bicyclic α-amino acid core), α-hydroxy-*p*-hydroxyphenyllactic acid (D-Hpla), and L-argininol (L-Argol). The L-Choi (**10**), L-Argol (**15**), and D-Hpla-D-Leu (**22**) segments were synthesized separately, and then assembled all together to frame the structure **1** as depicted in Schemes 1–4.

Synthesis of L-Choi segment (10):

Synthesis of this segment was initiated with L-Cbz-tyrosine (**3**) (Scheme 1). On oxidation with $PhI(OAc)_2$ followed by esterification with methanol, compound **3** afforded (2*S*,3a*R*,7a*R*)-1-benzyl 2-methyl 3a-hydroxy-6-oxo-3,3a,7,7a-tetrahydro-1*H*-indole-1,2(2*H*,6*H*)-dicarboxylate (**4**) in a diastereoselectivity of >98:2. Its *syn*-diastereomer **6** was achieved through a thermodynamic equilibration of the benzoyl-protected alcohol **5** via a retro-Michael/Michael addition sequence reaction. Compound **7** was obtained on debenzoylation of **6** in good yield using SmI_2 [3]. Hydrogenation of **7** followed by L-Selectride reduction and TBS protection afforded the L-Choi core (**10**).

Synthesis of L-Argol segment (15):

The L-Argol segment (**15**) was synthesized in six steps from L-arginine (**11**) (Scheme 2). Selective Alloc and Cbz protection followed by in situ sodium borohydride reduction [4] of the mixed anhydride formed with

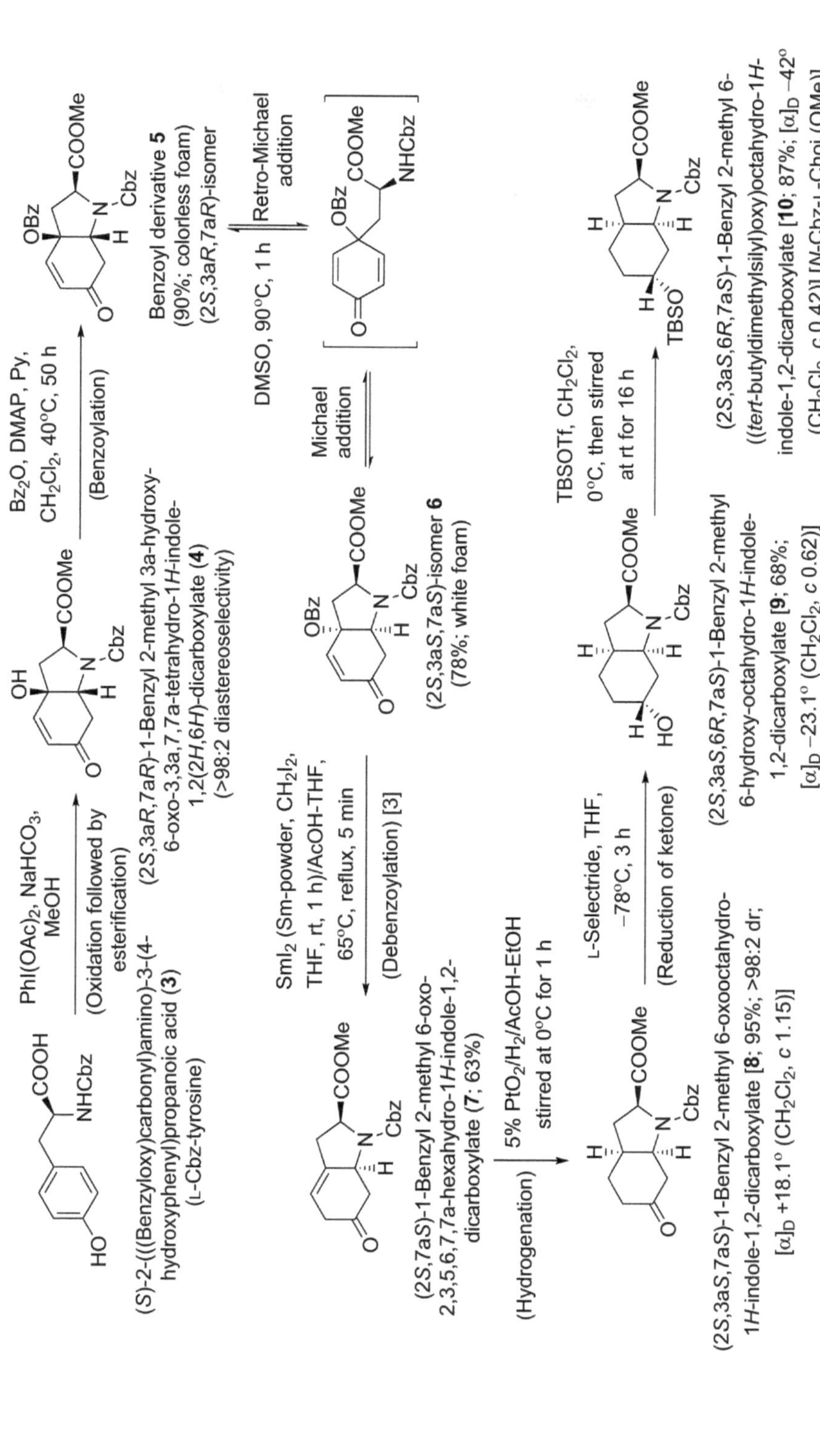

Scheme 1 Synthesis of L-Choi segment (**10**) [2].

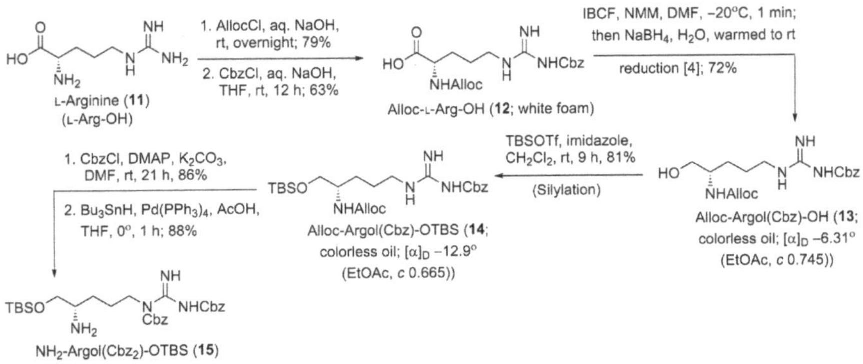

Scheme 2 Synthesis of L-Argol segment (**15**) [2].

isobutyl chloroformate (IBCF) yielded the primary alcohol Alloc-Argol (Cbz)-OH (**13**). Standard protection and deprotection techniques implemented in the next three steps afforded the L-Argol segment **15** in good yield.

Synthesis of D-Hpla-D-Leu segment (22):

The investigators started the synthesis of this segment using a $BF_3{\cdot}OEt_2$-catalyzed organocuprate addition of (*R*)-benzylglycidol **17** with the protected 4-bromophenol to generate the (*R*)-1-(benzyloxy)-3-(4-((triisopropylsilyl)oxy)phenyl)propan-2-ol (**18**) in 83% yield (Scheme 3). Standard protective group manipulations followed by Dess-Martin periodinane [5] and sodium perchlorate oxidations resulted in D-Hpla **20**. Finally, diethyl cyanophosphinate (DEPC)-mediated coupling to D-Leu-OBn (**21**) and hydrogenolysis provided the desired D-Hpla-D-Leu segment (D^{Leu}-**22**).

Assembly of L-Choi (10), L-Argol (15), and D-Hpla-D-Leu (22) segments: total synthesis of aeruginosin 298-A (1):

At this stage, the investigators assembled the respective segments to arrive at the tetrapeptide framework of the target molecule **1** (Scheme 4). The Cbz group of *N*-Cbz-L-Choi (OMe) (**10**) was exchanged with an Alloc group, and was then subjected to subsequent saponification and pentafluorophenyl ester-mediated coupling to L-Argol **15** [NH_2-Argol(Cbz_2)-OTBS] to afford the dipeptide Alloc-L-Choi(TBS)-L-Argol(Cbz_2)-OTBS (**23**). Alloc deprotection followed by 3-(diethoxyphosphoryloxy)-1,2,3-benzotriazin-4(3*H*)-one (DEPBT)-mediated coupling [7] to D^{Leu}-**22** resulted in the formation of the tetrapeptide D^{Leu}-**24** in 59% yield. The tetrapeptide was treated with HF (H_2O-MeCN 1:9) for 2 h, neutralized with aq. NaOH, and extracted into CH_2Cl_2/EtOAc. Finally, hydrogenolysis cleaved the Cbz-protective groups (34%; two steps) to furnish (+)-aeruginosin 298-A

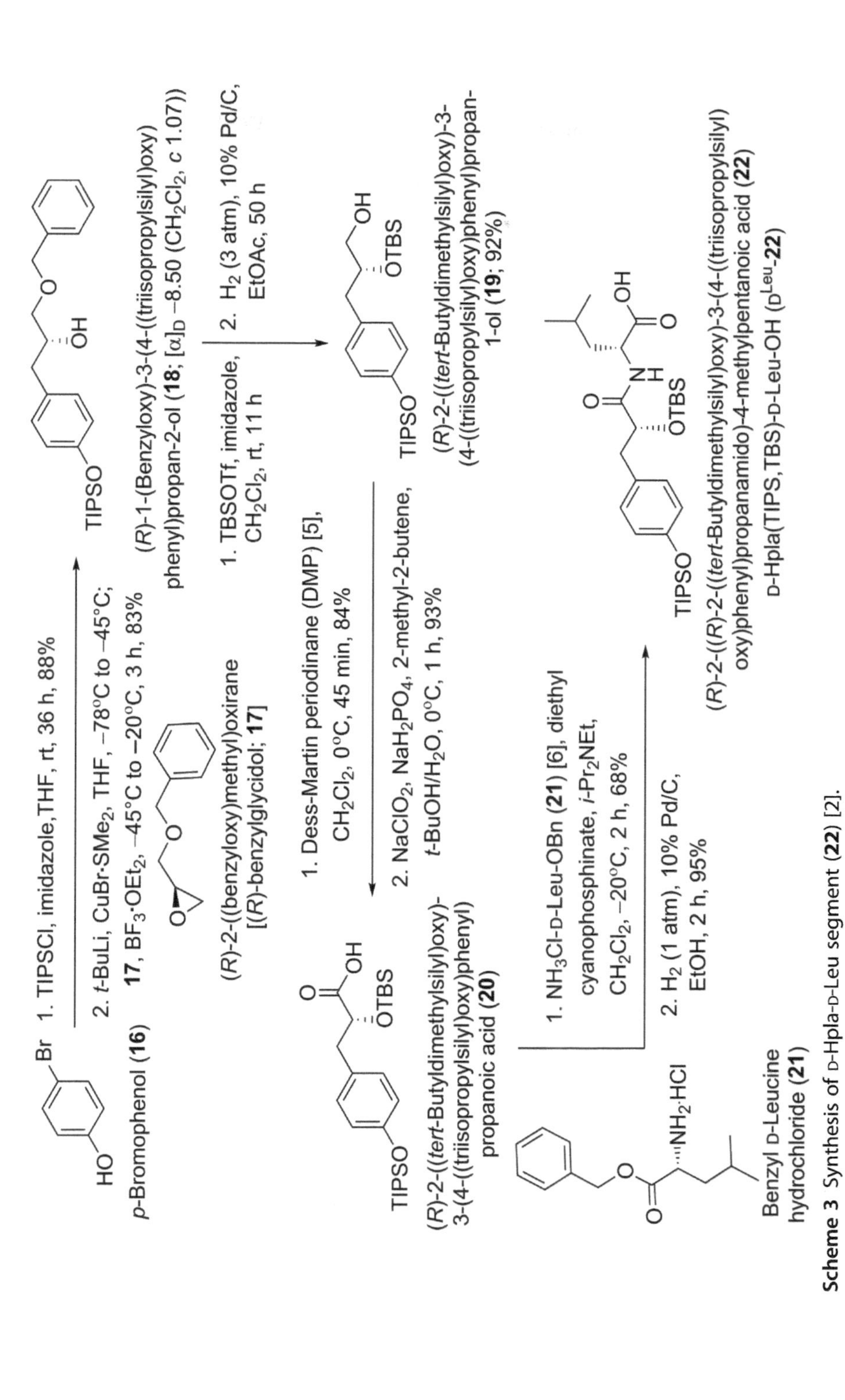

Scheme 3 Synthesis of D-Hpla-D-Leu segment (**22**) [2].

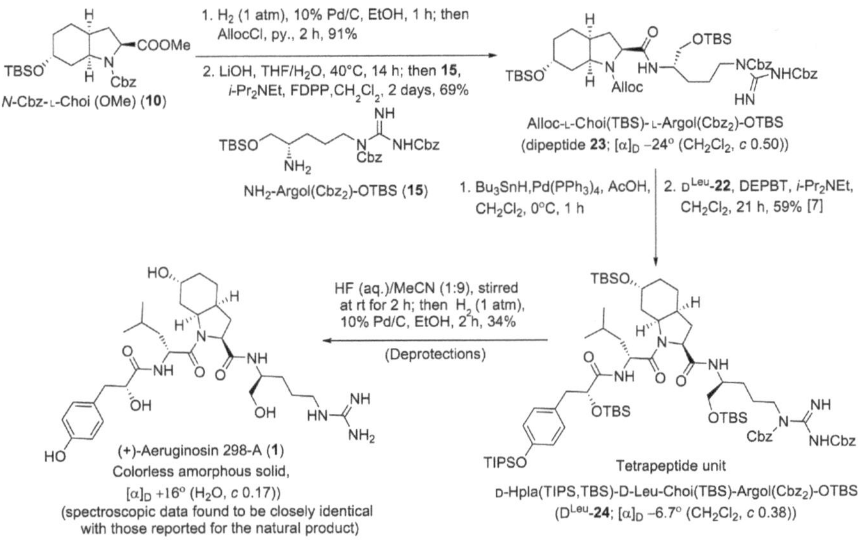

Scheme 4 Total synthesis of (+)-aeruginosin 298-A (**1**) [2].

(**1**) as white amorphous solid; spectroscopic data of this synthetic compound completely matched those reported for the natural product [1].

Almost in a parallel reporting, Bonjoch and his group [8] outlined another route for the total syntheses for both aeruginosin 298-A (**1**) and 298-B (**2**) (Schemes 5–7). The investigators first prepared L-Choi fragment **31** and D-Hpla fragment **36** as depicted below.

Synthesis of *N*-Boc-L-Choi(OMe) (31):

(2*S*,3a*S*,6*R*,7a*S*)-1-*tert*-Butyl 2-methyl 6-hydroxyoctahydro-1*H*-indole-1,2-dicarboxylate [*N*-Boc-L-Choi(OMe) (**31**)] was prepared starting from *N*-acetyl-L-tyrosine (**25**):

Synthesis of D-Hpla fragment (36):

The protected derivative of D-Hpla **36** was synthesized starting from β-(4-hydroxyphenyl)pyruvic acid (**32**) (Scheme 6); on enantioselective reduction following Wang's procedure [11], compound **32** furnished (*R*)-2-hydroxy-3-(4-hydroxyphenyl)propanoic acid (D-Hpla; **33**) in 92% yield with 86% *ee*. Then after sequential protection of both hydroxyl groups and recrystallization with (*R*)-1-phenylethylamine, the investigators obtained the desired **36** in >98% *ee*.

Assembly of the segments: total synthesis of aeruginosin 298-A (1) and aeruginosin 298-B (2):

The investigators [8] then assembled the corresponding fragments to achieve the target molecules aeruginosin 298-A (**1**) and aeruginosin 298-B (**2**) as described in the following Schemes 7 and 8.

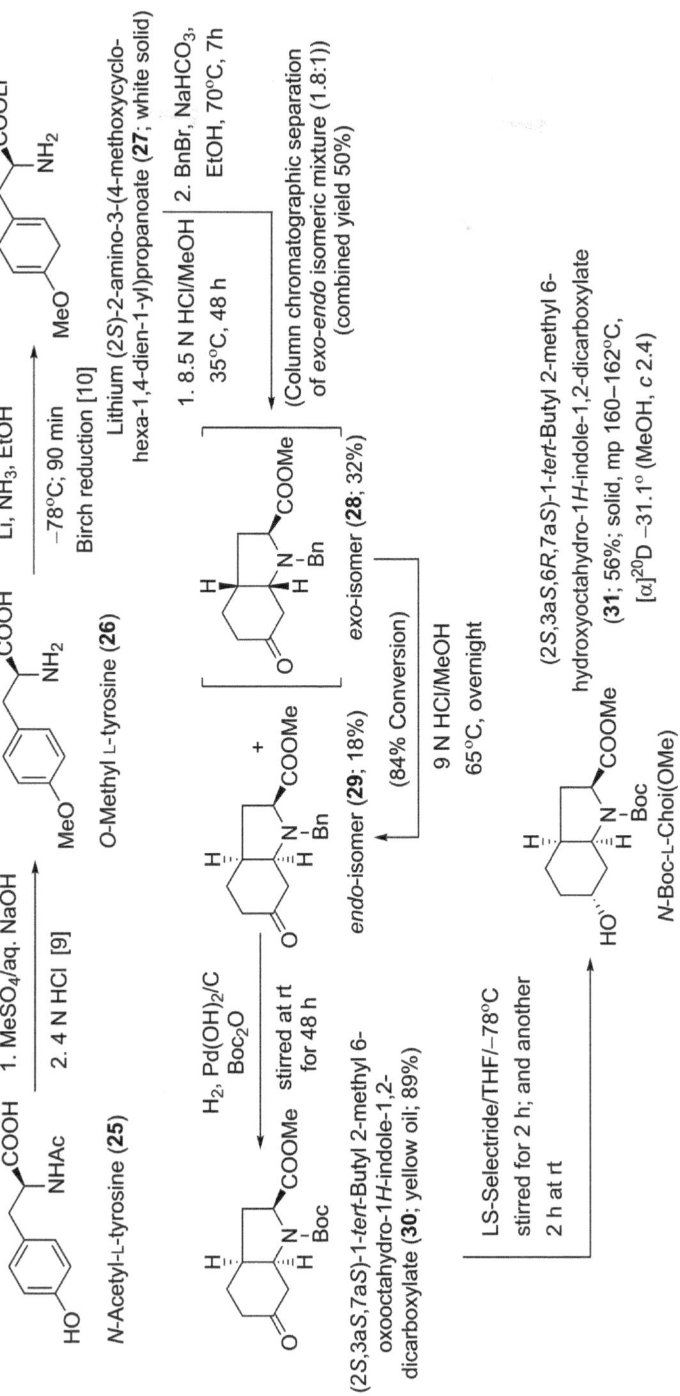

Scheme 5 Synthesis of *N*-Boc-L-Choi(OMe) (**31**) [8].

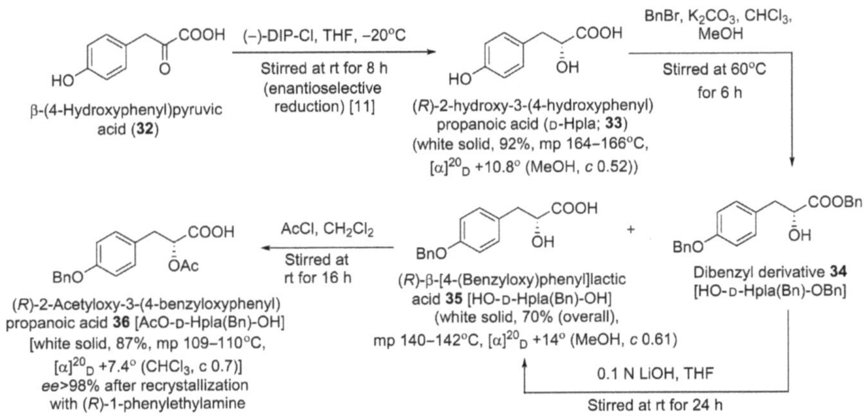

Scheme 6 Synthesis of protected derivative of D-Hpla (**36**) [8].

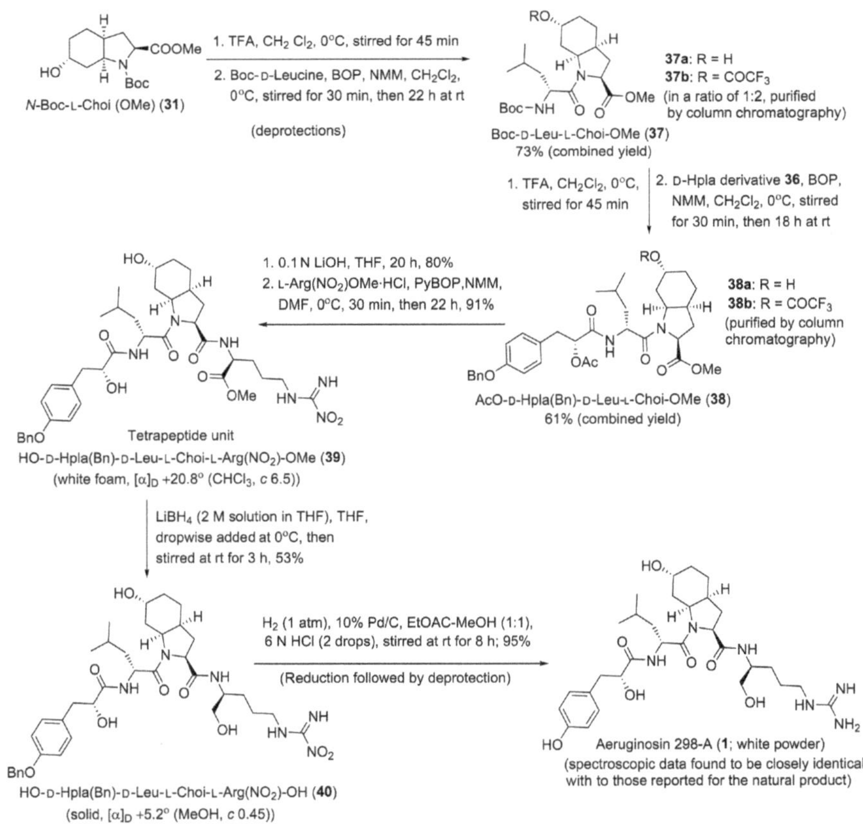

Scheme 7 Total synthesis of aeruginosin 298-A (**1**) [8].

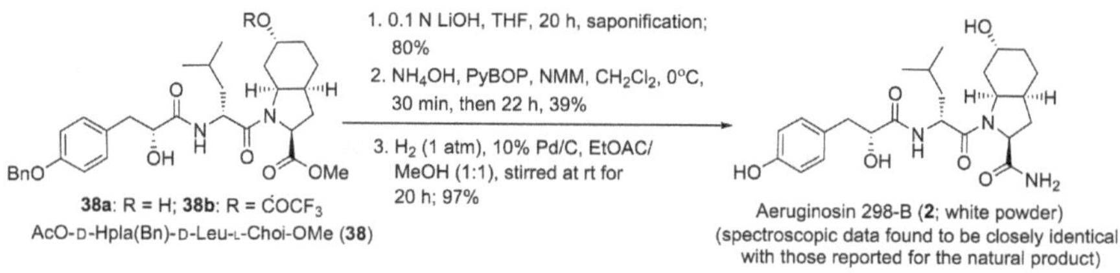

Scheme 8 Total synthesis of aeruginosin 298-B (**2**) [8].

Later on, Shibasaki and his group also developed a versatile and enantioselective synthetic process for aeruginosin 298-A (**1**) and its several attractive analogs using a catalytic asymmetric phase-transfer reaction and epoxidation, and evaluated their biological potentials as well [12, 13].

References

[1] (a) M. Murakami, Y. Okita, H. Matsuda, T. Okino, K. Yamaguchi, Tetrahedron Lett. 35 (1994) 3129; (b) M. Murakami, K. Ishida, T. Okino, H. Matsuda, K. Yamaguchi, Tetrahedron Lett. 36 (1995) 2785; (c) K. Ishida, H. Okita, H. Matsuda, T. Okino, M. Murakami, Tetrahedron 55 (1999) 10971.

[2] P. Wipf, J.-L. Methot, Org. Lett. 2 (2000) 4213.

[3] G.A. Molander, Org. React. 46 (1994) 211.

[4] M. Rodriguez, M. Llinares, S. Doulut, A. Heitz, J. Martinez, Tetrahedron Lett. 7 (1991) 923.

[5] (a) D.B. Dess, J.C. Martin, J. Am. Chem. Soc. 113 (1991) 7277; (b) R.E. Ireland, L. Liu, J. Org. Chem. 58 (1993) 2899.

[6] T. Wakamiya, M. Kamata, S. Kusumoto, H. Kobayashi, Y. Sai, Bull. Soc. Chem. Jpn. 71 (1998) 699.

[7] H. Li, X. Jiang, Y. Ye, C. Fan, T. Romoff, M. Goodman, Org. Lett. 1 (1999) 91.

[8] N. Valls, M. López-Canet, M. Vallribera, J. Bonjoch, Chem. Eur. J. 7 (2001) 3446.

[9] (a) B. Wünsch, M. Zott, Liebigs Ann. Chem. (1992) 39; (b) W. Siedel, K. Sturm, R. Geiger, Chem. Ber. 96 (1963) 1636.

[10] B.C. Laguzza, B. Ganem, Tetrahedron Lett. 22 (1981) 1483.

[11] Z. Wang, B. La, J.M. Fortunak, X.-J. Meng, G.W. Kabalka, Tetrahedron Lett. 39 (1998) 5501.

[12] T. Ohshima, V. Gnanadesikan, T. Shibuguchi, Y. Fukuta, T. Nemoto, M. Shibasak, J. Am. Chem. Soc. 125 (2003) 11206.

[13] Y. Fukuta, T. Ohshima, V. Gnanadesikan, T. Shibuguchi, T. Nemoto, T. Kisugi, T. Okino, M. Shibasaki, Proc. Natl. Acad. Sci. USA 101 (2004) 5433.

CHAPTER TWO

Ageladine A

Abbreviations

Biphenyl-PCy$_2$ 2-biphenyldicyclohexylphosphine
BOC *tert*-butoxycarbonyl
BOM benzyloxymethyl
DMF *N,N*-dimethylformamide
HPLC high performance liquid chromatography
MMPs matrix metalloproteinases
***n*-BuLi** *n*-butyllithium
Sc(OTf)$_3$ scandium triflate
THF tetrahydrofuran

Systematic name: 4-(4,5-Dibromo-1*H*-pyrrol-2-yl)-1*H*-imidazo[4,5-*c*]pyridin-2-amine

Compound class: Alkaloid (pyrrole-imidazole alkaloid)

Structure:

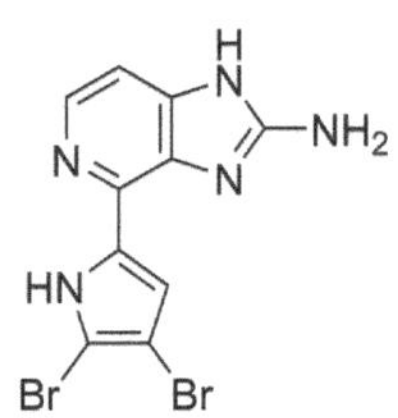

Ageladine A (**1**)

Natural source: *Agelas nakamurai* Hoshino (marine sponge collected near Kuchinoerabu-jima Island in southern Japan; family: Agelasidae) [1]

Pharmaceutical potential: MMPs inhibitor (antiangiogenic and anticancer) [1–10]; pH-sensitive membrane permeable dye [11, 12]

Synthetic routes:

Ageladine A (**1**), an anticancer marine pyrrole-imidazole alkaloid, was isolated from the marine sponge *Agelas nakamurai* by Fusetani and coworkers [1]; the brominated alkaloid metabolite is the first example of an imidazolopyridine marine natural product. Its promising MMPs' inhibitory

https://doi.org/10.1016/B978-0-08-102822-3.00002-X

activity and unusual compact structure motivated synthetic chemists toward its total synthesis; as a result at least five synthetic strategies for the total synthesis of ageladine A and its analogs have been reported so far [2–10]. In 2006, Meketa and Weinreb [2] reported the first total synthesis of **1** using a 6π-1-azaelectrocyclization and Suzuki-Miyaura coupling of *N*-Boc-pyrrole-2-boronic acid and a chloropyridine derivative as a key step. Soon after in the same year, Shengule and Karuso [3] reported the concise synthesis of **1** based on biomimetic principles involving a pivotal Pictet-Spengler-type condensation between 2-aminohistamine and *N*-Boc-4,5-dibromo-2-formyl pyrrole. Weinreb and his group further reported the third total synthesis using a variation of his original method employing a 6π-2-azatriene electrocyclization for the formation of the imidazolopyridine moiety [5–7]. Later on, Ando et al. reported a fourth synthesis of ageladine A and its analogs based on the Pictet-Spengler cyclization between the *N*-Boc-2-aminohistamine and *N*-protected 2-formyl pyrroles [8, 9]. Most recently, Karuso and his group reported the one-pot synthesis of **1** and a range of analogs from 2-aminohistamine and various heterocyclic aldehydes [10]. Herein the first two total synthetic strategies are discussed.

Meketa and Weinreb accomplished the total synthesis of the tricyclic heteroaromatic marine metabolite ageladine A (**1**) for the first time in 12 consecutive steps using a 6π-1-azaelectrocyclization and Suzuki-Miyaura coupling technique as the key reactions (Scheme 1) [2]. The readily available benzyloxymethyl (BOM)-protected tribromoimidazole (**2**) [13] was first metalated with *n*-BuLi at C-2, and a thiomethyl group was introduced thereon using dimethyl disulfide. A second equivalent of *n*-BuLi was then added to the reaction mixture to effect metalation at C-5, followed by treatment with DMF, leading to the formation of 3-benzyloxymethyl-5-bromo-2-methylsulfanylimidazole-4-carboxaldehyde (**3**) in 91% overall yield for the one-pot operation. The bromo vinylimidazole **4**, obtained from the imidazole aldehyde **3** via Wittig reaction, was then converted to BOM-protected vinylimidazole carboxylic acid **5** by means of lithiation of **4** at C-4 followed by treatment with carbon dioxide in 98% yield. Applying the methodology of Kikugawa et al. [14], the investigators transformed this BOM-protected compound **5** directly into *N*-methoxy imidoyl chloride **6** (87% yield), which then underwent 6π-1-azaelectrocyclization on refluxing in *o*-xylene at 150°C affording chloropyridine **7** in 84% isolated yield.

The investigators observed that Suzuki-Miyaura coupling of BOM-protected chloropyridine **7** with *N*-Boc-boronic acid **11** could not be effected, and thus they modified their strategy. The sulfide moiety in **7** was first oxidized

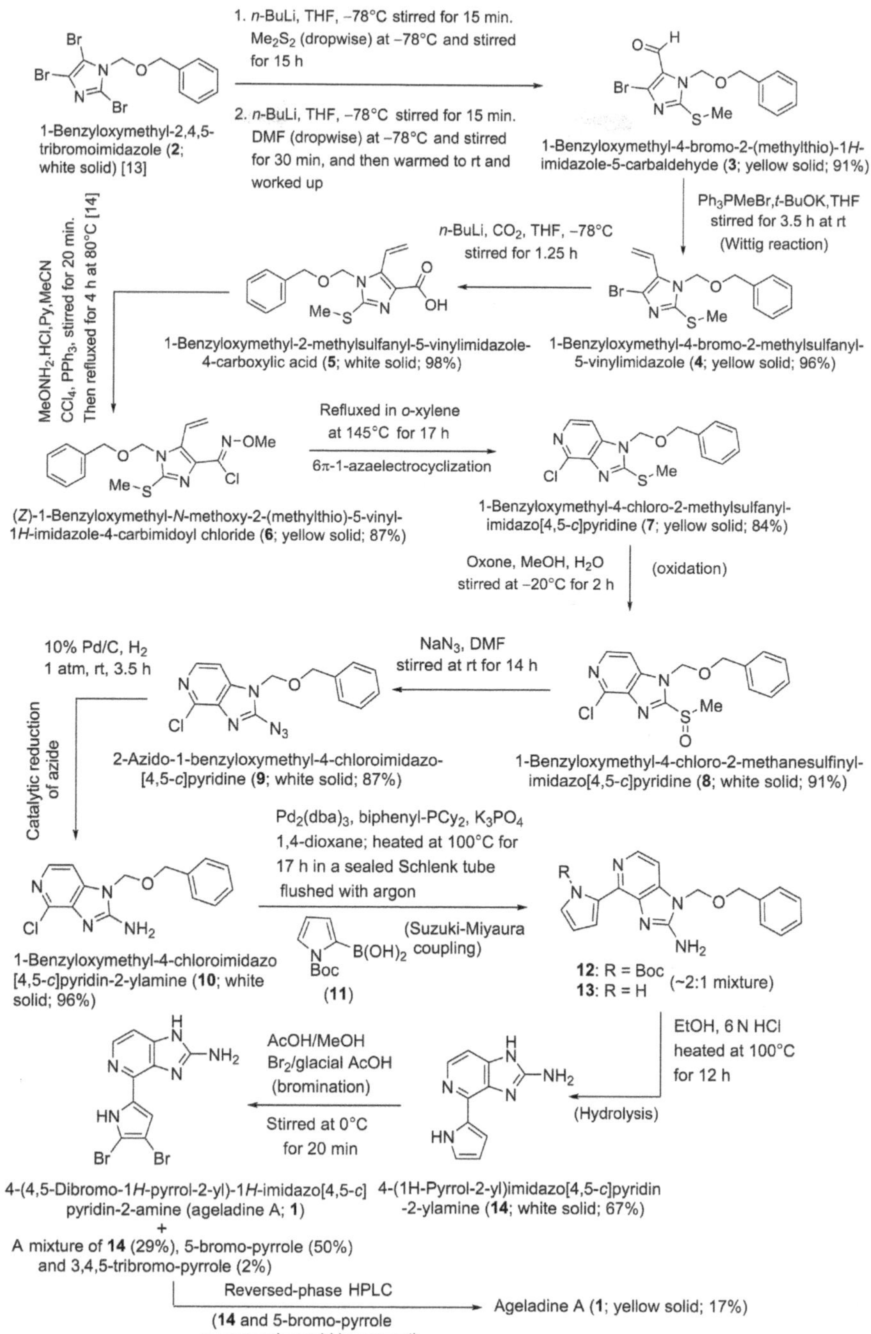

Scheme 1 Weinreb's total synthesis of ageladine A (**1**) [2].

with oxone to the sulfoxide **8**, followed by its displacement with sodium azide at room temperature affording thereby the azide **9** in 87% yield. Catalytic hydrogenation of this azide then cleanly produced 2-aminoimidazolopyridine **10**, which then underwent Suzuki-Miyaura coupling with **11** in the presence of Buchwald's 2-biphenyldicyclohexylphosphine ligand [15] to furnish a ~2:1 mixture of Boc-protected tricyclic pyrrole **12** and compound **13** (without the Boc group). This crude mixture was then hydrolyzed with 6 N HCl in ethanol to afford deprotected tricycle **14** in 67% overall yield from chloropyridine **10**. Finally, bromination of pyrrole **14** was effected upon treatment with bromine in cold acetic acid/methanol medium, which produced ageladine A (**1**) (17%), along with recovered starting material (**14**; 29%), the 5-bromo-pyrrole (50%), and a small amount of 3,4,5-tribromopyrrole (2%). These compounds were separable by reverse-phase HPLC. Both the monobromopyrrole and the starting material **14** could be recycled to the natural product. Synthetic ageladine A ($C_{10}H_8N_5Br_2$, yellow solid) was found to have identical physical and spectral properties to those reported for the natural material [1].

Soon after the first report of Meketa and Weinreb [2], a concise total synthesis of ageladine A in just two steps was published by Shengule and Karuso in the same year (Scheme 2) [3]. They achieved this shorter route for the synthesis of **1** by exploiting a Pictet-Spengler-type condensation between 2-aminohistamine (**15**) and *N*-Boc-4,5-dibromo-2-formylpyrrole (**16**) as the key step. The starting material **15** is commercially available or can be prepared from Boc-guanidine and β-alanine as per reported methods [16, 17]. Similarly, 4,5-dibromo-2-formylpyrrole (**16**) is also commercially available, and can be effected in near quantitative yield from 2-formylpyrrole in two steps [18]. Pictet-Spengler-type condensation between 2-aminohistamine (**15**) and *N*-Boc-4,5-dibromo-2-formylpyrrole (**16**) was carried out by stirring the reactants in ethanol at room temperature for 5 h in the presence of the Lewis acid catalyst $Sc(OTf)_3$ to furnish *N'*-Boc-5′-(2-amino-4,5,6,7-tetrahydroimidazo[4,5-*c*]pyridine-4-yl)-2′,3′-dibromopyrrole (**17**; yellow solid) in 44% yield. Dehydrogenation and deprotection of **17** were cleanly

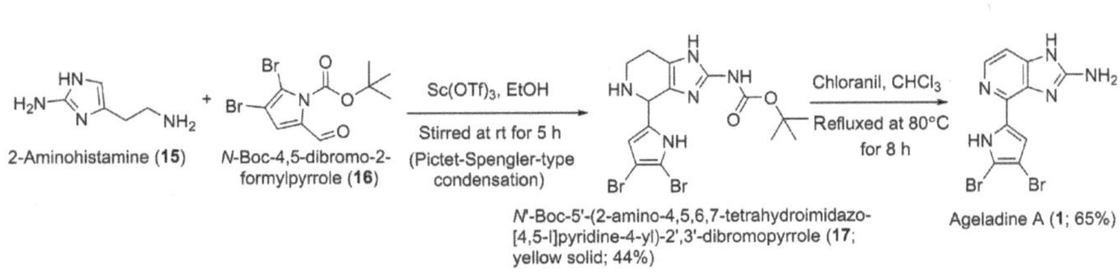

Scheme 2 Karuso's total synthesis of ageladine A (**1**) [3].

effected by refluxing it with chloranil in chloroform for 8 h to yield ageladine A (**1**) as a fluorescent yellow solid, identical in all respects to the natural product [1].

References

[1] M. Fujita, Y. Nakao, S. Matsunaga, M. Seiki, Y. Itoh, J. Yamashita, R.W.M. van Soest, N. Fusetani, J. Am. Chem. Soc. 125 (2003) 15700.
[2] M.L. Meketa, S.M. Weinreb, Org. Lett. 8 (2006) 1443.
[3] S.R. Shengule, P. Karuso, Org. Lett. 8 (2006) 4083.
[4] Y. Nakao, N. Fusetani, J. Nat. Prod. 70 (2007) 689.
[5] M.L. Meketa, S.M. Weinreb, Y. Nakao, N. Fusetani, J. Org. Chem. 72 (2007) 4892.
[6] M.L. Meketa, S.M. Weinreb, Tetrahedron 63 (2007) 9112.
[7] M.L. Meketa, S.M. Weinreb, Org. Lett. 9 (2007) 853.
[8] N. Ando, S. Terashima, Bioorg. Med. Chem. Lett. 17 (2007) 4495.
[9] N. Ando, S. Terashima, Bioorg. Med. Chem. Lett. 19 (2009) 5461.
[10] S.R. Shengule, W.L. Loa-Kum Cheung, C.R. Parish, M. Blairvacq, L. Meijer, Y. Nakao, P. Karuso, J. Med. Chem. 54 (2011) 2492.
[11] U. Bickmeyer, A. Grube, K.W. Klings, M. Köck, Biochem. Biophys. Res. Commun. 373 (2008) 419 (Erratum in: *Biochem. Biophys. Res. Commun.*, 2009, **383**, 519).
[12] U. Bickmeyer, M. Heine, I. Podbielski, D. Münd, M. Köck, P. Karuso, Biochem. Biophys. Res. Commun. 402 (2010) 489.
[13] R.W. Schumacher, B.S. Davidson, Tetrahedron 55 (1999) 935.
[14] Y. Kikugawa, L.H. Fu, T. Sakamoto, Synth. Commun. 23 (1993) 1061.
[15] T.E. Barder, S.D. Walker, J.R. Martinelli, S.L. Buchwald, J. Am. Chem. Soc. 127 (2005) 4685.
[16] R.G. Jones, E.C. Kornfeld, K.C. McLaughlin, J. Am. Chem. Soc. 72 (1950) 4526.
[17] T.L. Little, S.E. Webber, J. Org. Chem. 59 (1994) 7299.
[18] S.T. Handy, J.J. Sabatini, Y. Zhang, I. Vulfova, Tetrahedron Lett. 45 (2004) 5057.

CHAPTER THREE

(+)-Ainsliadimer A

Abbreviations

Ar argon
BINOL 1,1′-bi-2-naphthol
CH_2Cl_2 dichloromethane
DABCO 1,4-diazobicyclo[2.2.2]octane
DBU 1,8-diazabicyclo[5.4.0]undec-7-ene
EtOAc ethyl acetate
H_2O_2 hydrogen peroxide
HMPA hexamethylphosphoramide
***i*Pr** isopropyl
LDA lithium diisoproylamide
***m*-CPBA** *meta*-chloroperbenzoic acid
MsCl methanesulfonyl chloride
$NaBH_4$ sodium borohydride
NO nitric oxide
PhSe-SePh diphenyl diselenide
Py pyridine
$SOCl_2$ thionyl chloride
THF tetrahydrofuran

Systematic name: (3a*S*,3a′*S*,6a*R*,6a′*R*,7a′*S*,9a*R*,9b*S*,10a′*S*,10b′*S*,10c′*S*)-7a′,10a′-dihydroxy-3,3′,6,6′-tetramethylenehexadecahydro-2*H*-spiro[azuleno[4,5-*b*]furan-9,8′-cyclopenta[2,3]azuleno[4,5-*b*]furan]-2,2′,8(3*H*,9a*H*,9b*H*,9′*H*,10b′*H*,10c′*H*)-trione

Compound class: Guaianolide sesquiterpene dimer

https://doi.org/10.1016/B978-0-08-102822-3.00003-1

Structure:

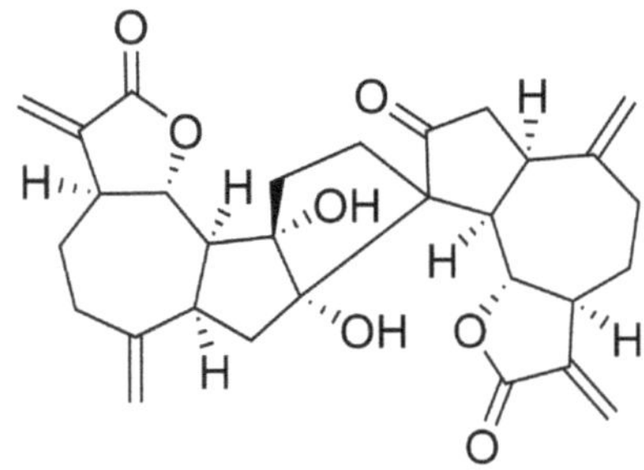

(+)-Ainsliadimer A (**1**)

Natural source: *Ainsliaea macrocephala* (whole plants; family: Compositae) [1]

Pharmaceutical potential: Inhibitor to NO production in RAW264.7 stimulated by lipopolysaccharide ($IC_{50}=2.41$ μg/mL) [1]

Synthetic route:

(+)-Ainsliadimer A (1), a guaianolide sesquiterpene dimer isolated from the Chinese medicinal plant *A. macrocephala*, bears a unique heptacyclic ring system having 11 contiguous stereogenic centers and a highly functionalized cyclopentane ring with three quaternary carbons connecting the two monomeric sesquiterpene lactone units. These unique architectural features combined with its potent inhibitory effect against nitric oxide (NO) production made this molecule as an ideal target for total synthesis. Lei and his group [2] reported the first total synthesis of (+)-ainsliadimer A starting from α-stantonin based on a concise and biomimetic approach involving 14 steps—a hydrogen bonding-promoted [4+2]-hetero-Diels-Alder dimerization is the key step of the synthetic strategy (Scheme 1). The synthetic approach demonstrated the feasibility of using nonenzymatic conditions to achieve the proposed biosynthesis path of the natural product **1** [1].

The investigators first synthesized dehydrozaluzanin C (**12**) from commercially available α-santonin (**2**). On photoirradiation with a high-pressure Hg lamp (500 W) in acetic acid, α-santonin (**2**) afforded *O*-acetylisophotosantonic lactone (**3**) as a colorless solid [3–5], which was subsequently converted to compound **4** in quantitative yield and excellent diastereoselectivity on hydrogenation of the double bond with Pd-C/H_2. Selective reduction of ketone **4**, followed by mesylation of the corresponding hydroxyl group and in situ antielimination, thereafter furnished alkene **6** (52% over two steps) [6]. Compound **6** underwent smooth saponification reaction with 5% aq.

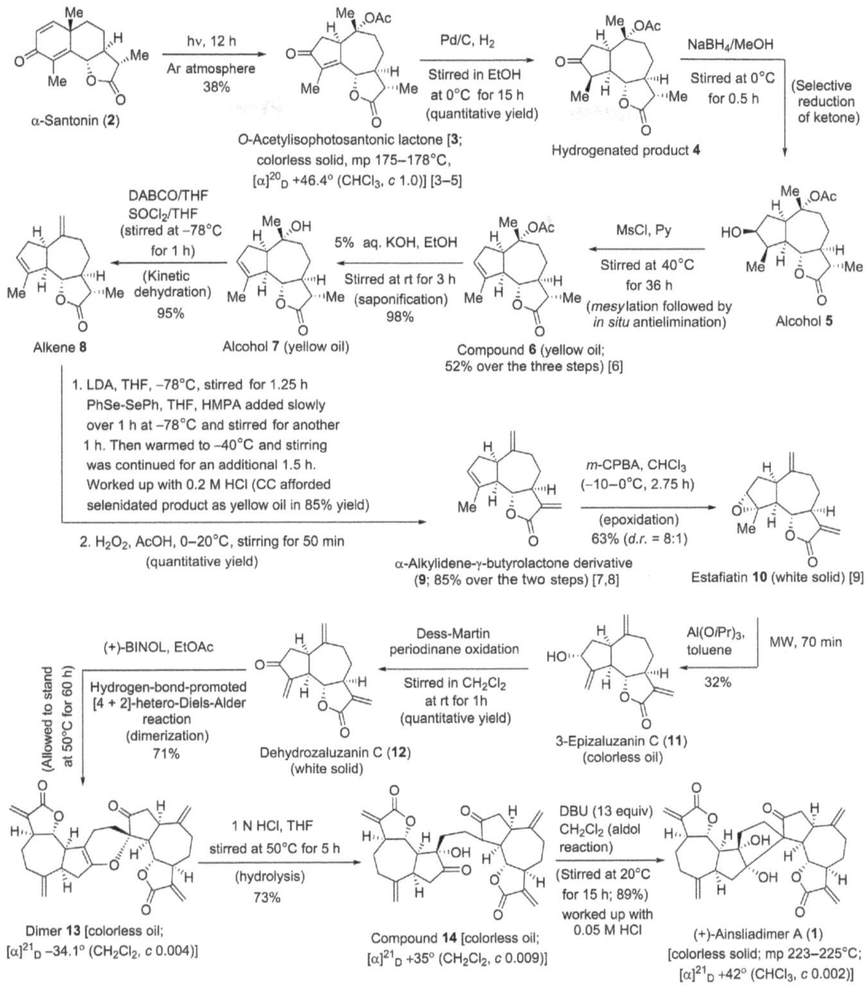

Scheme 1 Total synthesis of (+)-ainsliadimer A (**1**) [2].

KOH in ethanol to give alcohol **7** as yellow oil in 98% yield. Kinetically controlled dehydration of alcohol **7** to the disubstituted alkene **8** over tetrasubstituted alkene (in 12:1 ratio) was achieved on treating with $SOCl_2$ in the presence of DABCO as base at −78°C. Selenenylation of **8**, followed by oxidation of the resulting selenide and subsequent selenoxide elimination, produced the desired α-alkylidene-γ-butyrolactone **9** [7, 8], epoxidation of which with *m*-CPBA afforded estafiatin (**10**) as the major diastereomer in 63% yield [9]. Treatment of estafiatin (**10**) with aluminum isopropoxide in toluene under microwave irradiation generated 3-epizaluzanin C (**11**), which was further oxidized by Dess-Martin periodinane oxidation to afford dehydrozaluzanin C (**12**).

Dehydrozaluzanin C (**12**) in ethyl acetate solution was allowed to stand at 50°C for about 60 h in the presence of (+)-BINOL catalyst when the compound underwent hydrogen bonding-promoted [4 + 2]-hetero-Diels-Alder cycloaddition to generate the dimer **13** as colorless oil in 71% yield. Compound **13** was then hydrolyzed under mild acidic conditions to compound **14**, which in the final step of the synthesis underwent an efficient intramolecular aldol condensation on treatment with a large excess of DBU in CH_2Cl_2 at 20°C, to furnish the natural product (+)-ainsliadimer A (**1**; mf $C_{30}H_{34}O_7$) in 89% yield. The synthetic compound was found to have almost similar physical and spectral properties to those reported for natural compound [1].

References

[1] Z.-J. Wu, X.-K. Xu, Y.-H. Shen, J. Su, J.-M. Tian, S. Liang, H.-L. Li, R.-H. Liu, W.-D. Zhang, Org. Lett. 10 (2008) 2397.
[2] C. Li, X. Yu, X. Lei, Org. Lett. 12 (2010) 4284.
[3] W. Zhang, S. Luo, F. Fang, Q. Chen, H. Hu, X. Jia, H. Zhai, J. Am. Chem. Soc. 127 (2005) 18.
[4] G. Blay, L. Cardona, B. Garcia, L. Lahoz, J.R. Pedro, J. Org. Chem. 66 (2001) 7700.
[5] D.H.R. Barton, P. De Mayo, M. Shafiq, J. Chem. Soc. (1957) 929.
[6] E.B. Piers, K.F. Cheng, Chem. Commun. 562 (1959).
[7] M. Ando, K. Ibayashi, N. Minami, T. Nakamura, K. Isogai, J. Nat. Prod. 57 (1994) 433.
[8] M. Ando, H. Kusaka, H. Ohara, K. Takase, H. Ymaoka, Y. Yanagi, J. Org. Chem. 54 (1989) 1952.
[9] M.T. Edgar, A.E. Greene, P. Crabbe, J. Org. Chem. 44 (1979) 159.

CHAPTER FOUR

(−)-Aiphanol

Abbreviations

AcCl acetyl chloride
DIAD diisopropyl azodicarboxylate
DIBAL-H diisobutylaluminum hydride
DIPEA *N,N*-diisopropylethylamine (Hünig's base)
DMAP 4-dimethylaminopyridine
ee enantiomeric excess
HPLC high performance liquid chromatography
MOM methoxymethyl
MsCl mesyl chloride
PPh_3 triphenylphosphine
TEA triethylamine
THF tetrahydrofuran
TsCl *p*-toluenesulfonyl chloride

Systematic name: 5-((*E*)-2-((2*S*,3*S*)-3-(4-hydroxy-3,5-dimethoxyphenyl)-2-(hydroxymethyl)-2,3-dihydrobenzo[*b*][1,4]dioxin-6-yl)vinyl)benzene-1,3-diol

Compound class: Stilbenolignan

Structure:

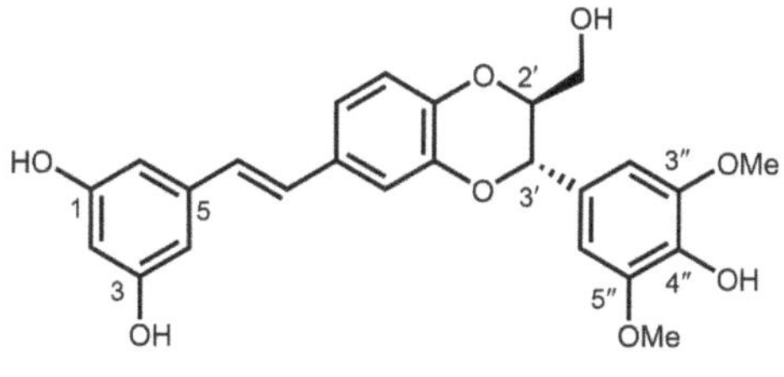

(−)-Aiphanol (**1**)

Natural source: *Aiphanes aculeata* Willd. (seeds; family: Arecaceae) [1]

Pharmaceutical potential: Cyclooxygenase-1 and -2 (COX-1/2) inhibitor [1, 2]

Total Synthesis of Bioactive Natural Products
https://doi.org/10.1016/B978-0-08-102822-3.00004-3

Synthetic routes:

Natural aiphanol (**1**), isolated by Kinghorn and coworkers from the seeds of *Aiphanes aculeata* Willd. (Arecaceae), is levorotatory (i.e., (−)-form; $[\alpha]_D = -21.8°$(MeOH, *c* 0.13)) [1]. This natural product not only possesses an unprecedented stilbenolignan skeleton in which a hydroxylated stilbene unit is connected to a phenylpropane moiety via a 1,4-dioxane bridge, but it also shows potent inhibition of COX-1 and -2 [1–3]. Both Pan and his group [3] and also Banwell et al. [2] reported total synthesis of (±)-aiphanol almost at the same time; the latter group of workers thereafter became successful in accomplishing the first enantioselective total synthesis of the stilbenolignan (−)-aiphanol (**1**) (Scheme 1) [4].

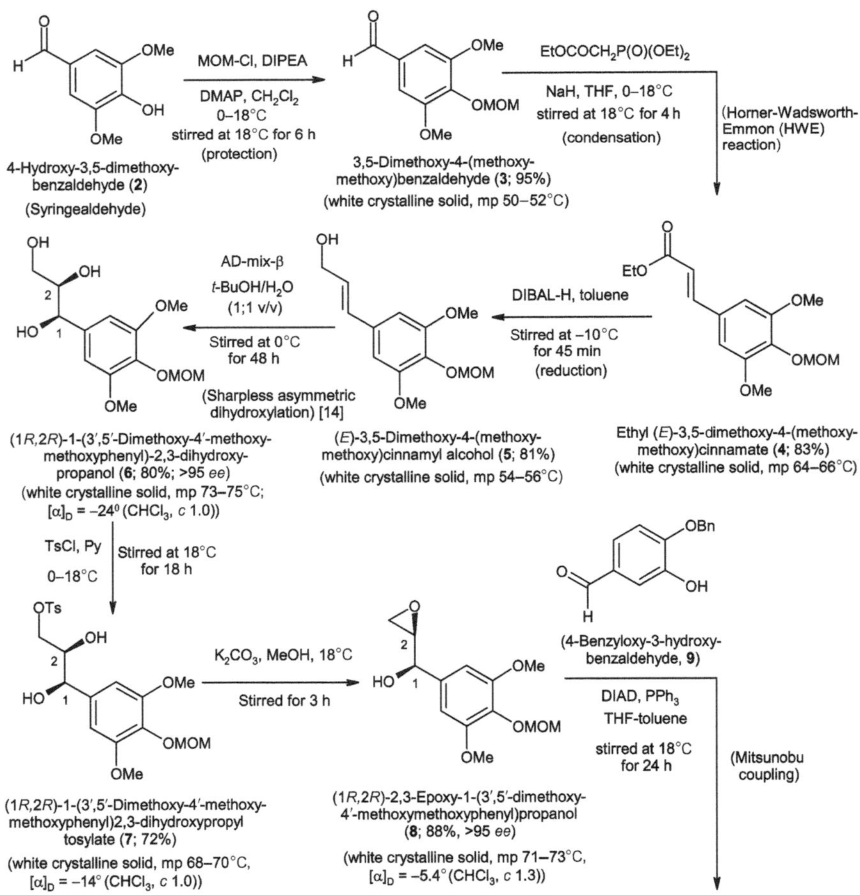

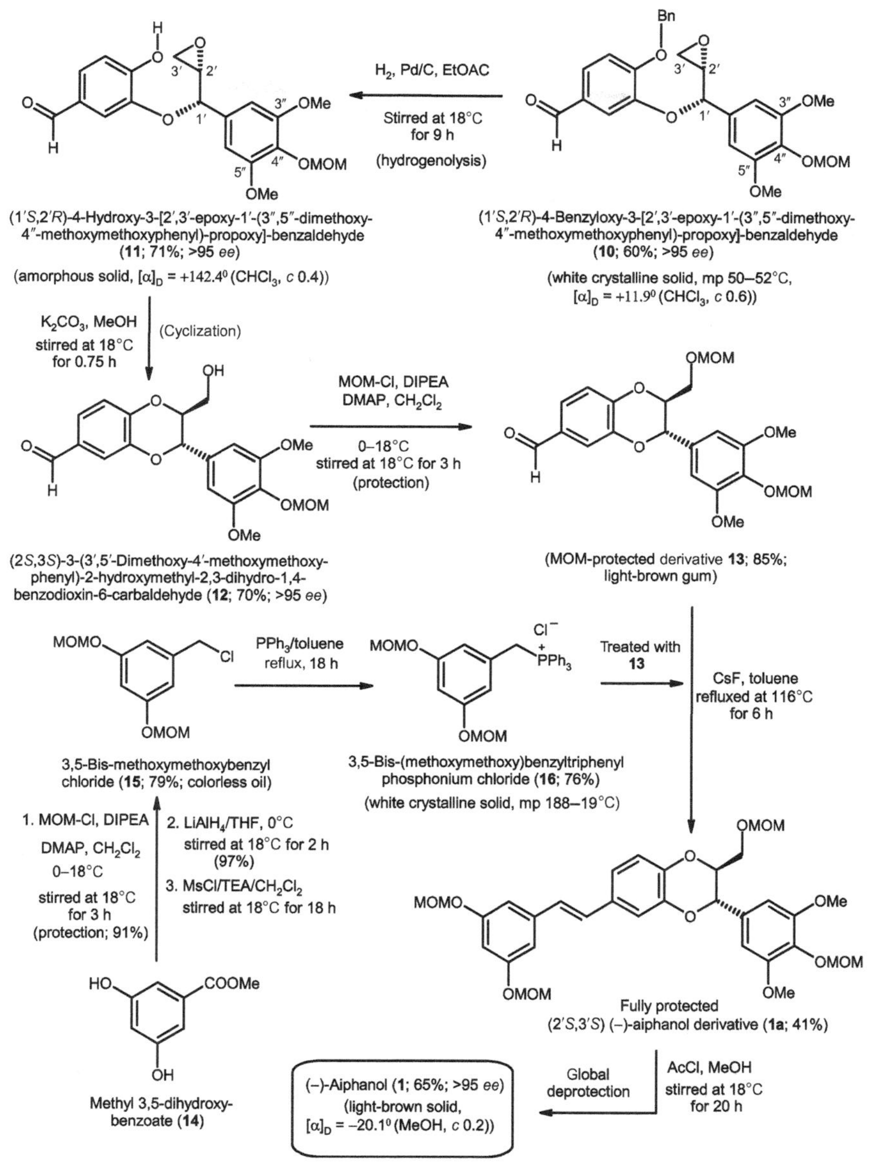

Scheme 1 Banwell's total synthesis of (−)-aiphanol (**1**) [4].

The MOM-ether derivative **3** of commercially available syringealdehyde (**2**) was first made to undergo Horner-Wadsworth-Emmon reaction [5] with triethyl phosphonoacetate in the presence of sodium hydride in THF to produce the *E*-configured α,β-unsaturated ester **4** in 79% yield over two steps. DIBAL-H reduction of the ester functionality in compound **4** furnished *E*-3,5-dimethoxy-4-(methoxymethoxy)cinnamyl alcohol **5** as white crystalline solid (81% yield), which on subsequent Sharpless asymmetric dihydroxylation [6] with AD mix-β afforded the (1*R*,2*R*)-triol derivative **6** (>95 *ee*) in 80% yield. This triol derivative **6** was then converted, via its tosylate **7**, into an epoxy-alcohol **8** in 63% yield and >95% *ee* as established by chiral HPLC analysis.

The investigators then carried out a Mitsunobu coupling [7] of the epoxy-alcohol **8** with 4-benzyloxy-3-hydroxybenzaldehyde (**9**) using DIAD and PPh_3 to obtain the (1′*S*,2′*R*)-adduct **10** as a white crystalline solid in 60% yield and >95% *ee*. Removal of the benzyl group in compound **10** was then carried out by hydrogenolysis over 5% Pd on C and using ethyl acetate as solvent to afford (1′*S*,2′*R*)-epoxide **11** (71% yield), which underwent smooth cyclization under the influence of K_2CO_3 in methanol to furnish compound (2*S*,3*S*)-3-(3′,5′-dimethoxy-4′-methoxymethoxyphenyl)-2-hydroxymethyl-2,3-dihydro-1,4-benzodioxin-6-carbaldehyde (**12**) in 70% yield, and in an enantiomeric purity (*ee*) of >95%. The MOM-ether derivative (**13**) of this 1,4-dioxane compound **12** was then refluxed with the phosphonium salt (**16**) [8] in the presence of cesium fluoride (CsF) in toluene to obtain the fully protected (2′*S*,3′*S*)-aiphanol derivative (**1a**) in 41% yield. Global removal of the MOM in **1a** was then achieved using MeOH and AcCl; the crude reaction product on preparative HPLC resolution afforded (2′*S*,3′*S*)-(−)-aiphanol (**1**) in 65% yield and >95% *ee*, as light-brown solid, $[\alpha]_D = -20.1°$ (MeOH, *c* 0.2) (Lit. $[\alpha]_D = -21.8°$ (MeOH, *c* 0.1)) [1]. Physical and spectral data of the synthesized compound were found to be consistent with the data reported for the natural product [1] and essentially identical to those obtained for the racemic material [2].

References

[1] D. Lee, M. Cuendet, J.S. Vigo, J.G. Graham, F. Cabieses, H.H.S. Fong, J.M. Pezzuto, A.D. Kinghorn, Org. Lett. 3 (2001) 2169.
[2] M.G. Banwell, A. Bezos, S. Chand, G. Dannhardt, W. Kiefer, U. Nowe, C.R. Parish, G.P. Savage, H. Ulbrich, Org. Biomol. Chem 1 (2003) 2427.
[3] X.L. Wang, J.P. Feng, X.G. Xie, X.P. Cao, X.F. Pan, Chin. Chem. Lett. 15 (2004) 1036.
[4] M.G. Banwell, S. Chand, G.P. Savage, Tetrahedron Asymmetry 16 (2005) 1645.
[5] B.E. Maryanoff, A.B. Reitz, Chem. Rev. 89 (1989) 863.
[6] H.C. Kolb, M.S. VanNiewenhze, K.B. Sharpless, Chem. Rev. 94 (1994) 2483.
[7] O. Mitsunobu, Synthesis (1981) 1 (and references cited therein).
[8] A. Kuboki, T. Yamamoto, S. Ohira, Chem. Lett. 32 (2003) 420.

CHAPTER FIVE

(+)-Amphidinolide T1

Abbreviations

Ar argon
$(ClCO)_2$ oxalyl chloride
Bn benzyl
BnOH benzylalcohol
CH_2Cl_2 dichloromethane
Cp_2TiMe_2 dimethyltitanocene (Petasis reagent)
DAMP 4-dimethylaminopyridine
DIBAL diisobutylaluminum hydride
DMSO dimethylsulfoxide
DTBMP 2,6-di-*tert*-butyl-4-methylpyridine
Et_3N triethylamine
EtOAc ethyl acetate
HMPA hexamethylphosphoramide
HSO_2Ph phenylsulfinic acid
***i*-Pr_2NEt** diisopropylethylamine
LAH lithium aluminum hydride
LiHMDS lithium hexamethyl disilazide
***n*-BuLi** *n*-butyllithium
NMO *N*-methylmorpholine-*N*-oxide
***p*-TsOH** *p*-toluenesulfonic acid
Py pyridine
TBS/TBDMS *tert*-butyldimethylsilyl
THF tetrahydrofuran
$TiCl_4$ titanium(IV) chloride
TIPS triisopropylsilyl
$TMS(CH_2)_2OH$ trimethylsilylethanol
TPAP tetra-*n*-propylammonium perruthenate
Tris-Cl/Tris-HCl tris(hydroxymethyl)aminomethane hydrochloride [$NH_2C(CH_2OH)_3 \cdot HCl$]
Ts 4-methylphenylsulfonyl (*p*-toluenesulfonyl)

Systematic name: (1*S*,6*S*,9*R*,13*R*,14*S*,17*R*,19*S*)-14-hydroxy-6,13,19-trimethyl-11-methylene-9-propyl-8,20-dioxabicyclo[15.2.1]icosane-7,15-dione

https://doi.org/10.1016/B978-0-08-102822-3.00005-5

Compound class: A 19-membered macrolide
Structure:

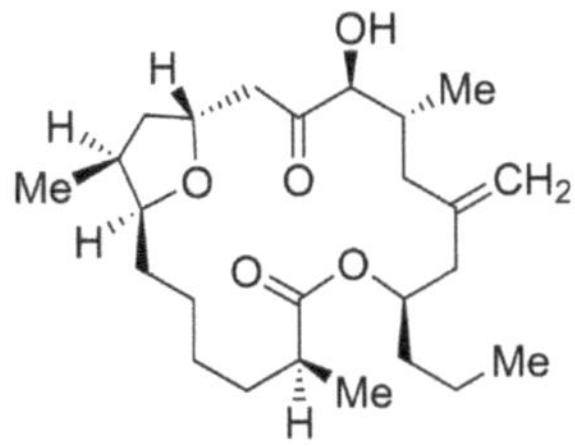

(+)-Amphidinolide T1 (**1**)

Natural source: Y-56 strain of the marine dinoflagellate *Amphidinium* sp. [1–3]

Pharmaceutical potential: Anticancer (showed potent activity against murine lymphoma L1210 and human epidermoid carcinoma KB cell lines) [2, 4, 5]

Synthetic route:

The significant biological activity, low abundance, and unique structural features of amphidinolide T1 (**1**) attracted the synthetic community for its total synthesis, and as a result four total syntheses have appeared so far in the literature. The first total synthesis of amphidinolide T1 was accomplished by Ghosh and Liu [6]; later on the other three total syntheses were reported by Fürstner et al. [7], Jamison et al. [8], and Yadav and Reddy [9] using different strategies to effect the macrocycle formation. The first total synthesis of amphidinolide T1 (**1**) reported by Ghosh and Liu is outlined herein.

The retrosynthetic approach conceived by the investigators is outlined in Scheme 1. Their stereocontrolled and convergent synthetic strategy for the macrolide **1** involves the assembly of the C_1–C_{10} sulfone segment **2** and the C_{11}–C_{21} segment **3** by an oxocarbenium ion-mediated alkylation and macrolactonization sequence. They planned to synthesize fragment **2** using cross-metathesis and a hydrogenation sequence between the terminal olefins of **5** and **6**, and enol ether **4** was designed to be the surrogate of fragment **3**.

The synthesis of C_1–C_{10} segment **2** is outlined in Scheme 2. Aldol condensation of (1*R*,2*S*)-(4-methylphenylsulfonamido)-2,3-dihydro-1*H*-inden-2-yl propionate (**7**) with 3-(benzyloxy)propanal (**8**) at −78°C furnished the aldol adduct **9** as a single diastereomer in 90% yield [10, 11].

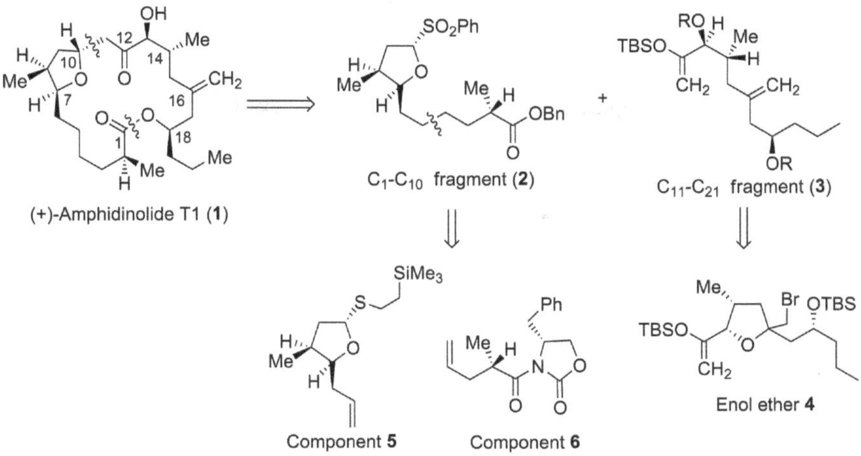

Scheme 1 Ghosh's retrosynthetic approach for (+)-amphidinolide T1 [6].

LAH reduction of **9** followed by selective sulfonylation of the primary hydroxyl group of diol **10**, displacement of the resulting sulfonate with cyanide, and acid-promoted lactonization ultimately afforded **11** in 81% overall yield as colorless oil. DIBAL reduction of the γ-lactone **11** followed by its reaction with trimethylsilylethanol and *p*-TsOH gave acetal **12** and its isomer as a 3.5:1 mixture, which was separated by column chromatography after removal of the benzyl group by hydrogenolysis. Swern oxidation of the resulting alcohol followed by Wittig olefination furnished alkene **5**, one of the substrates for cross-metathesis. Afterward the investigators carried out cross-metathesis reaction between the terminal alkenes **5** and **6** [12] to form the C_4–C_5 carbon–carbon bond [13] in the presence of 5 mol% of second-generation Grubbs' catalyst [14] whereby a 1:1 mixture (*E*:*Z*) of cross-metathesis product **13** was obtained. Catalytic hydrogenation of the alkene mixture followed by treatment of the saturated derivative with lithium phenylmethoxide furnished benzyl ester **14** in 85% yield. Exposure of **14** to phenylsulfinic acid and $CaCl_2$ afforded sulfone derivative **2** and its isomer as a 7:1 mixture in 95% yield [15].

Synthesis of the C_{11}–C_{21} segment is outlined in Scheme 3. Aldol reaction of (1*S*,2*R*)-(4-methylphenylsulfonamido)-2,3-dihydro-1*H*-inden-2-yl propionate (**15**) with benzyloxyacetaldehyde (**16**) afforded a single *syn*-diastereomer (**17**) in 95% yield [10, 11]. Protection of the resulting alcohol **17** as a TIPS ether followed by DIBAL-H reduction produced alcohol **18**, which was readily converted into the corresponding iodide (**19**). Preparation of aldehyde **21** was achieved from glycidyl tosylate **20** [14]. Treatment of

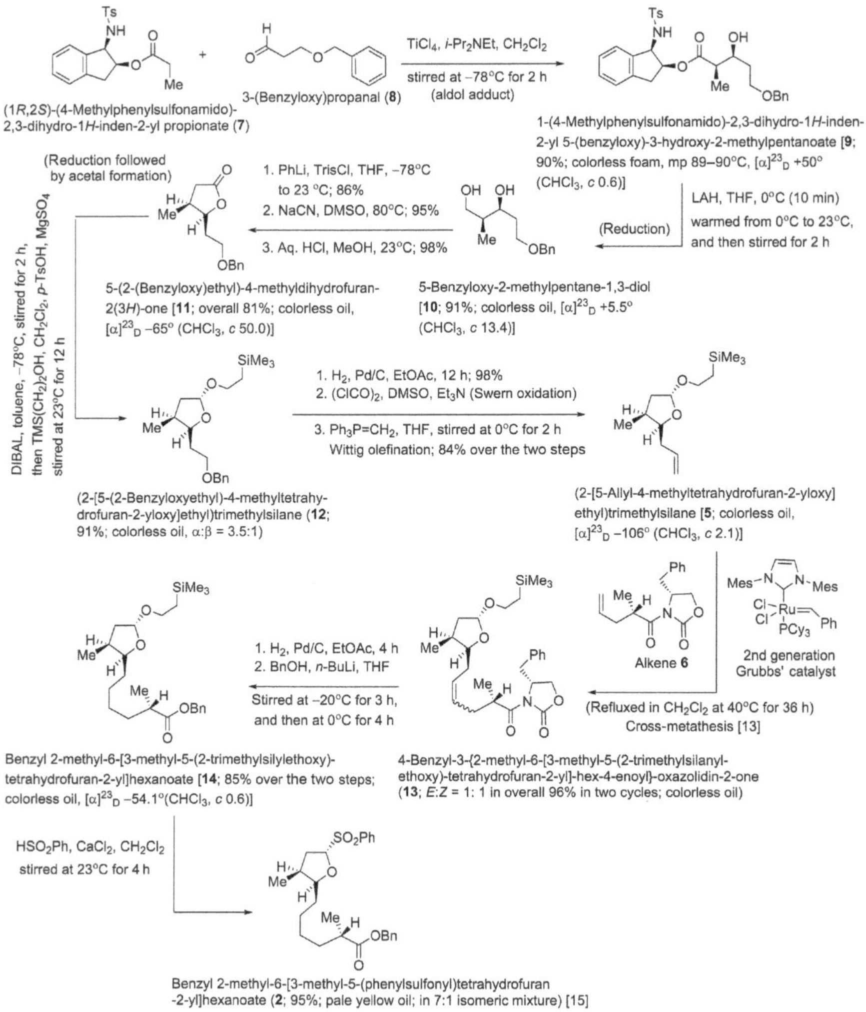

Scheme 2 Synthesis of the C_1–C_{10} segment (**2**) [6].

iodide **19** with *t*-BuLi generated the corresponding alkanyllithium, which was reacted with aldehyde **21** to obtain a 1:1 diastereomeric mixture of alcohols; this mixture upon TPAP oxidation yielded ketone **22** [16]. Olefination of **22** utilizing Petasis conditions [17] furnished alkene **23**. Reductive removal of the benzyl and TIPS ethers [18] followed by reaction of the resulting diol with NBS eventually resulted in the formation of bromotetrahydrofuran **25** as a 3:1 diastereomeric mixture. The investigators prepared this bromolactone to protect the C_{13}-alcohol as well as to protect

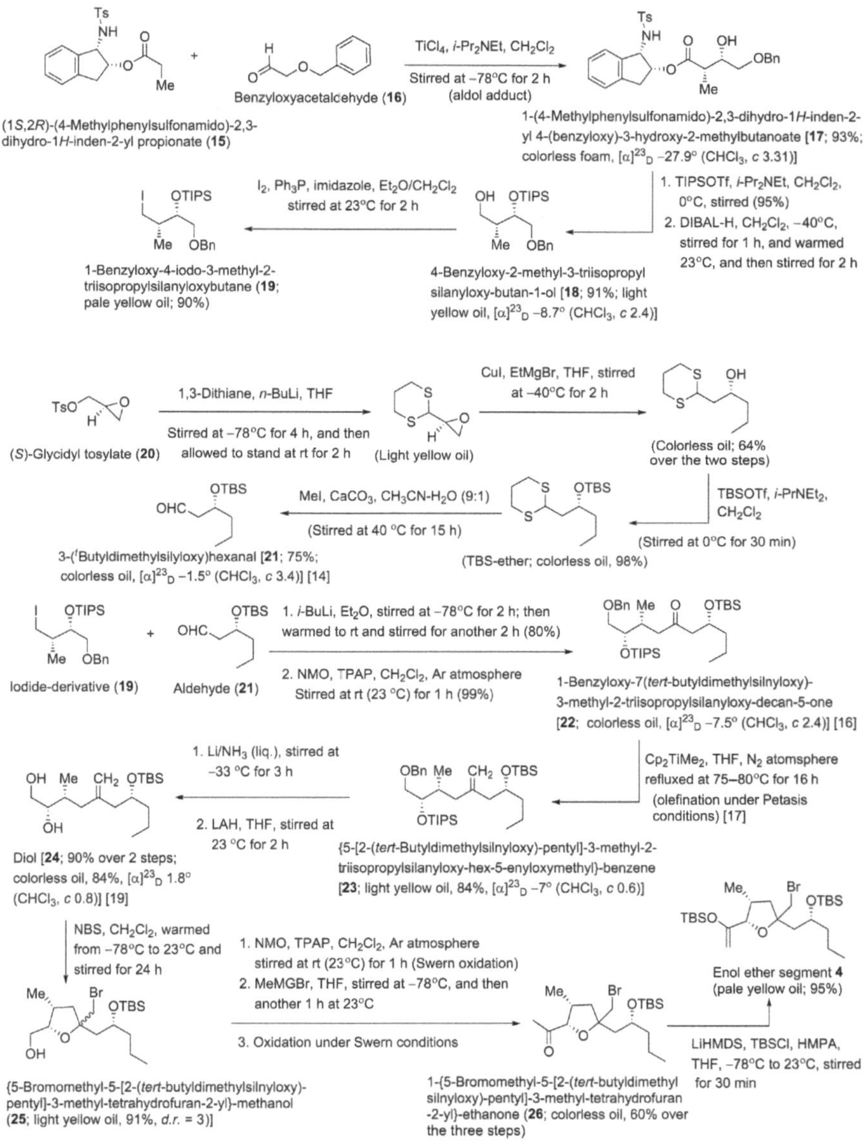

Scheme 3 Synthesis of the C_{11}–C_{21} segment (**4**) [6].

the sensitive *exo*-methylene group during the oxocarbenium ion-mediated alkylation process. The hydroxymethyl group of **25** was then converted to methyl ketone **26** in 60% overall yield. Treatment of **26** with LiHMDS followed by reaction of the resulting enolate with TBSCl afforded the vinyl ether segment (**4**) as pale yellow oil.

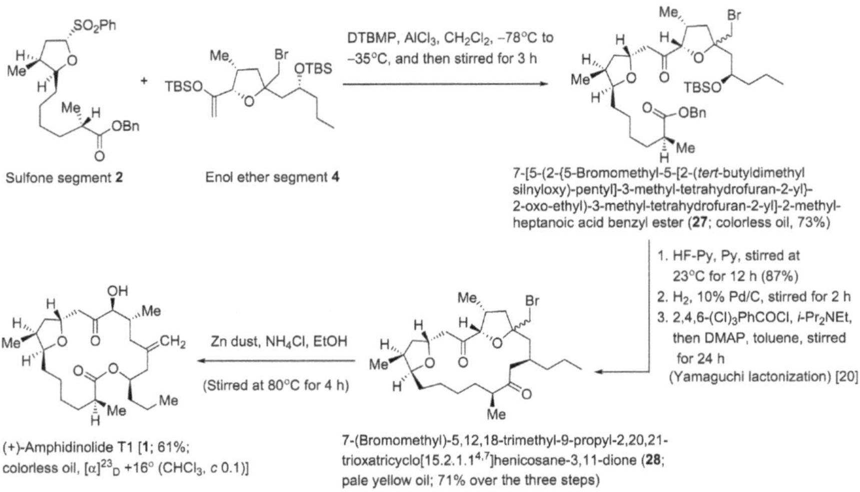

Scheme 4 Assembly of segments **2** and **4** to form (+)-amphidinolide T1 (**1**) [6].

The assembly of the sulfone segment **2** and enol ether segment **4** is depicted in Scheme 4. The investigators achieved this assembly of fragments **2** and **4** by an oxocarbenium ion-mediated alkylation reaction using modified Ley's protocol [19]. Treatment of **2** and **4** in the presence of excess $AlCl_3$ (6 equiv.) and DTBMP (1.2 equiv.) at −35°C resulted in the coupling product **27** as colorless oil in 73% yield as a single isomer. The C_{18}-silyl ether was then removed by exposure to HF-Py, and subsequent hydrogenolysis removed the benzylester. Macrolactonization of the resulting hydroxy acid under Yamaguchi conditions [20] afforded macrolactone **28** in 71% yield over three steps. Reductive unmasking of the bromoether with Zn dust and NH_4Cl in ethanol provided amphidinolide T1 (**1**), whose spectral data were found to be in agreement with those of natural compound [1–3].

References

[1] M. Tsuda, T. Endo, J. Kobayashi, J. Org. Chem. 65 (2000) 1349.
[2] J. Kobayashi, T. Kubota, T. Endo, M. Tsuda, J. Org. Chem. 66 (2001) 134.
[3] T. Kubota, T. Endo, M. Tsuda, M. Shiro, J. Kobayashi, Tetrahedron 57 (2001) 6175.
[4] T.K. Chakraborty, S. Das, Curr. Med. Chem. Anti-Cancer Agents 1 (2001) 131 (Review).
[5] J. Kobayashi, M. Tsuda, Nat. Prod. Rep. 21 (2004) 77 (Review).
[6] A.K. Ghosh, C. Liu, J. Am. Chem. Soc. 125 (2003) 2374.
[7] A. Fürstner, C. Aissa, J. Ragot, J. Am. Chem. Soc. 125 (2003) 15512.
[8] E.A. Colby, K.C. O'Brien, T.F. Jamison, J. Am. Chem. Soc. 127 (2005) 4297.
[9] Y.S. Yadav, C.S. Reddy, Org. Lett. 11 (2009) 1705.

[10] A.K. Ghosh, S. Fidanze, M. Onishi, K.A. Hussain, Tetrahedron Lett. 38 (1997) 7171.
[11] A.K. Ghosh, S. Fidanze, C.H. Senanayake, Synthesis (1998) 937.
[12] D. Schinzer, A. Bauer, J. Schiber, Chem. Eur. J. 5 (1999) 2492.
[13] H.E. Blackwell, D.J. O'Leary, A.K. Chatterjee, R.A. Washenfelder, D.A. Bussmann, R.H. Grubbs, J. Am. Chem. Soc. 122 (2000) 58 (and references therein).
[14] F. Yokokawa, T. Asano, T. Shioiri, Org. Lett. 26 (2000) 4169.
[15] L.A. Paquette, J. Tae, J. Org. Chem. 61 (1996) 7860.
[16] E.J. Corey, D.C. Ha, Tetrahedron Lett. 29 (1998) 3171.
[17] N.A. Petasis, E.I. Bzowej, J. Am. Chem. Soc. 112 (1990) 6394.
[18] E.J. Corey, G.B. Jones, J. Org. Chem. 57 (1992) 1028.
[19] S.V. Ley, B. Lygo, A. Wonnacott, Tetrahedron Lett. 26 (1989) 535.
[20] J. Inanaga, K. Hirata, H. Saeki, T. Katsuki, M. Yamaguchi, Bull. Chem. Soc. Jpn. 52 (1979) 1989.

CHAPTER SIX

Ancistroealaine A and Ancistrotanzanine B

Abbreviations

Ar argon
$B(OMe)_3$ trimethyl borate
CH_2Cl_2 dichloromethane
LAH lithium aluminum hydride
***n*-BuLi** *n*-butyllithium
$Pd_2(dba)_3$ tris(dibenzylideneacetone)dipalladium(0)
Ph_3P triphenylphosphine
$POCl_3$ phosphoryl chloride/phosphorus oxychloride
PTC phase-transfer catalyst
THF tetrahydrofuran

Systematic name: (3*S*)-5-(4,5-Dimethoxy-7-methylnaphthalen-1-yl)-6,8-dimethoxy-1,3-dimethyl-3,4-dihydroisoquinoline (Ancistroealaine A: *P*-isomer; Ancistrotanzanine B: *M*-isomer)

Compound class: Naphthylisoquinoline alkaloids

Structure:

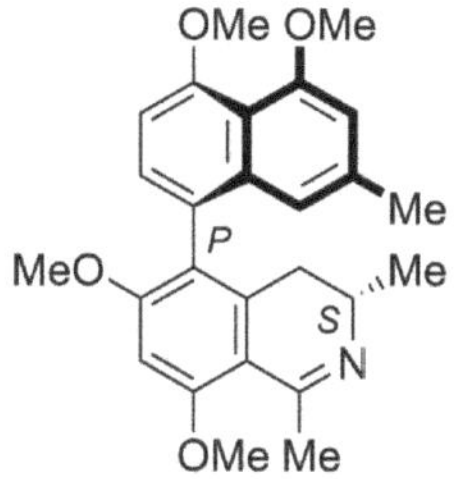

Ancistroealaine A (**1**)

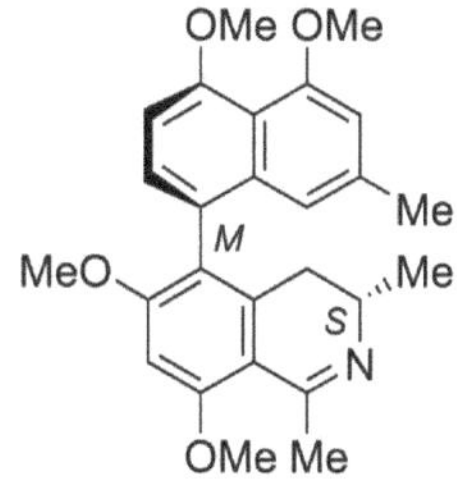

Ancistrotanzanine B (**2**)

Natural sources:

Ancistroealaine A (**1**): *Ancistrocladus ealaensis* (the Central African liana; family: Ancistrocladaceae) [1]

https://doi.org/10.1016/B978-0-08-102822-3.00006-7

Ancistrotanzanine B (**2**): *A. tanzaniensis* (the East African species; leaves) [2]

Pharmaceutical potential: Antileishmanials [1, 2]

Synthetic route:

The first total synthesis of the naphthylisoquinoline alkaloid ancistroealaine A (**1**) and its atropo-diastereomer ancistrotanzanine B (**2**) was reported by Bringmann et al. in 2003 (Schemes 1 and 2) [3]. The key step is to construct the rotationally hindered and thus stereogenic biaryl

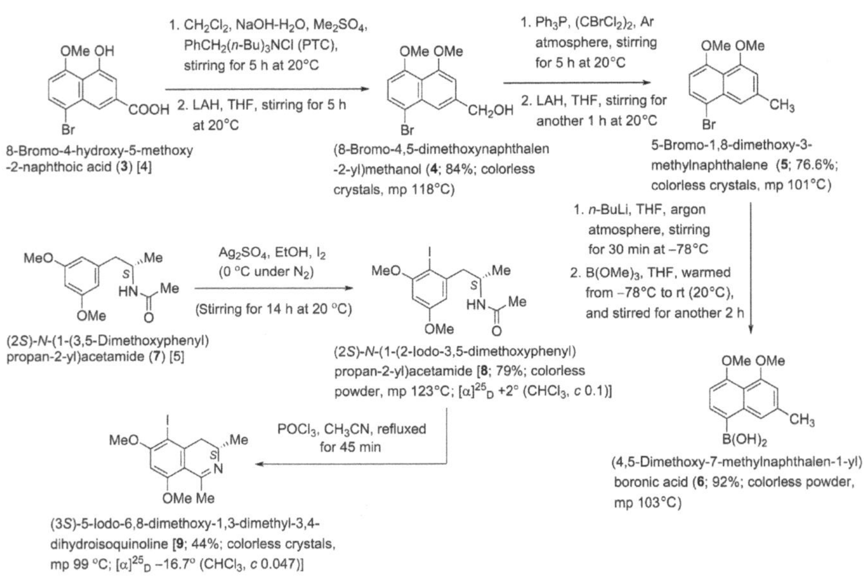

Scheme 1 Preparation of Suzuki precursors **6** and **9** [3].

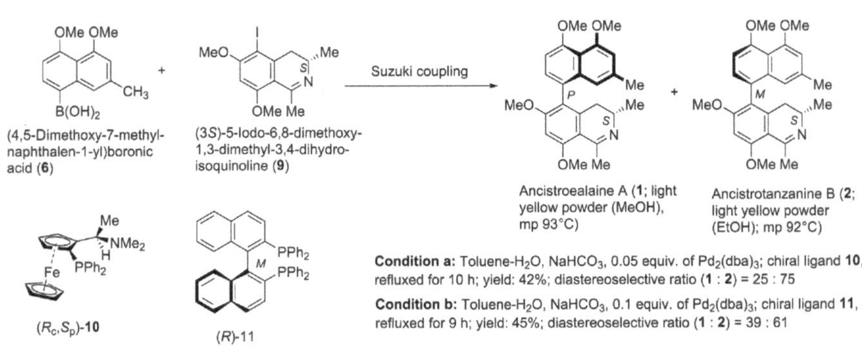

Scheme 2 Suzuki coupling between precursors **6** and **9** to produce ancistroealaine A (**1**) and ancistrotanzanine B (**2**) [3].

axis by Suzuki coupling between the naphthalene building block **6** and the halogen-substituted dihydroisoquinoline **9** (as heterocyclic portion).

Naphthylboronic acid (**6**) was prepared from 8-bromo-4-hydroxy-5-methoxy-2-naphthoic acid (**3**) by phase-transfer-catalyzed *O*-methylation of the phenolic OH-function, LAH reduction of the carboxylate, and deoxygenation of the resulting primary alcohol **4** (by hydroxy-halogen exchange and subsequent renewed alanate reduction), followed by lithiation and subsequent reaction with trimethyl borate. The brominated dihydroisoquinoline (**9**) was prepared from the known acetamide **7** [5], which was iodinated and cyclized using the Bischler-Napieralski reaction (Scheme 1). In the next step, the Suzuki coupling reaction between precursors **6** and **9** was carried out under various conditions using chiral palladium catalyst and chiral ligands. Under condition "a" (optimized condition) in the presence of chiral ligand **10**, ancistroealaine A (**1**) and ancistrotanzanine B (**2**) were obtained in overall 42% yield with a diastereoselective ratio of 25:75; conversely, under condition "b" (another optimized condition) in the presence of chiral ligand **11**, the naphthylisoquinoline alkaloids (**1** and **2**) were obtained in overall 45% yield with a diastereoselective ratio of 39:61 (Scheme 2). The physical and spectral data of the synthetic compounds **1** and **2** exactly matched those of the natural alkaloids [1, 2].

References

[1] G. Bringmann, A. Hamm, C. Günther, M. Michel, R. Brun, V. Mudogo, J. Nat. Prod. 63 (2000) 1465.

[2] G. Bringmann, M. Dreyer, J. Faber, P.W. Dalsgaard, D. Stærk, J. Jaroszewski, H. Ndangalasi, F. Mbago, R. Brun, M. Reichert, K. Maksimenka, S.B. Christensen, J. Nat. Prod. 66 (2003) 1159.

[3] G. Bringmann, A. Hamm, M. Schraut, Org. Lett. 16 (2003) 2805.

[4] G. Bringmann, R. Götz, S. Harmsen, J. Holenz, R. Walter, Liebigs Ann. Chem. (1996) 2045.

[5] G. Bringmann, R. Weirich, H. Reuscher, J.R. Jansen, L. Kinzinger, T. Ortmann, Liebigs Ann. Chem. (1993) 877.

CHAPTER SEVEN

(−)-Andrographolide

Abbreviations

DDQ 2,3-dichloro-5,6-dicyanobenzoquinone
DMF *N,N*-dimethylformamide
DMSO dimethylsulfoxide
HMPA hexamethylphosphoramide
LDA lithium diisopropylamide
$NaBH_4$ sodium borohydride
Ph_3P triphenylphosphine
PPTs pyridinium *p*-toluenesulfonate
rt room temperature
TBAF tetra-*n*-butylammonium fluoride
TBSCl *tert*-butyldimethylsilyl chloride
THF tetrahydrofuran

Systematic name: (*S*,*E*)-4-Hydroxy-3-(2-((1*R*,4a*S*,5*R*,6*R*,8a*S*)-6-hydroxy-5-(hydroxymethyl)-5,8a-dimethyl-2-methylenedecahydronaphthalen-1-yl)ethylidene)dihydro-furan-2(3*H*)-one

Compound class: *ent*-Labdane diterpenoid

Structure:

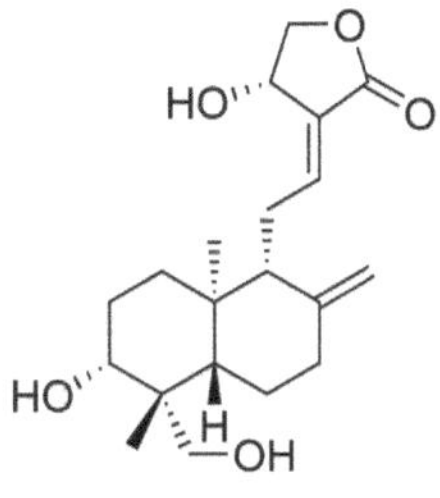

(−)-Andrographolide (**1**)

Natural source: *Andrographis paniculata* Nees (mainly in leaves; family: Acanthaceae) [1–5]

Pharmaceutical potentials: Antiinflammatory [6–17], antioxidant [10, 18, 19], anticancer and antitumor [20–25], immunomodulatory [26],

https://doi.org/10.1016/B978-0-08-102822-3.00007-9

hepatoprotective [27–30], antiplatelet aggregation [31, 32], antidiabetic [33], antiviral [34], and a few other activities also [35–41].

Synthetic route:

The first total synthesis of (−)-andrographolide (**1**) was achieved by Li and coworkers [42] via the biomimetic cyclization (cation-olefin annulation) of an epoxy homoiodo allylsilane precursor **14** prepared starting from geraniol (**2**) (Scheme 1). 3,7-Dimethyl-8-((tetrahydro-2*H*-pyran-2-yl)oxy) octa-2,6-dien-1-ol (**8**) was first prepared from geraniol (**2**), in an overall

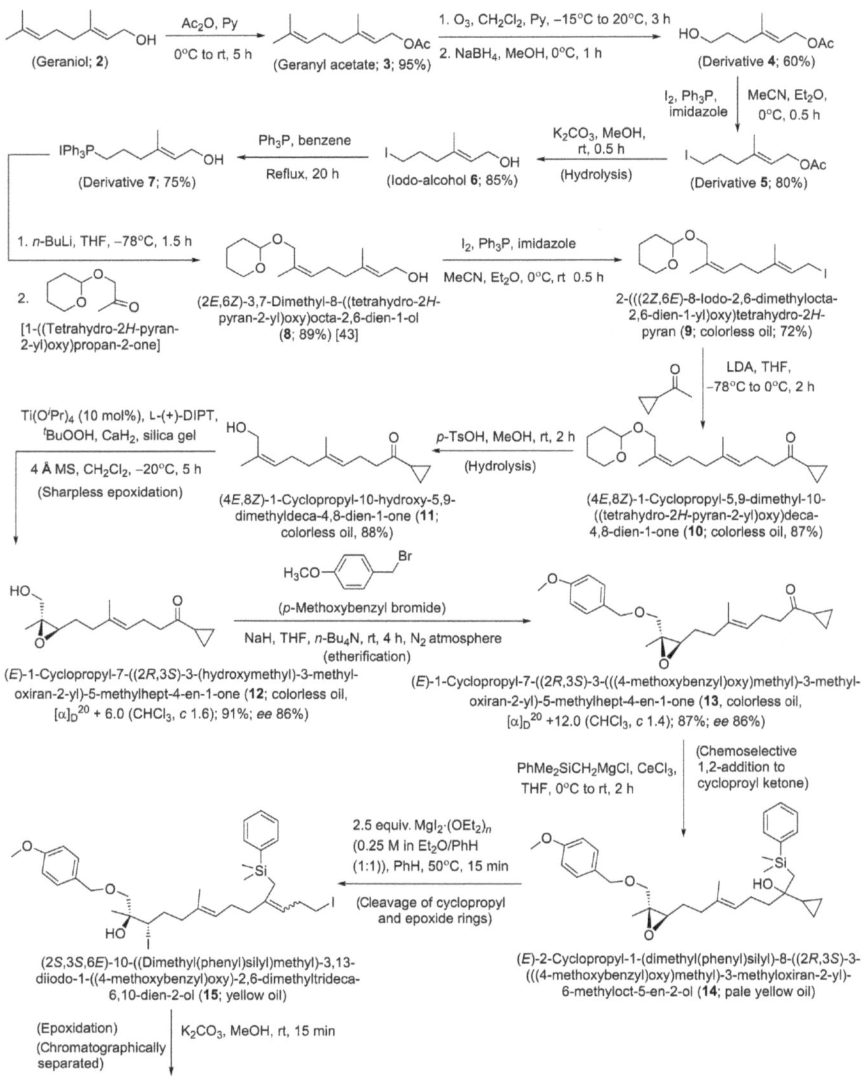

Scheme 1—Cont'd

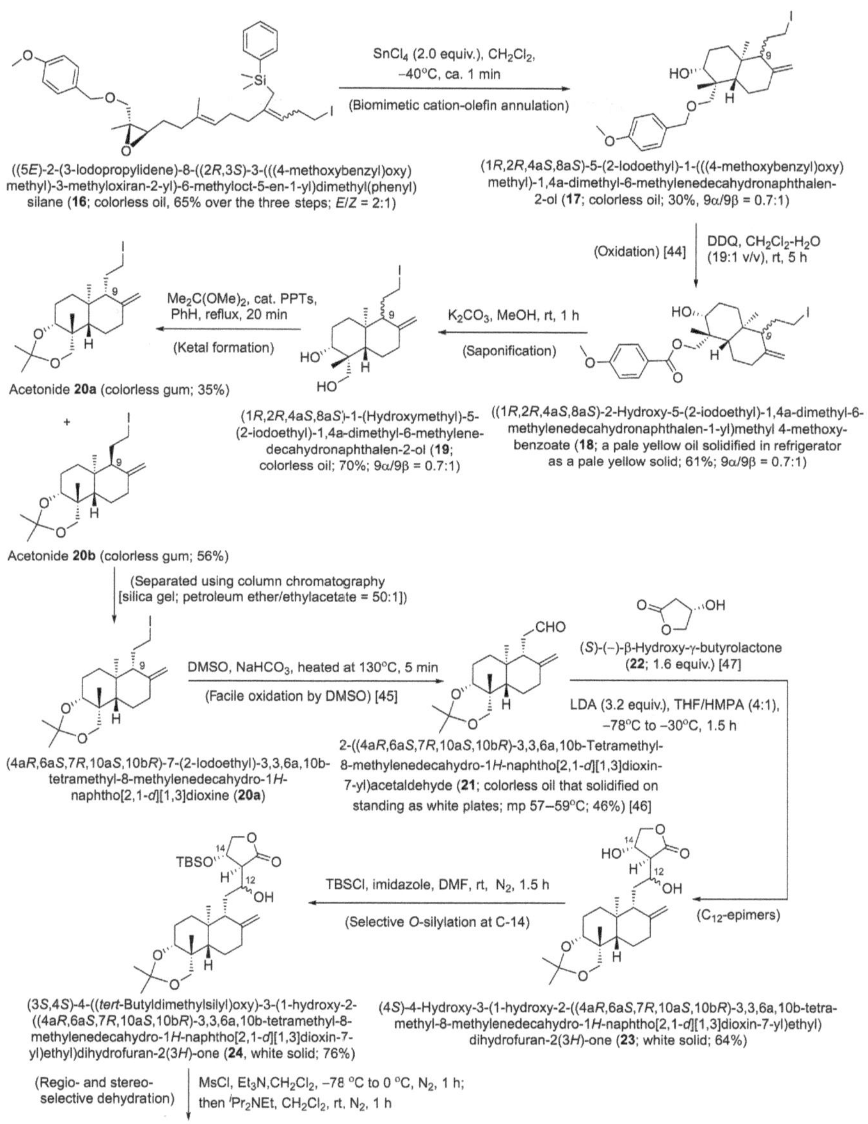

Scheme 1—Cont'd

(Continued)

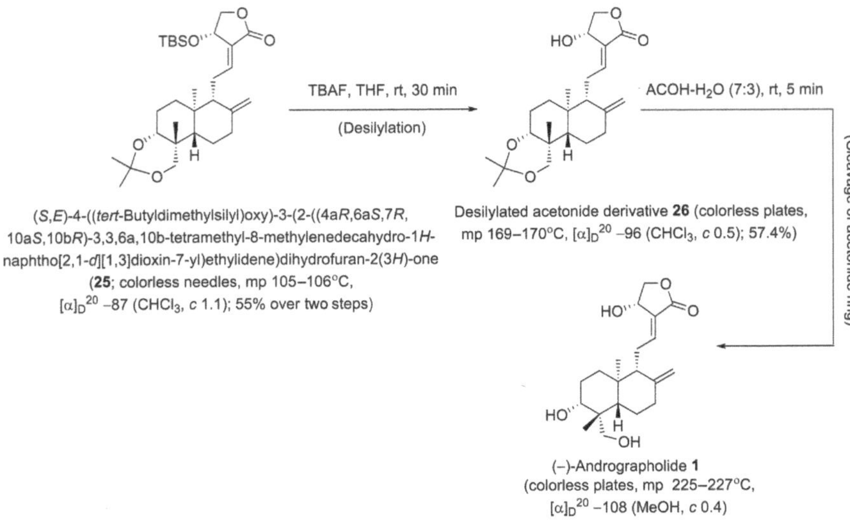

Scheme 1 Li's total synthesis of (−)-andrographolide (**1**) [42].

yield of 26% over eight steps, following standard procedure [43]. The allylic alcohol **11**, obtained from **8**, was then subjected to Sharpless epoxidation to generate the epoxide **12** (colorless oil), which underwent etherification with *p*-methoxybenzyl bromide to provide epoxide **13** (86% *ee*) in 79% yield over the two steps. The investigators developed the optimized three-step route to synthesize the homoiodo allylsilane **16** from cyclopropyl ketone **13**, involving (1) chemoselective 1,2-addition of (phenyldimethylsilyl)methylcerium chloride to **13** in THF at 0°C, (2) exposure of the resulting cyclopropyl carbinol **14** to freshly prepared MgI_2 etherate to cleave both the cyclopropyl and epoxide rings to have bis-iodo allylsilane intermediate **15**, and (3) treatment of the crude intermediate **15** with potassium carbonate in methanol to furnish the epoxy homoiodo allylsilane **16** in 65% yield from **13** as a mixture of geometric isomers (*E*/*Z* 2:1).

In the next step, this epoxy homoiodo allylsilane precursor **16** was subjected to undergo the biomimetic cation-olefin annulation on treating with freshly distilled $SnCl_4$ in dichloromethane and subsequent aqueous work-up to form a bicyclic alcohol **17** as a mixture of C_9 epimers ($\alpha/\beta = 0.7:1$) as colorless oil in 30% yield. This crude product was then converted into the readily separable acetonides **20a** and **20b** by a sequence of reactions that involved: (1) DDQ oxidation in aqueous dichloromethane at room temperature conditions to give the ester **18** in 61% yield [44], (2) saponification of the resulting ester derivative **18** to form the iodo-diol, 1-(hydroxymethyl)-5-(2-iodoethyl)-1,4a-dimethyl-6-methylene-decahydronaphthalen-2-ol

(**19**; colorless oil; 70%; 9α/9β=0.7:1), and (3) reacetonization of the corresponding diol **19**, followed by column chromatographic separation of the acetonide **20a**. Compound **20a** underwent facile oxidation with DMSO on heating at 130°C [45] to afford 2-((4a*R*,6a*S*,7*R*,10a*S*,10b*R*)-3,3,6a,10b-tetramethyl-8-methylenedecahydro-1*H*-naphtho[2,1-*d*][1,3] dioxin-7-yl)acetaldehyde (**21**; colorless oil that solidified on standing as white plates; mp 57–59°C; 46%) [46], which served as the key intermediate for the attachment of the B-ring lactone side chain. This aldehydic derivative **21** on aldol condensation with the corresponding lithium enolate of (*S*)-(−)-β-hydroxybutyrolactone (**22**) [47] gave dihydroxy lactone **23** (as a mixture of C_{12} epimers), which was selectively *O*-silyated at C-14 and dehydrated regio- and stereoselectively via the corresponding mesylate intermediate to give the *E*-configurated lactone **25** in 55% yield over the two steps as colorless needles. Finally, the investigators were successful in isolating (−)-andrographolide (**1**) on standard desilylation and acetonide cleavage of **25**, as colorless crystals (mp 225–227°C) from methanol, which was found to be identical spectroscopically with natural andrographolide [3].

References

[1] M.K. Gorter, Rec. Trav. Chim. 30 (1911) 151.
[2] R.N. Chakravarti, D. Chakravarti, Indian Med. Gaz. 86 (1951) 96.
[3] T. Fujita, R. Fujitani, Y. Takeda, Y. Takaishi, T. Yamada, M. Kido, I. Miura, Chem. Pharm. Bull. 32 (1984) 2117.
[4] M. Rajani, N. Shrivastava, M.N. Ravishankara, Pharm. Biol. 38 (2000) 204.
[5] P. Kulyal, U.K. Tiwari, A. Shukla, A.K. Gaur, Indian J. Chem. 49B (2010) 356.
[6] W.-F. Chiou, C.-F. Chen, J.-J. Lin, Br. J. Pharmacol. 129 (2000) 1553.
[7] J. Sinha, S. Mukhopadhyay, N. Das, M.K. Basu, Drug Deliv. 7 (2000) 209.
[8] Y.-C. Shen, C.-F. Chen, W.-F. Chiou, Br. J. Pharmacol. 135 (2002) 399.
[9] T. Wang, B. Liu, W. Zhang, B. Wilson, J.-S. Hong, J. Pharmacol. Exp. Ther. 308 (2004) 975.
[10] K. Sheeja, P.K. Shihab, G. Kuttan, Immunopharmacol. Immunotoxicol. 28 (2006) 129.
[11] A.A. Abu-Ghefreh, H. Canatan, C.I. Ezeamuzie, Int. Immunopharmacol. 9 (2009) 313.
[12] Z. Bao, S. Guan, C. Cheng, S. Wu, S.H. Wong, D.M. Kemeny, B.P. Leung, W.S.F. Wong, Am. J. Respir. Crit. Care Med. 179 (2009) 657.
[13] J. Li, L. Luo, X. Wang, B. Liao, G. Li, Cell Mol. Immunol. 6 (2009) 381.
[14] S. Suebsasama, P. Pongnaratorn, J. Sattayasai, T. Arkaravichien, S. Tiamkao, C. Aromdee, Arch. Pharm. Res. 32 (2009) 1191.
[15] J. Levita, A. Nawawi, A. Mutholib, S. Ibrahim, J. Appl. Sci. 10 (2010) 1481.
[16] C.V. Chandrasekaran, A. Gupta, A. Agarwal, J. Ethnopharmacol. 129 (2010) 203.
[17] W.W. Chao, Y.H. Kuo, B.F. Lin, J. Agric. Food Chem. 58 (2010) 2505.
[18] G.A. Akowuah, I. Zharik, A. Mariam, Food Chem. Toxicol. 46 (2008) 3616.
[19] F.L. Lin, S.J. Wu, S.C. Lee, L.T. Ng, Phytother. Res. 23 (2009) 958.
[20] S. Rajagopal, R.A. Kumar, D.S. Deevi, C. Satyanarayana, R. Rajagopalan, J. Exp. Ther. Oncol. 3 (2003) 147.

[21] C. Satyanarayana, D.S. Deevi, R. Rajagopalan, N. Srinivas, S. Rajagopal, BMC Cancer 4 (1) (2004).
[22] K.K. Shen, T.Y. Liu, C. Xu, L.L. Ji, Z.T. Wang, Yao. Xue. Xue. Bao. 44 (2009) 973.
[23] Y.C. Lee, H.H. Lin, C.H. Hsu, C.J. Wang, T.A. Chiang, J.H. Chen, Eur. J. Pharmacol. 632 (2010) 23.
[24] Y. Tan, K.H. Chiow, D. Huang, S.H. Wong, Br. J. Pharmacol. 159 (2010) 1497.
[25] J. Zhou, C.N. Ong, G.M. Hur, H.M. Shen, Biochem. Pharmacol. 79 (2010) 1242.
[26] W. Wang, J. Wang, S.F. Dong, C.H. Liu, P. Italiani, S.H. Sun, J. Xu, D. Boraschi, S.P. Ma, D. Qu, Acta Pharmacol. Sinica 31 (2010) 191.
[27] S.S. Handa, A. Sharma, Indian J. Med. Res. 92 (1990) 276.
[28] S.S. Handa, A. Sharma, Indian J. Med. Res. 92 (1990) 284.
[29] P.K.S. Visen, B. Shukla, G.K. Patnaik, B.N. Dhawan, J. Ethnopharmacol. 40 (1993) 131.
[30] I.B. Koul, A. Kapil, Indian Aust. J. Pharm. 26 (1994) 296.
[31] C.Y. Zhang, B.K. Tan, J. Ethnopharmacol. 56 (1997) 97.
[32] E. Amroyan, E. Gabrielian, A. Panossian, G. Wikman, H. Wagner, Phytomedicine 6 (1999) 27.
[33] Z. Zhang, J. Jiang, P. Yu, X. Zeng, J.W. Larrick, Y. Wang, J. Trans. Med. 7 (2009) 62–73.
[34] C. Calabrese, S.H. Berman, J.G. Babish, X. Ma, L. Shinto, M. Dorr, K. Wells, C.A. Wenner, L.J. Standish, Phytother. Res. 14 (2000) 333.
[35] K. Maiti, A. Gantait, K. Mukherjee, B.P. Saha, P.K. Mukherjee, J. Nat. Remed. 6 (2006) 1.
[36] X. Suo, H. Zhang, Y. Wang, Biomed. Chromatogr. 21 (2007) 730.
[37] F. Zhao, E.Q. He, L. Wang, K. Liu, J. Asian Nat. Prod. Res. 10 (2008) 473.
[38] A.P. Raina, A. Kumar, S.K. Pareek, Indian J. Pharm. Sci. 69 (2007) 473.
[39] Jada, S.R., Subur, G.S., Mattews, C., Hamzah, A.S., Lajis, N.H., and Saad, M.S., Phytochemistry, 68, 904.
[40] Y. Wang, J. Wang, Q. Fan, J. Geng, Cell Res. 17 (2007) 933.
[41] G. Brahmachari, Handbook of Pharmaceutical Natural Products, Wiley-VCH, Weinheim, Germany, 2010, pp. 34–38.
[42] H.-T. Gao, B.-L. Wang, W.-D.Z. Li, Tetrahedron 70 (2014) 9436.
[43] (a) K. Tago, M. Arai, H.J. Kogen, Chem. Soc., Perkin Trans. 1 (2000) 2073; (b) S. Amslinger, K. Kis, S. Hecht, P. Adam, F. Rohdich, D. Arigoni, A. Bacher, W. Eisenreich, J. Org. Chem. 67 (2002) 4590; (c) J.P. Malerich, D. Trauner, J. Am. Chem. Soc. 125 (2003) 9554.
[44] (a) S.J. Mickel, G.H. Sedelmeier, D. Niederer, F. Schuerch, G. Koch, E. Kuesters, R. Daeffler, A. Osmani, M. Seeger-Weibel, E. Schmid, A. Hirni, K. Schaer, R. Gamboni, A. Bach, S. Chen, W. Chen, P. Geng, C.T. Jagoe, F.R. Kinder Jr., G.T. Lee, J. McKenna, T.M. Ramsey, O. Repic, L. Rogers, W.-C. Shieh, R.-M. Wang, L. Waykole, Org. Process. Res. Dev. 8 (2004) 107; (b) J.A. Marshall, S.-P. Xie, J. Org. Chem. 60 (1995) 7230.
[45] (a) N. Kornblum, W.J. Jones, G.J. Anderson, J. Am. Chem. Soc. 81 (1959) 4113; (b) A. P. Johnson, A. Pelter, J. Chem. Soc. 520 (1964); (c) T. Oishi, H. Tsuchikawa, M. Murata, M. Yoshida, M. Morisawa, Tetrahedron Lett. 44 (2003) 6387; (d) T. Kai, X.-L. Sun, K.M. Faucher, R.P. Apkarian, E.L. Chaikof, J. Org. Chem. 70 (2005) 2606; (e) G.E. Besong, J.M. Bostock, Angew. Chem. Int. Ed. 44 (2005) 6403; (f) K. Furuta, M. Maeda, Y. Hirata, S. Shibata, K. Kiuchi, M. Suzuki, Bioorg. Med. Chem. Lett. 17 (2007) 5487.
[46] S. Nanduri, V.K. Nyavanandi, S.S.R. Thunuguntla, M. Velisoju, S. Kasu, S. Rajagopal, R.A. Kumar, R. Rajagopalan, J. Iqbal, Tetrahedron Lett. 45 (2004) 4883.
[47] (a) H.-M. Shieh, G.D. Prestwich, J. Org. Chem. 46 (1981) 4319; (b) D.J. Edmonds, K. W. Muir, D.J. Procter, J. Org. Chem. 68 (2003) 3190.

CHAPTER EIGHT

Anolignan A

Abbreviations

Bu_3P tributylphosphine
Bu_4NF tetrabutylammonium fluoride
CH_2Cl_2 dichloromethane
DMAP 4-dimethylaminopyridine
Et_3N triethylamine
Et_3SiH triethylsilane
HCOOH formic acid
Me_3SiCH_2MgCl trimethylsilylmethylmagnesium chloride
Ms methanesulfonyl
***n*-BuLi** *n*-butyllithium
$Pd_2(dba)_3$ tris(dibenzylideneacetone)dipalladium(0)
$PdCl_2(PPh_3)_2$ bis(triphenylphosphine)palladium(II) dichloride
PhLi phenyl lithium
TFA trifluoroacetic acid
THF tetrahydrofuran
$TiCl_4$ titanium (IV) chloride
TMS trimethylsilyl
TsCl *p*-toluene sulfonyl chloride

Systematic name: 4-(3-(Benzo[*d*][1,3]dioxol-5-ylmethyl)-2-methylenebut-3-en-1-yl)benzene-1,3-diol

Compound class: Lignan (dibenzylbutadiene type)

Structure:

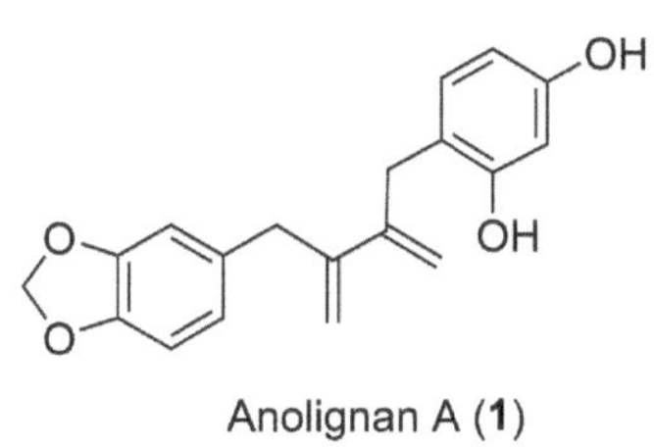

Anolignan A (**1**)

Natural source: *Anogeissus acuminata* (Roxb. ex DC.) Guill. & Perr. var. *lanceolata* Wall ex C.B. Clarke (ground stems; family: Combretaceae) [1]

Total Synthesis of Bioactive Natural Products
https://doi.org/10.1016/B978-0-08-102822-3.00008-0

Pharmaceutical potential: Anti-HIV (HIV-1 reverse transcriptase inhibitor) [1]

Synthetic routes:

The characteristic unsymmetrical 2,3-disubstituted 1,3-butadiene structure within an anolignan A molecule coupled with its interesting biological activity prompted synthetic chemists to investigate the total synthesis of this natural lignoid **1**. The first total synthesis of anolignan A (**1**) was reported by Hatakeyama and his group in eight steps in 30% overall yield from piperonal employing $TiCl_4$-mediated addition of a 1-trimethylsilyl-2,3-butadiene derivative to an aldehyde as the key step (Scheme 1).

Propargyl alcohol **5** was first prepared from piperonal (**2**) via acetate **4** by a three-step sequence involving propargylation using tetrahydro-2-(2-propynyloxy)-2*H*-pyran (**3**), acetylation, and removal of the benzylic acetoxy group with TFA-Et_3SiH in 65% overall yield. The tosylate derivative **6** on reaction with trimethylsilylmethylmagnesium chloride (Me_3SiCH_2MgCl) in the presence of cuprous cyanide and lithium chloride produced 1-trimethylsilyl-2,3-butadiene derivative **7** selectively in good yield [3], which then underwent a smooth $TiCl_4$-mediated addition to 2,4-diacetoxybenzaldehyde (**8**) in the presence of acetonitrile with concomitant acyl migration to give 1,3-butadiene derivative **9**. Upon treatment of compound **9** with HCOOH-Et_3N in the presence of $PdCl_2(PPh_3)_2$ in boiling dioxane [4], the reductive removal of the benzylic acetoxy group took place with complete regioselectivity to give **10** exclusively. Finally, saponification of **10** afforded anolignan A (**1**) as colorless needles, mp 97–98°C, which exhibited spectral properties (^1H and ^{13}C NMR, IR, MS) in accord with those reported for the natural molecule [1].

Later on, Mori and his group synthesized anolignan A using ruthenium-catalyzed cross-enyne metathesis as the key step; the 1,3-diene moiety was constructed by the introduction of the methylene part of ethylene into alkyne using Grubbs' catalyst (Scheme 2). The starting alkyne diacetate (**15**) was first prepared from piperonal (**2**) and ethynyltrimethylsilane (**11**). Piperonal was added to a THF solution of lithium trimethylsilylacetylide, and after complete exhaustion of the aldehyde, tetrabutylammonium fluoride in THF was added to the mixture. After the usual work-up, alkyne **12** was obtained in 95% yield. Treatment of **12** with 2 equiv. of BuLi and then 2,4-dimesyloxybenzaldehyde (**23**) afforded the diol derivative **14** followed by the protection of its hydroxy groups via acetylation to furnish the alkyne diacetate (**15**) as colorless sticky oil in overall 62% yield. To carry out enyne cross-metathesis reaction for the construction of 1,3-diene moiety, the

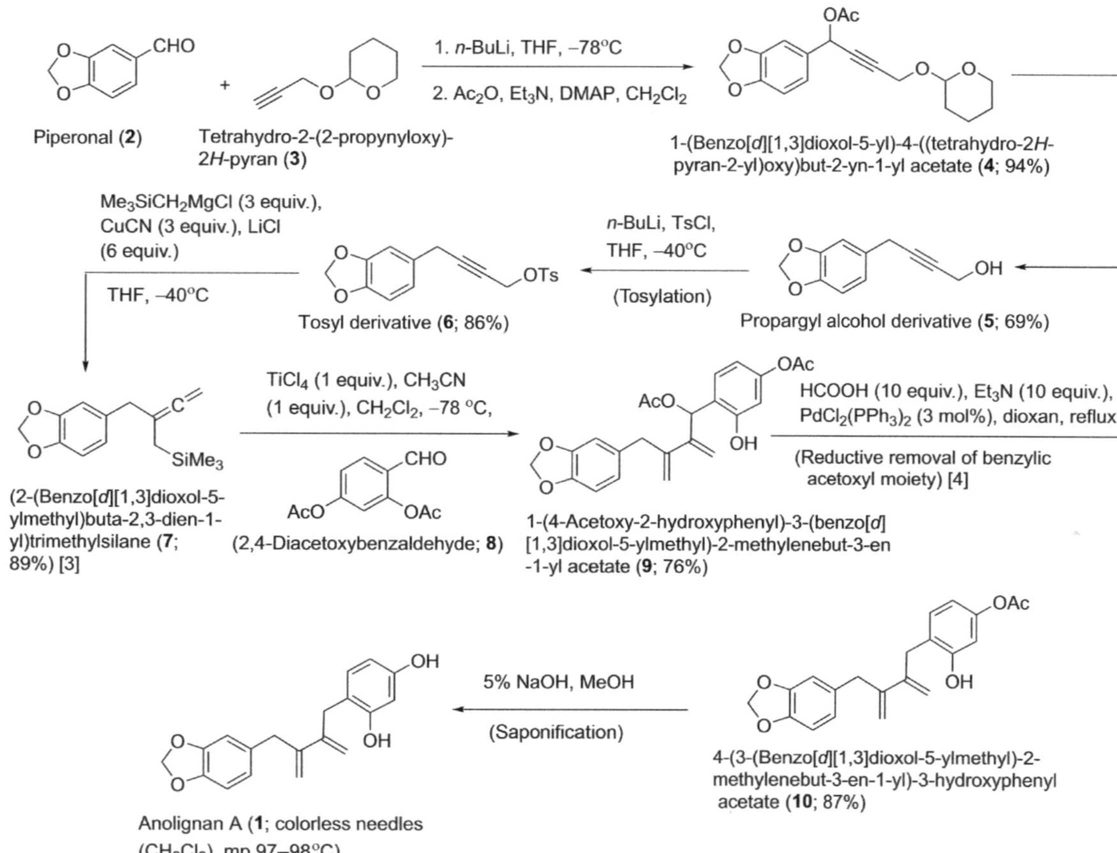

Scheme 1 Hatakeyama's total synthesis of anolignan A (**1**) [2].

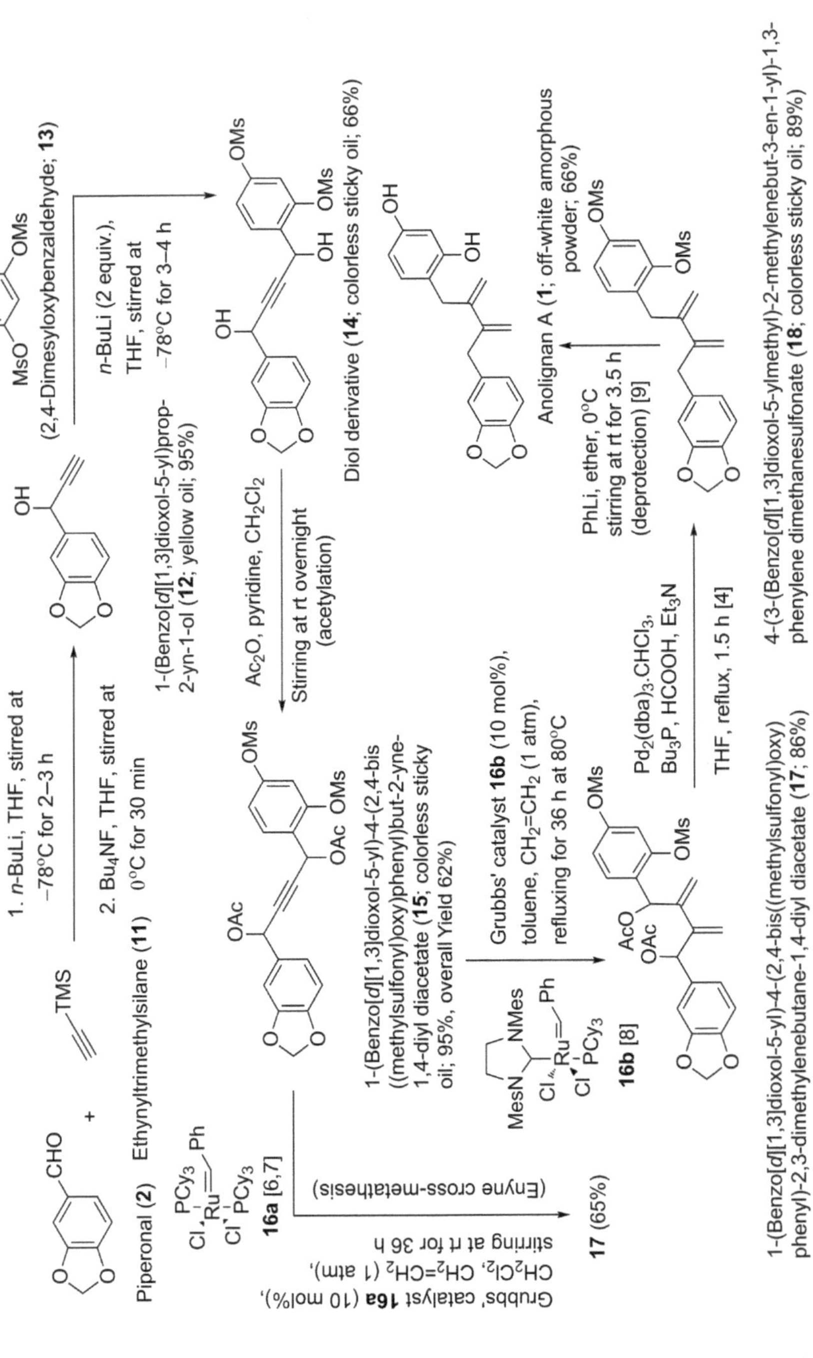

Scheme 2 Mori's total synthesis of anolignan A (**1**) [5].

investigators investigated the efficacy of both the Grubbs' catalysts **16a** [6, 7] and **16b** [8]. A CH_2Cl_2 solution of **15** on stirring for 36 h in the presence of 10 mol% of **16a** under ethylene gas (1 atm) at room temperature produced the 1,3-diene diacetate **17** in 65% yield along with the starting material **15** in 15% yield. On the other hand, when a new-generation ruthenium carbene complex **16b** containing *N*-heterocyclic carbene ligand was used in toluene under reflux conditions, the yield for this reaction increased to 86%. Hydrogenolysis of two acetoxy groups in **17** was successfully achieved by treatment with $Pd_2(dba)_3.CHCl_3$ and tributylphosphine in the presence of formic acid and triethylamine [4] to give **18**, which on deprotection with PhLi in ether at room temperature [9] afforded anolignan A (**1**) as off-white amorphous powder in 66% yield. The overall yield of **1** under this scheme was estimated as 30%, and spectral data of the synthetic compound were in agreement with those of anolignan A reported in the literature [1].

References

[1] A.M. Rimando, J.M. Pezzuto, N.R. Fransworth, T. Santisuk, V. Reutrakul, K. Kawanishi, J. Nat. Prod. 57 (1994) 896.
[2] M. Luo, A. Matsui, T. Esumi, Y. Iwabuchi, S. Hatakeyama, Tetrahedron Lett. 41 (2000) 4401.
[3] T. Nishiyama, T. Esumi, Y. Iwabuchi, H. Irie, S. Hatakeyama, Tetrahedron Lett. 39 (1998) 43.
[4] J. Tsuji, I. Minami, I. Shimizu, Synthesis (1986) 623.
[5] M. Mori, K. Tonogaki, N. Nishiguchi, J. Org. Chem. 67 (2002) 224.
[6] A. Kinoshita, N. Sakakibara, M. Mori, J. Am. Chem. Soc. 119 (1997) 12388.
[7] A. Kinoshita, N. Sakakibara, M. Mori, Tetrahedron 55 (1999) 8155.
[8] M. Scholl, S. Ding, C.W. Lee, R.H. Grubbs, Org. Lett. 1 (1999) 953.
[9] J.E. Baldwin, D.H.R. Barton, I. Dainis, J.L.C. Pereira, J. Chem. Soc. C (1965) 2283.

CHAPTER NINE

Anolignan B

Abbreviations

Ac_2O acetic anhydride
Bu_3P tributylphosphine
CH_2Cl_2 dichloromethane
Et_3N triethylamine
***n*-BuLi** *n*-butyllithium
$Pd_2(dba)_3$ tris(dibenzylideneacetone)dipalladium(0)
THF tetrahydrofuran

Systematic name: 4,4′-(2,3-Dimethylenebutane-1,4-diyl)diphenol
Compound class: Lignan (dibenzylbutadiene type)
Structure:

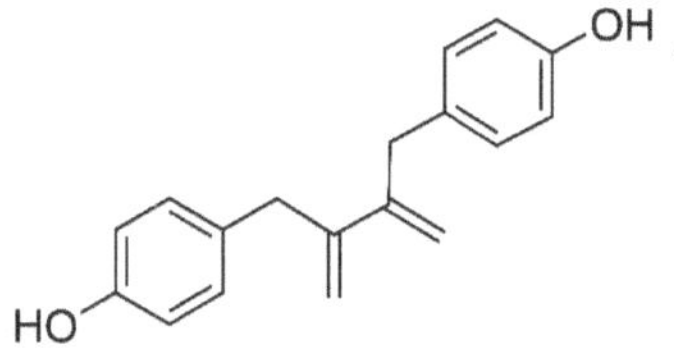

Anolignan B (**1**)

Natural source: *Anogeissus acuminata* (Roxb. ex DC.) Guill. & Perr. var. *lanceolata* Wall ex C.B. Clarke (ground stems; family: Combretaceae) [1]

Pharmaceutical potential: Anti-HIV (HIV-1 reverse transcriptase inhibitor) [1]

Synthetic route:

Total synthesis of the anti-HIV natural lignoid anolignan B was first reported by Mori and his group using ruthenium-catalyzed cross-enyne metathesis as the key step; the 1,3-diene moiety was constructed by the introduction of the methylene part of ethylene into alkyne using Grubbs' catalyst (Scheme 1). Condensation of 4-benzenesulfonyloxybenzaldehyde (**3**) with lithium acetylide produced 4-(1-hydroxyprop-2-yn-1-yl)phenyl-benzenesulfonate (**4**) as colorless oil, which was treated with BuLi (2 equiv.) and then another molecule of **3** followed by acetylation to afford alkyne

https://doi.org/10.1016/B978-0-08-102822-3.00009-2

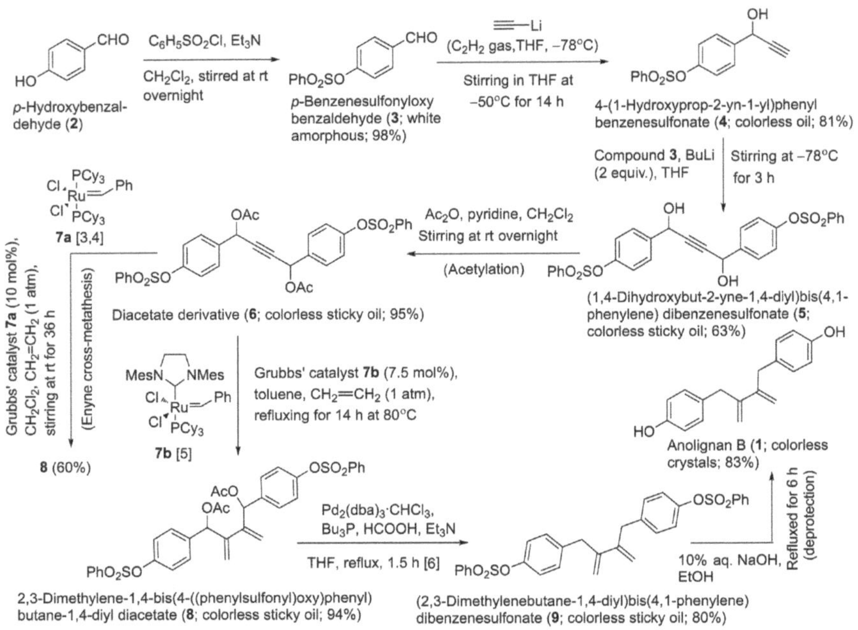

Scheme 1 Mori's total synthesis of anolignan B (**1**) [2].

diacetate derivative (**6**; 95%). To carry out an enyne cross-metathesis reaction for the construction of 1,3-diene moiety, the investigators investigated the efficacy of both the Grubbs' catalysts **7a** [3, 4] and **7b** [5]. A CH_2Cl_2 solution of **6** on stirring for 36 h in the presence of 10 mol% of **7a** under ethylene gas (1 atm) at room temperature produced the 1,3-diene diacetate **8** in 60% yield along with the starting material **6** in 32% yield. However, treatment of **6** with 7.5 mol% of **7b** as a catalyst in toluene at 80°C for 14 h under ethylene gas furnished 1,3-diene **8** in 94% yield. Removal of the two acetoxy groups with a palladium catalyst followed by deprotection of the benzenesulfonyl group with aqueous NaOH afforded anolignan B as colorless crystals in 83% yield. The spectral data of the synthetic compound were in agreement with those of anolignan B reported in the literature [1].

References

[1] A.M. Rimando, J.M. Pezzuto, N.R. Fransworth, T. Santisuk, V. Reutrakul, K. Kawanishi, J. Nat. Prod. 57 (1994) 896.
[2] M. Mori, K. Tonogaki, N. Nishiguchi, J. Org. Chem. 67 (2002) 224.
[3] A. Kinoshita, N. Sakakibara, M. Mori, J. Am. Chem. Soc. 119 (1997) 12388.
[4] A. Kinoshita, N. Sakakibara, M. Mori, Tetrahedron 55 (1999) 8155.
[5] M. Scholl, S. Ding, C.W. Lee, R.H. Grubbs, Org. Lett. 1 (1999) 953.
[6] J. Tsuji, I. Minami, I. Shimizu, Synthesis (1986) 623.

CHAPTER TEN

Antrocamphin A

Abbreviations

Et_3N triethyl amine
$(KSO_3)_2NO$ potassium nitrosodisulfonate (Frémy's salt)
$Pd(PPh_3)_4$ tetrakis(triphenylphosphine)palladium(0)
THF tetrahydrofuran
$TiCl_3$ titanium trichloride

Systematic name: 1,2,5-Trimethoxy-3-methyl-4(3-methyl-but-3-en-1-ynyl)benzene

Compound class: Benzenoid

Structure:

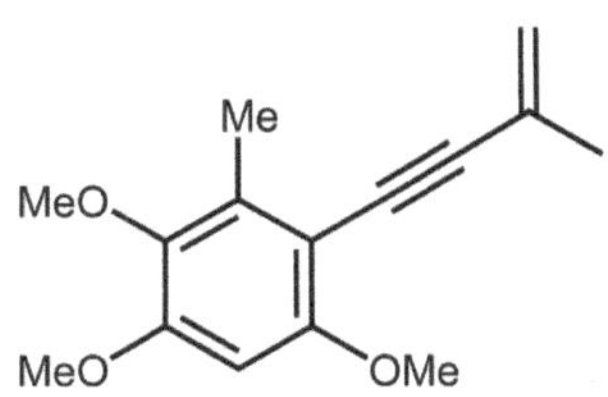

Antrocamphin A (**1**)

Natural source: *Antrodia camphorata* Wu, Ryvarden & Chang (syn. *Taiwanofungus camphorates*; family: Polyporaceae, Aphyllophorales), a parasitic fungus on the inner heartwood wall of the endemic species *Cinnamomum kanehirai* Hay (Lauraceae) in Taiwan [1, 2]

Pharmaceutical potential: Antiinflammatory [1–3]

Synthetic route:

The first total synthesis of antrocamphin A was accomplished by Wu and coworkers in six steps starting from *o*-vanillin (**2**) (Scheme 1) [3]. Compound **2** on hydrogenation with 10% palladium-on-carbon formed 2-hydroxy-3-methoxytoluene (**3**; 75%) followed by its conversion to the 1,4-quinone derivative **4** as yellow solid (78% yield) on oxidation with potassium nitrosodisulfonate (Fremy's salt). Quinone **4** was then reduced

https://doi.org/10.1016/B978-0-08-102822-3.00010-9

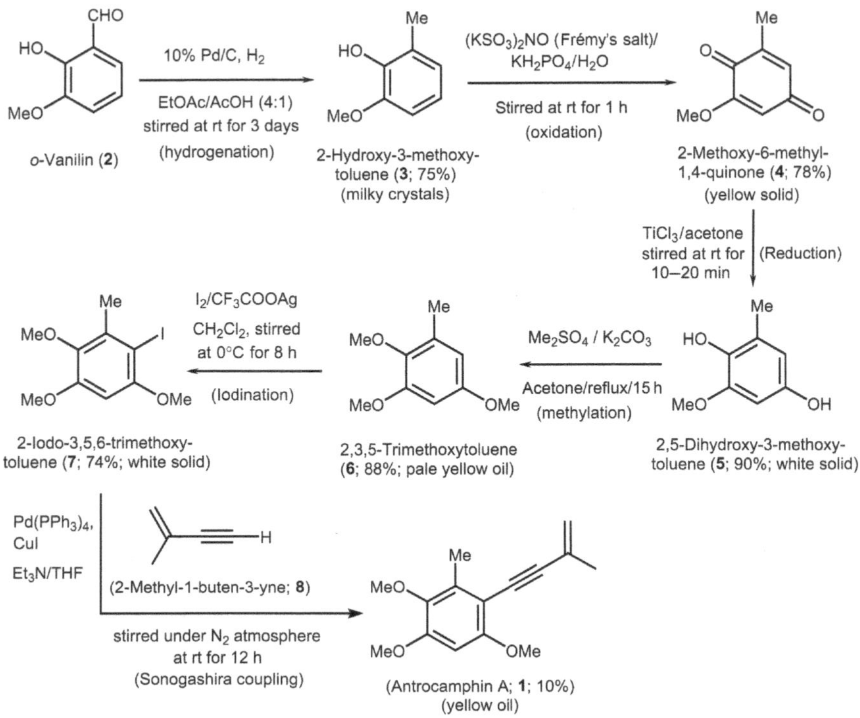

Scheme 1 Total synthesis of antrocamphin A (**1**).

to the hydroquinone derivative **5** (white solid; 90% yield) using $TiCl_3$, and thereafter the phenolic hydroxyl groups in **5** were methylated with dimethyl sulfate in the presence of K_2CO_3 resulting in the formation of 2,3,5-trimethoxytoluene (**6**) as pale yellow oil in 88% yield. Compound (**7**), 2-iodo-3,5,6-trimethoxytoluene, a key synthon was then prepared as white solid in 74% yield via iodination of **7** using I_2 and CF_3COOAg. In the final step, the investigators utilized the Sonogashira reaction to couple iodo **7** with 2-methyl-1-buten-3-yne (**8**) in the presence of the catalysts Pd $(PPh_3)_4$ and CuI to produce the desired compound antrocamphin A (**1**) as yellow oil in 10% yield. The Sonogashira coupling reaction also enabled the present investigators to synthesize a series of 14 different antrocamphin A analogs, and they carried out detailed studies with antrocamphin A and its synthetic analogs to evaluate their comparative antiinflammatory and anti-platelet aggregation potentials [3].

References

[1] J.J. Chen, W.J. Lin, C.H. Liao, P.C. Shieh, J. Nat. Prod. 70 (2007) 989.
[2] Y.H. Hsieh, F.H. Chu, Y.S. Wang, S.C. Chien, S.T. Chang, J.F. Shaw, C.Y. Chen, W.W. Hsiao, Y.H. Kuo, S.Y. Wang, J. Agric. Food Chem. 58 (2010) 3153.
[3] C.-L. Lee, C.-H. Huang, H.-C. Wang, D.-W. Chuang, M.-J. Wu, S.-Y. Wang, T.-L. Hwang, C.-C. Wu, Y.-L. Chen, F.-R. Chang, Y.-C. Wu, Org. Biol. Chem. 9 (2011) 70.

CHAPTER ELEVEN

Arenamide A

Abbreviations

Ala alanine
BAIB [bis(acetoxy)iodo]benzene
Bn benzyl
CH_2Cl_2 dichloromethane
(+)-DET (*R*,*R*)-diethyltartarate
DIPEA *N*,*N*-diisopropylethylamine
DMAP 4-dimethylaminopyridine
EDC 1-ethyl-3-(3-dimethylaminopropyl)carbodiimide
Et_3N triethylamine
Gly glycine
HMDA 3-hydroxy-4-methyl decanoic acid
HOBt hydroxybenzotriazole
Leu leucine
Phe phenylalanine
***p*-TsCl** *p*-toluene sulfonyl chloride
SAE Sharpless asymmetric epoxidation
TBS tetrabutylsilyl
TFA trifluoroacetic acid
TEMPO 2,2,6,6-tetramethyl-1-piperidinyloxyl
THF tetrahydrofuran
Val valine

Systematic name: (3*S*,6*S*,9*S*,12*S*,19*S*)-3-Benzyl-9-isobutyl-12-isopropyl-6-methyl-19-((*S*)-octan-2-yl)-1-oxa-4,7,10,13,16-pentaazacyclononadecane-2,5,8,11,14,17-hexaone

Compound class: Cyclohexadepsipeptide

Structure:

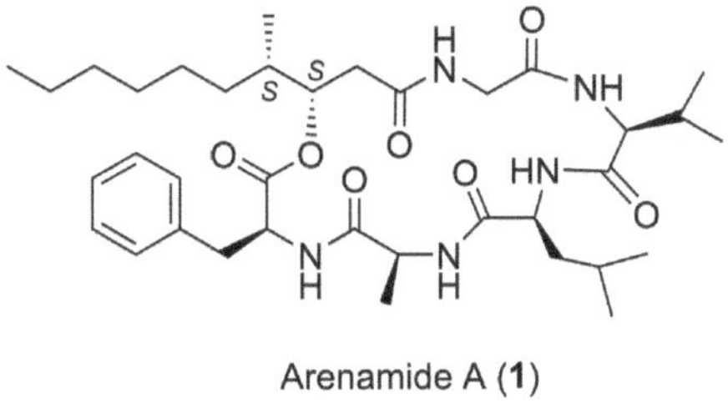

Arenamide A (**1**)

https://doi.org/10.1016/B978-0-08-102822-3.00011-0

Natural source: *Salinispora arenicola* strain CNT-088 (marine actinomycete; fermentation broth) [1]

Pharmaceutical potential: Antitumor (NFκB inhibitor); inhibitor to NO production; cytotoxic [1]

Synthetic route:

Chandrasekhar and his group [2] reported a convenient approach for the total synthesis of the marine cyclic depsipeptide arenamide A (**1**) and its (*R*,*R*) diastereomer for the first time. Their retrosynthetic approach for the molecule is shown in Scheme 1.

Preparation of (5*S*,6*S*)-5-hydroxy-6-methyl-1-(4-nitrophenyl) dodecan-3-one (5):

Fragment **5** was first prepared starting from the known allyl alcohol (*E*)-5-(benzyloxy)-2-methyl pent-2-en-1-ol (**6**) [3] (Scheme 2). Alcohol **6** on Sharpless asymmetric epoxidation [4] using (+)-DET, Ti(O*i*Pr)$_4$, and *t*BHP followed by reductive opening with $NaCNBH_4$ under $BF_3{\cdot}Et_2O$ catalysis [5] furnished exclusively the 1,3-diol derivative **7** in good yield. Selective

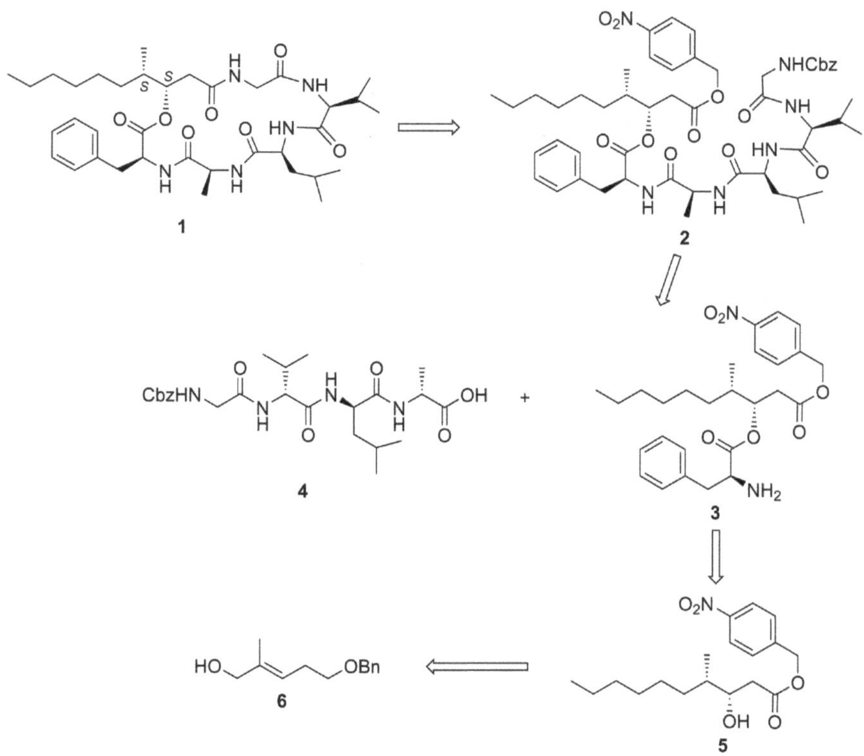

Scheme 1 Retrosynthetic approach for arenamide A (**1**) [2].

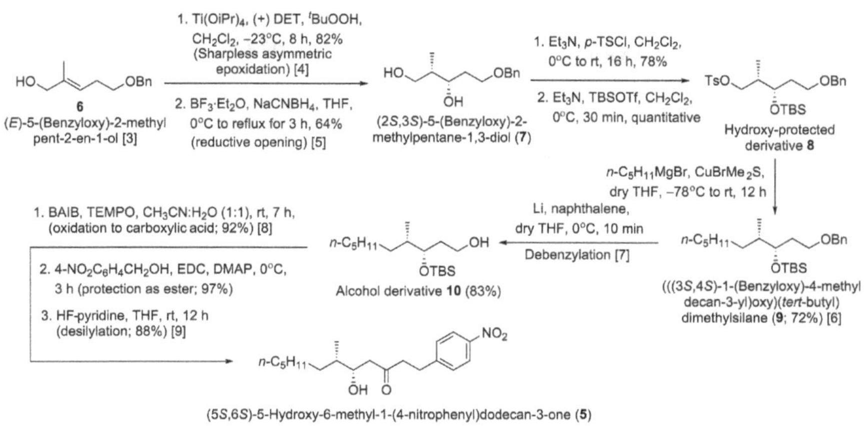

Scheme 2 Preparation of fragment **5** [2].

tosylation of 1° alcohol and silylation of 2° alcohol gave the hydroxyl-protected derivative **8** in over 75% for two steps. A copper-mediated C—C bond formation on tosylate **8** with pentyl magnesium bromide afforded compound **9** in 72% yield [6], which on debenzylation gave 1° alcohol derivative **10** [7]. This, in turn, underwent a smooth oxidation [8] with TEMPO to carboxylic acid followed by its subsequent protection as *p*-nitro benzyl ester and desilylation of 2° alcohol [9] provided the fragment **5** in good overall yield (Scheme 2).

Preparation of tetrapeptide fragment 4:

Tetrapeptide fragment **4** was synthesized from the commercially available protected (L)-amino acids (Scheme 3). The condensation of alanine

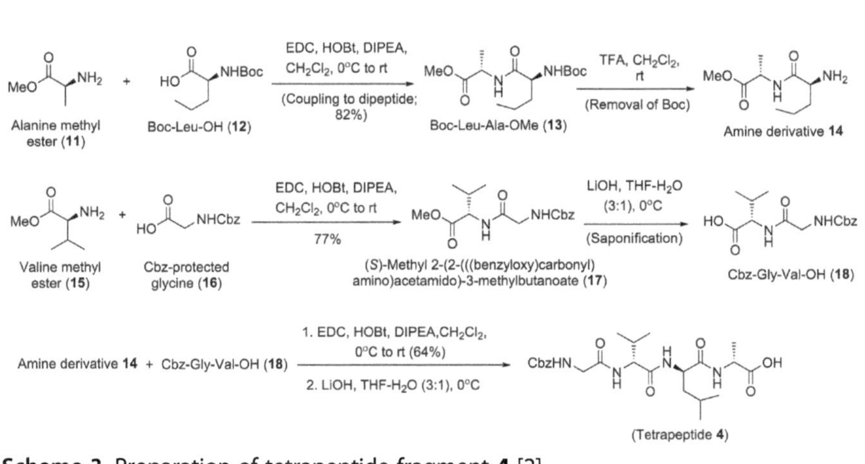

Scheme 3 Preparation of tetrapeptide fragment **4** [2].

methyl ester **11** and Boc-Leu-OH **12** in the presence of EDC and HOBt as coupling reagents produced dipeptide (Boc-Leu-Ala-OMe) **13** in 82% yield, which on removal of the Boc group gave amine derivative **14**. The protected dipeptide Cbz-Gly-Val-OH **18** was then prepared by coupling of valine methyl ester **15** with Cbz-protected glycine **16** followed by hydrolysis using lithium hydroxide. Coupling between both of these dipeptides **14** and **18** followed by hydrolysis of the ester functionality ultimately furnished the tetrapeptide fragment **4** in good yield (Scheme 3).

Synthesis of arenamide A (1):

The investigators then prepared another key intermediate Phe-HMDA **3** from the coupling reaction between fragment **5** and Boc-protected phenyl alanine (Boc-Phe-OH) **19** under EDC and DMAP conditions followed by treating with TFA in dichloromethane. The free amine derivative **3** underwent further coupling smoothly with tetrapeptide acid **4** under the standard coupling conditions for amide bond formation to afford hexadepsipeptide **2** in 74% yield. Finally, deprotection of *p*-nitro benzyl and Cbz group by hydrogenation using Pd/C in isopropyl alcohol and THF mixture, followed by cyclization of linear hexadepsipeptide under high dilution in dichloromethane using EDC and HOBt as coupling reagents furnished arenamide A (**1**) in 58% yield (Scheme 4).

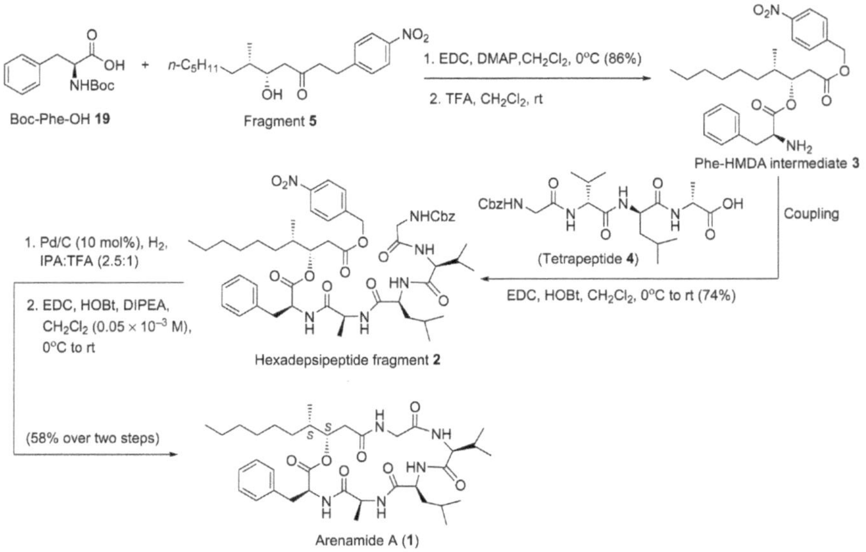

Scheme 4 Synthesis of arenamide A (**1**) [2].

References

[1] R.N. Asolkar, K.C. Freel, P.R. Jensen, W. Fenical, T.P. Kondratyuk, E.-J. Park, J.M. Pezzuto, J. Nat. Prod. 72 (2009) 396.
[2] S. Chandrasekhar, G. Pavankumarreddy, K. Sathish, Tetrahedron Lett. 50 (2009) 6851.
[3] K.B. Sawant, F. Ding, M.P. Jennings, Tetrahedron Lett. 48 (2007) 5177.
[4] T. Katsuki, K.B. Sharpless, J. Am. Chem. Soc. 102 (1980) 5974.
[5] S.D. Hiscock, P.B. Hitchcock, P.J. Parsons, Tetrahedron 54 (1998) 11567.
[6] K. Mori, Tetrahedron 37 (1981) 1341.
[7] H.-J. Liu, J. Yip, K.-S. Shia, Tetrahedron Lett. 38 (1997) 2253.
[8] J.B. Epp, T.S. Widlanski, J. Org. Chem. 64 (1999) 293.
[9] K.C. Nicolaou, S.E. Webber, Synthesis (1986) 453.

CHAPTER TWELVE

Atroviridin

Abbreviations

Ar argon gas
BBr_3 tribromoborate
Bn benzyl
Bu_4NI tetrabutylammonium iodide
CAN ceric(IV) ammonium nitrate
CH_2Cl_2 dichloromethane
CH_3CN acetonitrile
$CHCl_3$ chloroform
DBU 1,8-diazabicyclo[5.4.0]undec-7-ene
DMF *N,N*-dimethylformamide
DMSO dimethylsulfoxide
IBX 2-iodoxybenzoic acid
***m*-CPBA** *meta*-chloroperbenzoic acid
MEMCl β-methoxyethoxymethyl
MnO_2 manganese dioxide
$NaBH_4$ sodium borohydride
NaH sodium hydride
***n*-BuLi** *n*-butyl lithium
NHC *N*-heterocyclic carbene
$Pd(PPh_3)_4$ tetrakis(triphenylphosphine)palladium
TBAF tetrabutylammonium fluoride
TBSCl *tert*-butylchlorodimethylsilane
THF tetrahydrofuran

Systematic name: 8-Hydroxy-2,2-dimethylpyrano[3,2-*b*]xanthene-5,6,12(2*H*)-trione

Compound class: Xanthonoid (tetracyclic polyoxygenated xanthone)

Structure:

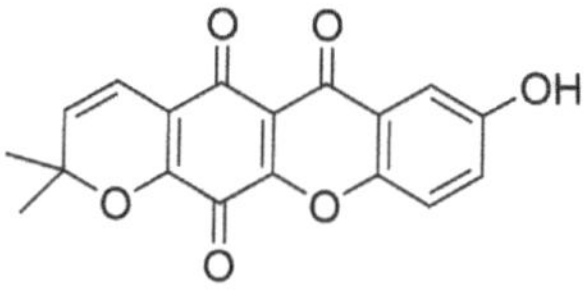

Atroviridin (**1**)

https://doi.org/10.1016/B978-0-08-102822-3.00012-2

Natural source: *Garcinia atroviridis* (stem bark; family: Guttiferae) [1, 2]

Pharmaceutical potential: *Garcinia* xanthones are reported to possess a wide variety of bioactivities [3–8]

Synthetic routes:

Theodorakis and his group accomplished the first total synthesis of atroviridin (**1**) using a coupling reaction between an aryl bromide **9** and an aldehyde **11** followed by subsequent intramolecular conjugate addition on a quinone precursor in 14 steps (Scheme 1) [9]. Aryl bromide **9** was synthesized from commercially available 2,5-dimethoxybenzaldehyde (**2**). Regioselective bromination of **2** with a slight excess of bromine in acetic acid produced 4-bromo-2,5-dimethoxybenzaldehyde (**3**) in 45% yield [10]. Baeyer-Villiger oxidation of **3** followed by hydrolysis of the resulting formate ester under basic conditions afforded phenol **4** in 99% yield [11]. On *O*-alkylation with 1,1-dimethylprop-2-ynyl methyl carbonate (**5**) using DBU and catalytic $CuCl_2$ in acetonitrile, compound **4** afforded alkyne **6** (84% yield) [12]. The methyl hydroquinone **6** was then treated with CAN in 40% aqueous acetonitrile, thereby transiently producing quinone **7**, which was extracted from the orange reaction mixture and, without further purification, was heated at 40°C in toluene when a red-colored chromenequinone **8** was produced in 77% yield over two steps [13]. Reduction of quinone **8** with $NaBH_4$ and AcOH in THF yielded the resulting hydroquinone, which was immediately converted to its MEM ether **9** using MEMCl and DIPEA (73% yield over two steps) [14, 15].

The commercially available 2,5-dihydroxy benzaldehyde (**10**) was then protected as the corresponding silyl ether **11** and subsequently alkylated with the lithium salt of **9** to afford benzylic alcohol **12** in 52% yield. Oxidation of **12** with Dess-Martin periodinane in dichloromethane furnished benzophenone **13** in 66% yield [16]. Deprotection of the MEM ethers of **13** was effected by using an excess of $ZnBr_2$ in dichloromethane to give hydroquinone **14** in 70% yield [15]. Now the task was to oxidize the hydroquinone ring and subsequent conjugate addition. The investigators accomplished the oxidation of **14** with IBX in 5% DMF/$CHCl_3$ at ambient temperature to produce the quinone **15** in 31% yield [17–19]. Treatment of **20** with TBAF in THF resulted in deprotection of TBS ethers to generate quinone **16**, which ultimately underwent in situ intramolecular cyclization to afford atroviridin (**1**). The investigators assumed this final deprotection protocol with concomitant annulations to give the target compound might be a proposed biosynthetic pathway for xanthone and xanthonoid natural products [9].

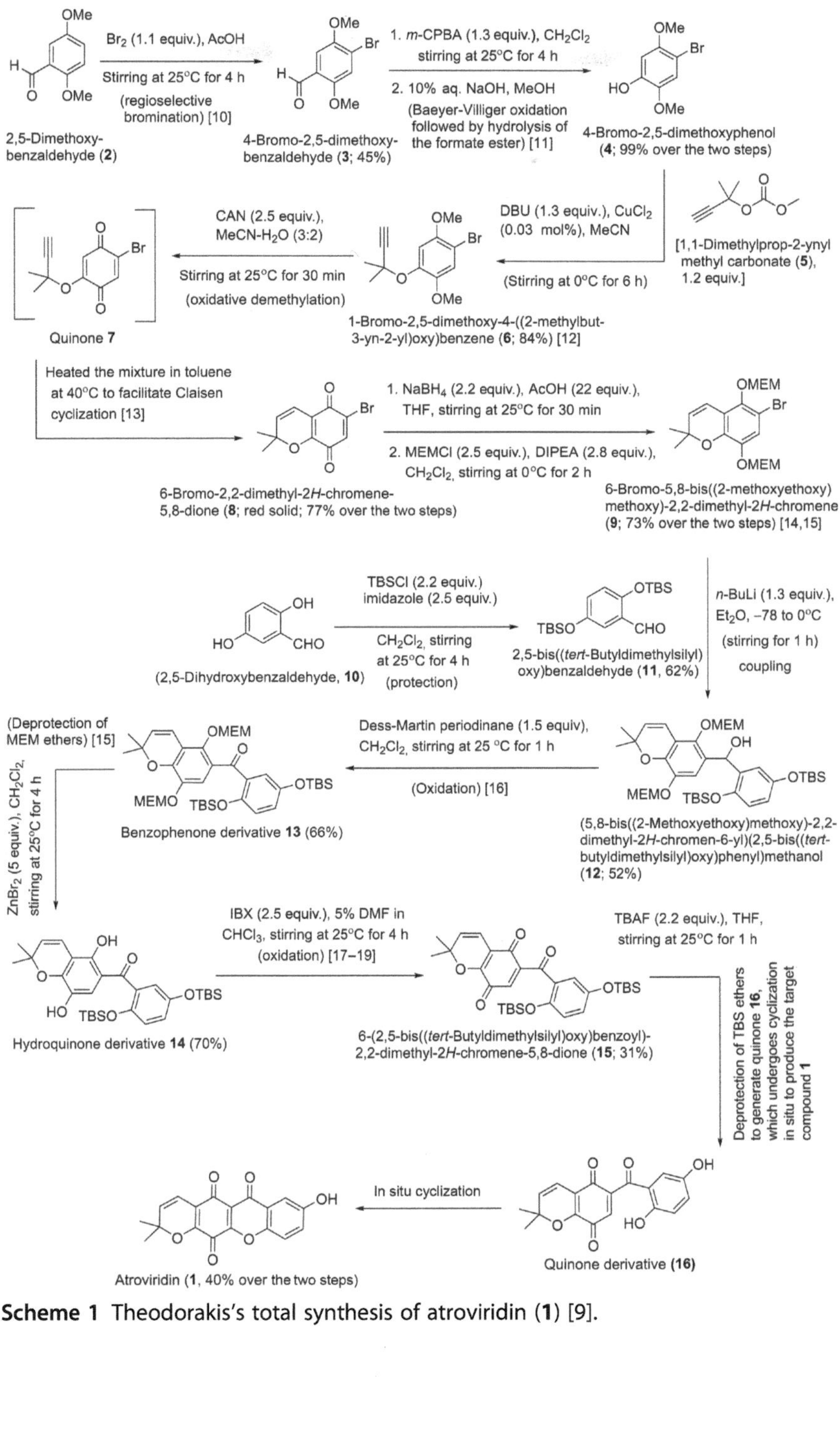

Scheme 1 Theodorakis's total synthesis of atroviridin (**1**) [9].

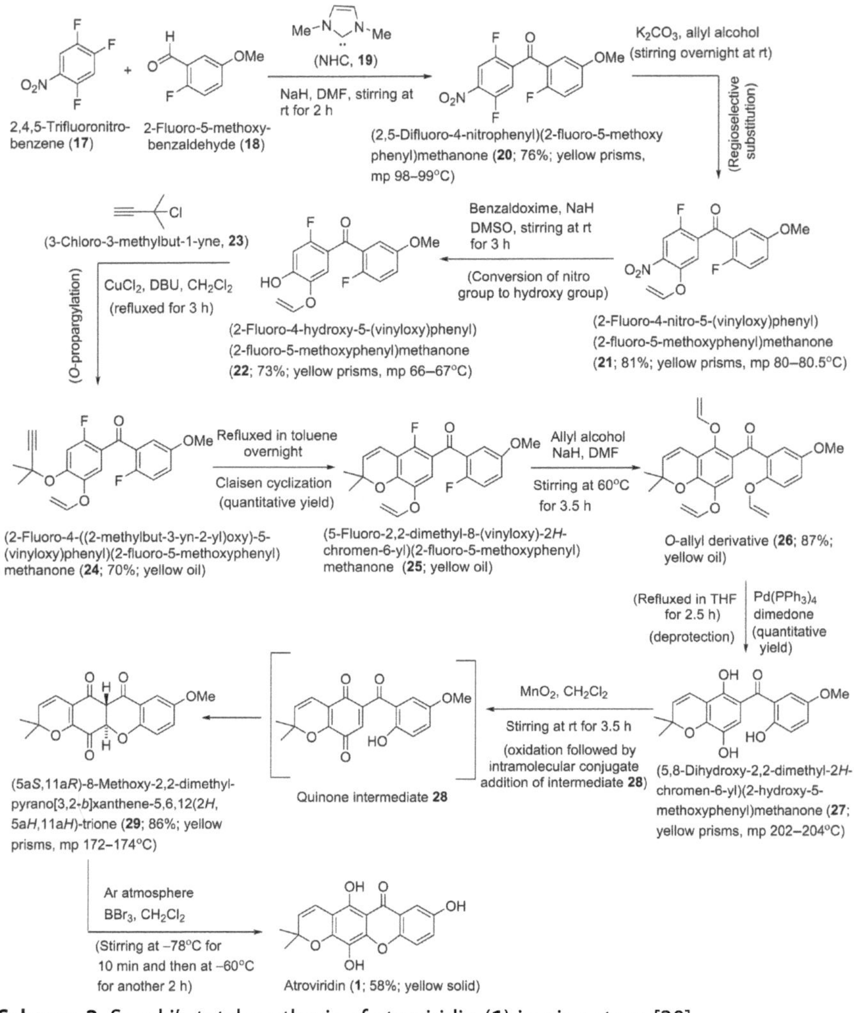

Scheme 2 Suzuki's total synthesis of atroviridin (**1**) in nine steps [20].

Suzuki and his group reported the total synthesis of atroviridin (**1**) in 9 (Scheme 2) and 11 steps (Scheme 3) by a linear route involving the NHC-catalyzed aroylation of the fluorobenzene derivative, Claisen cyclization of the *O*-propargylated benzophenones, and intramolecular 1,4-addition of the quinone intermediates [20]. The investigators first prepared (2,5-difluoro-4-nitrophenyl)(2-fluoro-5-methoxyphenyl)methanone (**20**) from 2,4,5-trifluoronitrobenzene (**17**) and 2-fluoro-6-methoxybenzaldehyde (**18**) by NHC catalysis as described in Scheme 2. The C-4 fluorine within **20** was regioselectively substituted by an allyloxy group in the reaction with

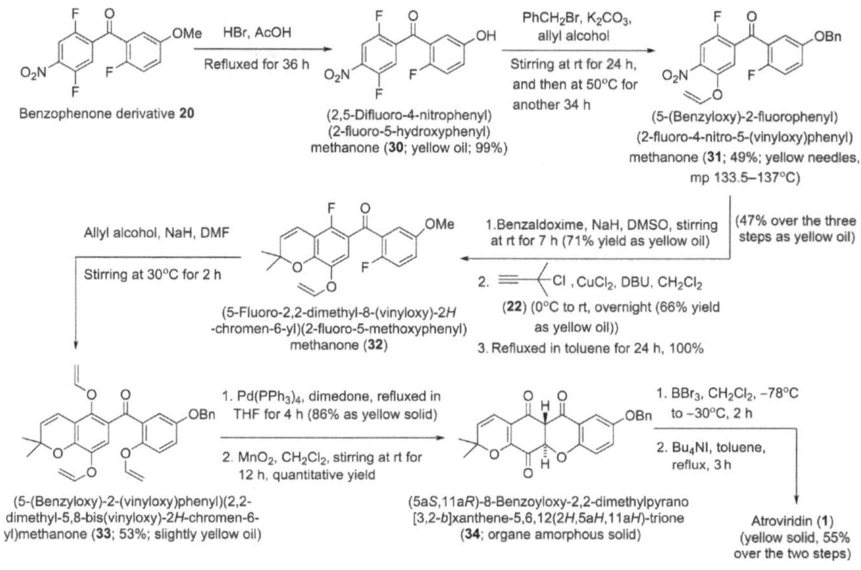

Scheme 3 Suzuki's alternative total synthesis of atroviridin (**1**) in 11 steps [20].

K_2CO_3/allyl alcohol at room temperature to afford the intermediate **21** in 81% yield. The nitro group was converted to hydroxyl function by treating **21** with the anion of benzaldoxime in DMSO at room temperature for 3 h (73% yield). The resulting compound **22** was *O*-propargylated with 3-chloro-3-methylbutyne (**23**) using DBU and catalytic $CuCl_2$ in dichloromethane to produce derivative **24**, which then underwent Claisen cyclization in refluxing toluene to furnish the chromene **25**. The remaining fluoro groups of **25** were replaced by allyloxy groups in the reaction with the alkoxide of allyl alcohol in DMF at 60°C for 3.5 h to afford the intermediate **26** in 87% yield as yellow oil. The removal of allyl groups using dimedone and $Pd(PPh_3)_4$ in THF quantitatively afforded the trihydroxy intermediate **27**; on oxidation by MnO_2 in CH_2Cl_2 at room temperature, compound **27** resulted in the formation of the tetracyclic triketone **29** (86% yield), which was supposed to be produced by the spontaneous intramolecular conjugate addition of **28**. Removal of the methyl group from **29** using BBr_3 in dichloromethane eventually afforded atroviridin (**1**) as yellow solid in 58% yield.

In the same report [20], Suzuki et al. outlined an alternative route for the total synthesis of the xanthone **1** in 11 steps (Scheme 3). They found that the last step in Scheme 2 involving the removal of the methyl group showed low reproducibility due to difficulties both in preventing undesired products and in purifying **1**; to overcome these difficulties they developed an alternative protecting group for the 7-hydroxy group (Scheme 3). Thus the benzyl ether **31** was prepared from **30** and then subjected to the conversion of the nitro group to the hydroxyl group, followed by *O*-propargylation and Claisen cyclization to produce 5-fluoro-2,2-dimethyl-8-(vinyloxy)-2*H*-chromen-6-yl)(2-fluoro-5-methoxyphenyl) methanone (**32**). The remaining fluoro groups were substituted to allyloxy groups cautiously to avoid the co-occurring Claisen rearrangement of the 10a-allyloxy group of the product **33**. After removal of the allyl groups, oxidation with MnO_2 afforded the trihydroxy intermediate **34** in good yield. Deprotection of the benzyl ether using BBr_3 followed by aromatization with *n*-Bu_4NI in refluxing toluene ultimately afforded the xanthone atroviridin (**1**).

References

[1] J. Kosin, N. Ruangrungsi, C. Ito, H. Furukawa, Phytochemistry 47 (1988) 1167.
[2] D. Parmana, N.H. Lajis, M.M. Mackeen, A.M. Ali, N. Aimi, M. Kitajima, H. Takayama, J. Nat. Prod. 64 (2001) 976.
[3] J. Asano, K. Chiba, M. Tada, T. Yoshii, Phytochemistry 41 (1996) 815.
[4] M. Iinuma, T. Ito, R. Miyake, H. Tosa, T. Tanaka, V. Chelladura, Phytochemistry 47 (1998) 1169.
[5] C. Okudaira, Y. Ikeda, S. Konndo, S. Furuya, Y. Hirabayashi, T. Koyano, Y. Saito, K. Umezawa, J. Enzym. Inhib. 15 (2000) 129.
[6] S. Suksamraran, N. Suwannapoch, W. Phakhodee, J. Thanuhiranlert, P. Ratananukul, N. Chimnoi, A. Suksamrarn, Chem. Pharm. Bull. 51 (2003) 857.
[7] M. Hamada, K. Iikubo, Y. Ishikawa, A. Ikeda, K. Umezawa, S. Nishiyama, Bioorg. Med. Chem. Lett. 13 (2003) 3151.
[8] Y. Sukpondma, V. Rukachaisirikul, S. Phongpaichit, Chem. Pharm. Bull. 53 (2005) 850.
[9] E.J. Tisdale, D.A. Kochman, E.A. Theodorakis, Tetrahedron Lett. 44 (2003) 3281.
[10] M.S. Sardessai, H.N. Abramson, Org. Prep. Proced. Int. 23 (1991) 419.
[11] U. Wriede, M. Fernandez, K.F. West, D. Harcourt, H.W. Moore, J. Org. Chem. 52 (1987) 4485.
[12] J.D. Godfrey Jr., R.H. Mueller, T.C. Sedergran, N. Soundararajan, V.J. Colandrea, Tetrahedron Lett. 35 (1994) 6405.
[13] P.E. Brown, R.A. Lewis, M.A. Waring, J. Chem. Soc. Perkin Trans. 1 (1990) 2979.
[14] C. Pugh, Org. Lett. 2 (2000) 1329.
[15] E.J. Corey, J.-L. Gras, P. Ulrich, Tetrahedron Lett. 17 (1976) 809.
[16] D.B. Dess, J.C. Martin, J. Am. Chem. Soc. 113 (1991) 7277.
[17] T. Wirth, Angew. Chem. Int. Ed. 40 (2001) 2812.
[18] K.C. Nicolaou, T. Montagnon, P.S. Baran, Y.-L. Zhong, J. Am. Chem. Soc. 124 (2002) 2245.
[19] J.D. More, N.S. Finney, Org. Lett. 4 (2002) 3001.
[20] Y. Suzuki, Y. Fukuta, S. Ota, M. Kamiya, M. Sato, J. Org. Chem. 76 (2011) 3960.

CHAPTER THIRTEEN

Bauhinoxepin J

Abbreviations

$(NH_4)_2S_2O_8$ ammonium persulfate
CH_2Cl_2 dichloromethane
CH_3CN acetonitrile
DBU 1,8-diazabicyclo[5.4.0]undec-7-ene
DMAP 4-dimethylaminopyridine
DMF *N*,*N*-dimethylformamide
Et_3N triethylamine
LAH lithium aluminum hydride
MOM methoxymethyl
***n*-BuLi** *n*-butyl lithium
rt room temperature
Salcomine *N*,*N*-bis(salicylidene)ethylenediaminocobalt(II)
TFAA trifluoroacetic anhydride [$(CF_3CO)_2O$]
THF tetrahydrofuran

Systematic name: 2-Methoxy-10,11-dihydrodibenzo[*b*,*f*]oxepine-1,4-dione

Compound class: Dibenzoxepin derivative

Structure:

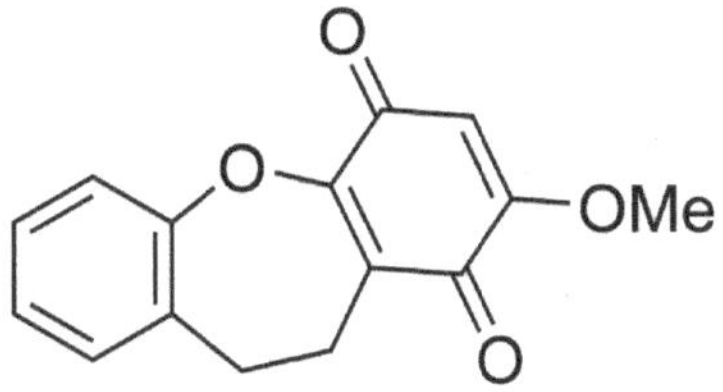

Bauhinoxepin J (**1**)

Natural source: *Bauhinia purpurea* L. (roots; family Leguminosae) [1]

Pharmaceutical potential: Antiinflammatory, cytotoxic, and antimalarial [1]

https://doi.org/10.1016/B978-0-08-102822-3.00013-4

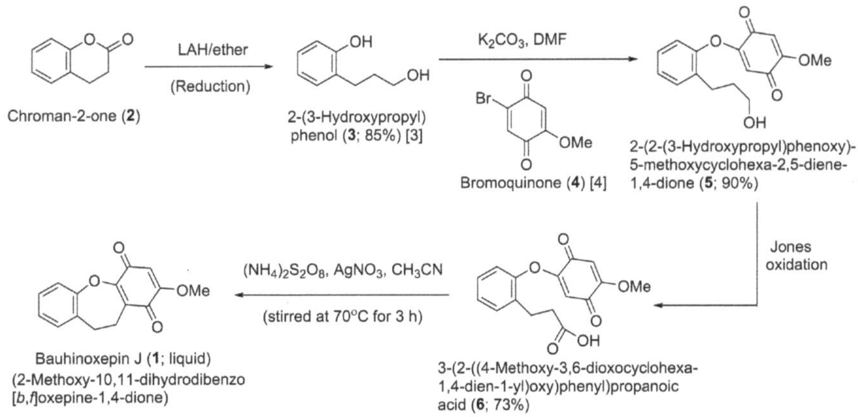

Scheme 1 Kraus's total synthesis of bauhinoxepin J (**1**) [2].

Synthetic route:

Bauhinoxepin J (**1**), a dihydrodibenz[*b,f*]oxepin isolated from *Bauhinia purpurea* [1], was found to be potent bioactive, and the first total synthesis of this natural product was achieved by Kraus et al. in four steps using an intramolecular persulfate-mediated radical addition to a quinone as the key step (Scheme 1) [2]. The starting material, chroman-2-one (**2**), was reduced by LAH to obtain 2-(3-hydroxypropyl)phenol (**3**) in 85% yield [3], which was then treated with 2-bromo-5-methoxycyclohexa-2,5-diene-1,4-dione (**4**) in the presence of potassium carbonate in DMF to produce the 2-(2-(3-hydroxypropyl)phenoxy)-5-methoxycyclohexa-2,5-diene-1,4-dione (**5**; 90%) [4]. The resulting alcohol **5** on Jones oxidation afforded the carboxylic acid derivative **6** in 73% yield. In the final step, compound **6** was treated with silver nitrate and ammonium persulfate in acetonitrile followed by heating the reaction mixture at 70°C for 3 h when there occurred an intramolecular radical addition to the quinine moiety, thereby furnishing the desired product **1** in 40% isolated yield. The overall yield for bauhinoxepin J was calculated as 25%. The physical and spectral (IR, ^{1}H NMR, ^{13}C NMR, LRMS, and HRMS) data of synthetic compound was found to be comparable to those published for natural bauhinoxepin J [1].

Later on, Yoshida and coworkers reported an approach to the dibenzo[*b,f*] oxepin core of **1** by using an oxidative dearomatization/cyclization protocol (13% overall yield in seven steps) [5]. More recently, Katoh and his group [6] accomplished the total synthesis of bauhinoxepin J (**1**) in 31% overall yield in six steps from the starting materials 1,4-dimethoxy-2-(methoxymethoxy)

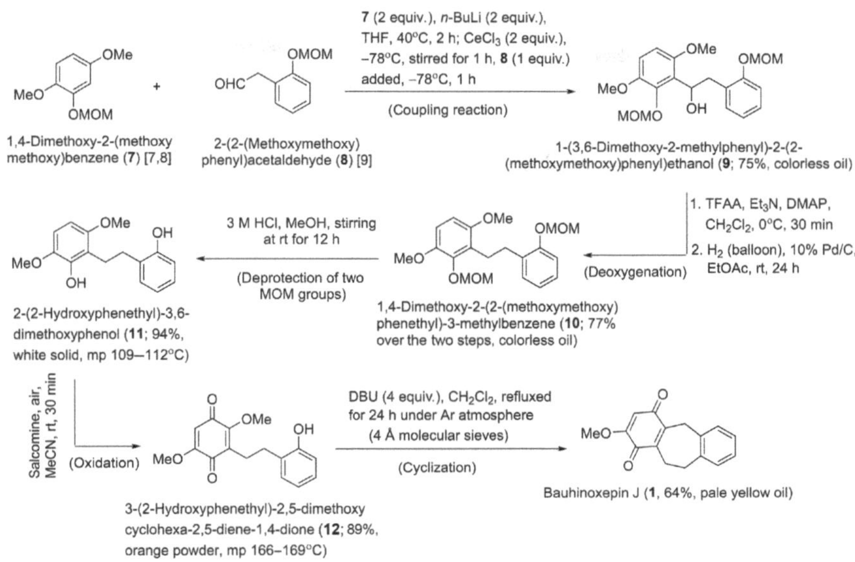

Scheme 2 Katoh's total synthesis of bauhinoxepin J (**1**) [6].

benzene (**7**) [7, 8] and 2-(2-(methoxymethoxy)phenyl)acetaldehyde (**8**) [9], involving a coupling reaction between the aromatic moieties to construct the requisite carbon framework as a key step (Scheme 2). 1-(3,6-Dimethoxy-2-methylphenyl)-2-(2-(methoxymethoxy)phenyl)ethanol (**9**) was thus obtained in 75% yield as colorless oil on coupling reaction under optimized conditions. On removal of the hydroxy group in **9** via trifluoroacetylation followed by hydrogenolysis of the resulting trifluoroacetate, the deoxygenated product, 1,4-dimethoxy-2-(2-(methoxymethoxy)phenethyl)-3-methylbenzene (**10**, colorless oil) in 77% overall yield over two steps was obtained. Deprotection of the two MOM groups in **10** under acidic conditions produced bisphenol **11** in 94% yield. The subsequent crucial chemoselective quinone formation reaction was successfully achieved by exposure of **11** to air (atmospheric oxygen) in the presence of Salcomine (0.5 equiv.) in acetonitrile at room temperature for 30 min, which resulted in a clean formation of the desired quinine, 3-(2-hydroxyphenethyl)-2,5-dimethoxycyclohexa-2,5-diene-1,4-dione (**12**) as an orange powder in high yield (89%). In the final step, compound **12** underwent smooth intramolecular ether cyclization under optimized conditions (a highly dilute solution (0.5 mm) of CH_2Cl_2 in the presence ofDBU (4 equiv.) and 4 Å molecular sieves under reflux conditions in argon atmosphere) furnishing bauhinoxepin J (**1**) in 64% yield as pale yellow oil. The spectroscopic properties (IR, ^1H and

^{13}C NMR, and HRMS) of synthetic sample **1** were found to be identical to those of natural compound [1].

References

[1] S. Boonphong, P. Puangsombat, A. Baramee, C. Mahidol, S. Ruchirawat, P. Kittakoop, J. Nat. Prod. 70 (2007) 795.
[2] G.A. Kraus, A. Thite, F. Liu, Tetrahedron Lett. 50 (2009) 5303.
[3] M. Botta, S. Quici, G. Pozzi, G. Marzanni, R. Pagliarin, S. Barra, S.G. Crich, Org. Biomol. Chem. 2 (2004) 570.
[4] H. Tohma, H. Morioka, Y. Harayama, M. Hashizume, Y. Kita, Tetrahedron Lett. 42 (2001) 6899.
[5] M. Yoshida, Y. Maeyama, K. Shishido, Heterocycles 80 (2010) 623.
[6] K. Narita, K. Nakamura, Y. Abe, T. Katoh, Eur. J. Org. Chem. (2011) 4985.
[7] T. Capecchi, C.B. de Koning, J.P. Michael, J. Chem. Soc. Perkin Trans. 1 (2000) 2681.
[8] U. Wriede, M. Fernandez, K.F. West, D. Harcourt, H.W. Moore, J. Org. Chem. 52 (1987) 4485.
[9] H.-S. Moon, S.-I. Nam, S.-D. Kim, D.Y. Kim, B.J. Gwag, Y.A. Lee, S.-H. Yoon, J. Pharm. Pharmacol. 54 (2002) 935.

CHAPTER FOURTEEN

Beta-Lapachone

Abbreviations

Ac_2O acetic anhydride
BCl_3 boron trichloride
CH_2Cl_2 dichloromethane
CrO_3 chromic trioxide
EDDA ethylenediamine diacetate
Et_3N triethylamine
HCOOH formic acid
LAH lithium aluminum hydride
THF tetrahydrofuran

Systematic name: 3,4-Dihydro-2,2-dimethyl-2*H*-naphtho[1,2-*b*]pyran-5,6-dione

Compound class: Pyranonaphthoquinone

Structure:

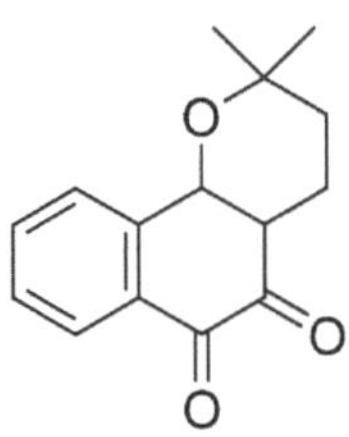

β-Lapachone (**1**)

Natural source: *Austroplenckia populnea* (Reiss.) Lundell. (root wood; family: Celastraceae) [1]; *Tabebuia avellanedae* (the lapacho tree; barks; family: Bignoniaceae) [2]

Pharmaceutical potentials: Anticancer, antitumor, antiproliferative, antiviral, antibacterial, antifungal, antiparasitic, antiinflammatory [3–25]

Total Synthesis of Bioactive Natural Products
https://doi.org/10.1016/B978-0-08-102822-3.00014-6

Synthetic routes:

In view of the great pharmaceutical importance of β-lapachone (**1**), a couple of synthetic methods have already been reported so far [26–29]. In an earlier investigation [26], β-lapachone was synthesized using initially a photochemical reaction between 3-methylcrotonaldehyde and 1,4-naphthoquinone; however, there was no report on the yield for the last step. Later on, Amaral and Barnes [27] accomplished the total synthesis of β-lapachone in eight steps with an overall yield of 23% starting from α-naphthol (Scheme 1). The hydroxyketone **4** was first prepared on treating the ester derivative naphthalene-1-yl 3-methylbut-2-enolate (obtained by esterification of α-naphthol with 3-methylcrotonic acid chloride) with boron trichloride in 75% yield [30]. Compound **4** was then cyclized with a mixture of formic and hydrochloric acids to furnish 2,2-dimethyl-2*H*-benzo[*h*]chromen-4(3*H*)-one (**5**; 84%), which gave the corresponding alcohol **6** on LAH reduction in 82% yield. The resulting alcohol **6** was easily dehydrated under acidic conditions to yield olefin 2,2-dimethyl-2*H*-benzo[*h*]chromene (**7**; 83%). The reduction of only the double bond within **7** was effected by hydrogenation using rather large amounts of catalyst and a low hydrogen pressure so that the naphthalene ring remained intact to obtain 2,2-dimethyl-3,4-dihydro-2*H*-benzo[*h*]chromene (**8**) as colorless oil in quantitative yield. Both compounds **7** and **8** are natural products; **8** was first isolated from *Tectona granais* Linn. [31] and both **7** and **8** were found in various *Galium* species [32]. These compounds were also synthesized earlier in rather poor yield [33]. On nitration the chromene **8** produced the nitro compound **9** in excellent yield (95%), which on reduction with Sn/HCl followed by acetylation with acetic anhydride afforded the amide derivative **10** (84%). On treating with concentrated nitric acid the amide **10** in glacial acetic acid under ice-bath temperature furnished β-lapachone in quite reasonable yield (62%), characterized from detailed spectral studies.

In 2003, Perez-Sacau et al. described the preparation of β-lapachone by means of sulfuric acid-catalyzed cyclization of lapachol; however, this method afforded both α- and β-lapachones in 34% and 39% yields, respectively [28]. Lee and his group [29] reported a concise total synthesis of **1** in just three steps from commercially available 4-methoxy-1-naphthol (**11**) by benzopyran formation, catalytic hydrogenation, and Jones oxidation (Scheme 2). Treatment of **11** with 3-methyl-2-butenal in the presence of 20 mol% of EDDA in refluxing chloroform produced lapachenole (**12**) in 60% yield, a natural product isolated from *Tabebuia chrysantha* [32]. Catalytic

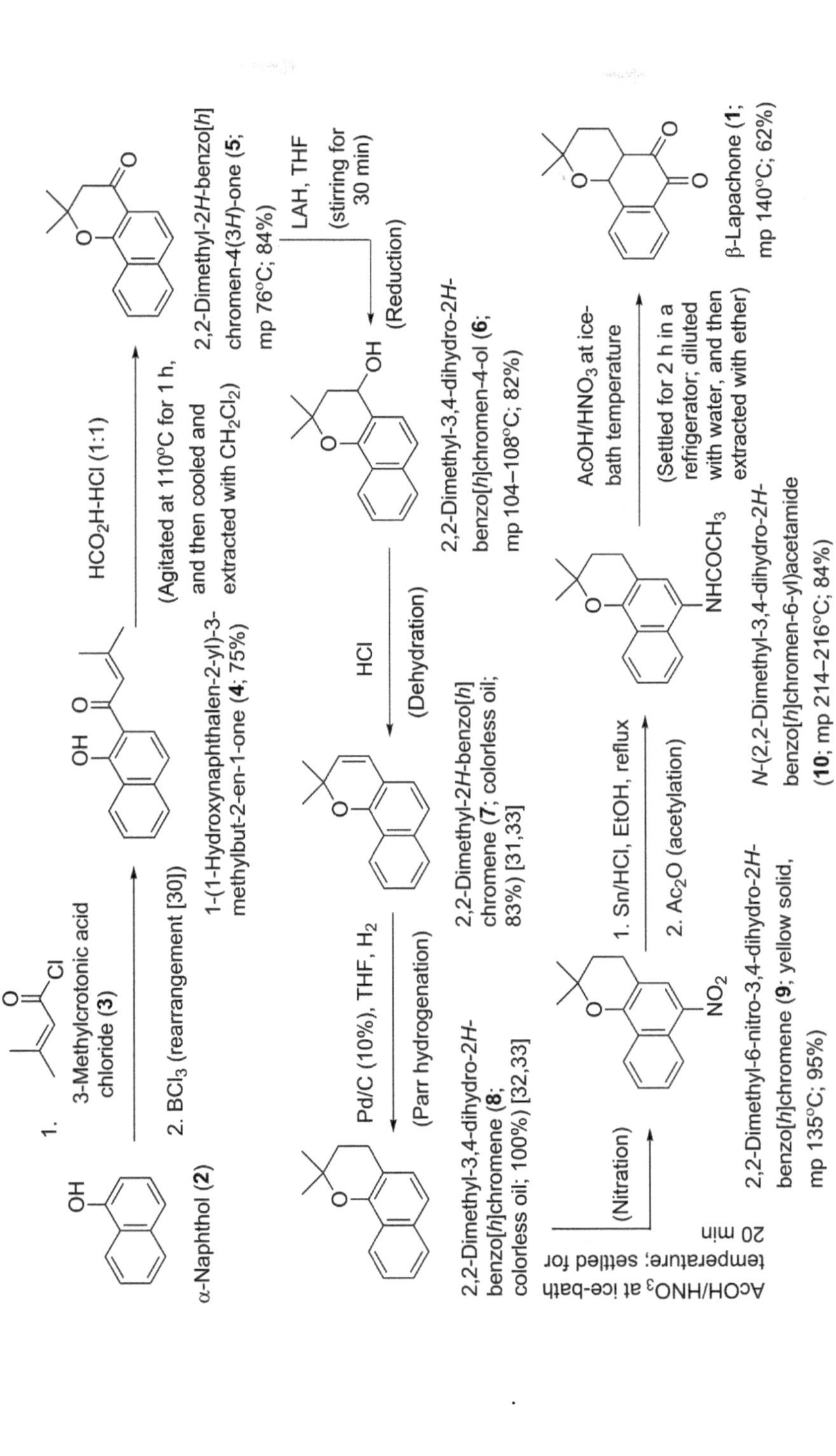

Scheme 1 Amaral and Barnes total synthesis of β-lapachone (**1**) [27].

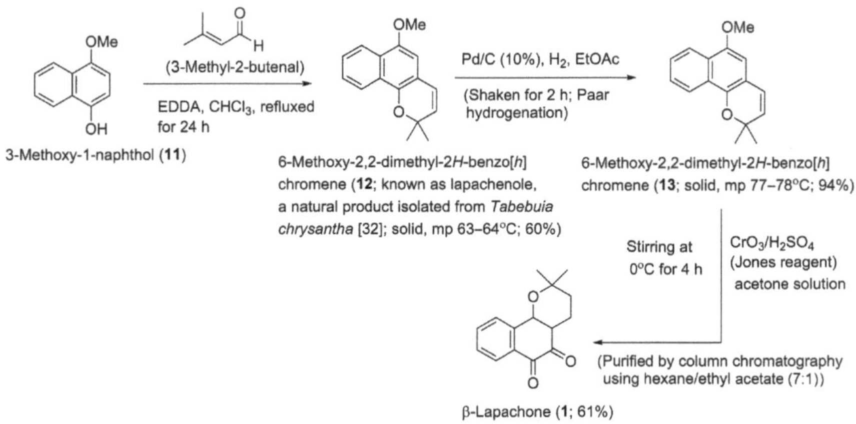

Scheme 2 Lee's total synthesis of β-lapachone (**1**) [29].

hydrogenation of **12** over Pd/C (20 psi) in ethyl acetate gave 6-methoxy-2,2-dimethyl-2*H*-benzo[*h*]-chromene (**13**) as solid product (94%), which on Jones oxidation afforded β-lapachone (**1**) in 61% yield. The structure was assigned by comparison with reported physical and spectral data for β-lapachone.

References

[1] J.R. de Sousa, G.D.F. Silva, T. Miyakoshi, C.-L. Chen, J. Nat. Prod. 69 (2006) 1225.
[2] H.J. Woo, K.Y. Park, C.H. Rhu, W.H. Lee, B.T. Choi, G.Y. Kim, Y.M. Park, Y.H. Choi, J. Med. Food 9 (2006) 161.
[3] A.M. Goncalves, M.E. Vasconcellos, R. Docampo, F.S. Cruz, W. De Souza, W. Leon, Mol. Biochem. Parasitol. 1 (1980) 167.
[4] K. Schaffner-Sabba, K.H. Schmidt-Ruppin, W. Wehrli, A.R. Schuerch, J.W. Wasley, J. Med. Chem. 27 (1984) 990.
[5] C.J. Li, C. Wang, A.B. Pardee, Cancer Res. 55 (1995) 3712.
[6] C.C. Lai, T.J. Liu, L.K. Ho, M.J. Don, Y.P. Chau, Histol. Histopathol. 13 (1998) 89.
[7] Y.P. Chau, S.G. Shiah, M.J. Don, M.L. Kuo, Free Radical. Biol. Med. 24 (1998) 660.
[8] A.G. Ravelo, A. Estévez-Braun, H. Chávez-Orellana, E. Pérez-Sacau, D. Mesa-Siverio, Curr. Top. Med. Chem. 4 (2004) 241.
[9] A.B. Pardee, Y.Z. Li, C.J. Li, Curr. Cancer Drug Targets, 2 (2002) 227.
[10] E.A. Bey, M.S. Bentle, K.E. Reinicke, Y. Dong, C.-R. Yang, L. Girard, J.D. Minna, W.G. Bornmann, J. Gao, D.A. Boothman, Proc. Natl. Acad. Sci. 104 (2007) 11832.
[11] S.M. Planchon, S.M. Wuerzberger-Davis, J.J. Pink, K.A. Robertson, W.G. Bornmann, D.A. Boothman, Oncol. Rep. 6 (1999) 485.
[12] S.G. Shiah, S.E. Chuang, Y.P. Chau, S.C. Shen, M.L. Kuo, Cancer Res. 59 (1999) 391.
[13] M.-J. Don, Y.-H. Chang, K.-K. Chen, L.-K. Ho, Y.-P. Chau, Mol. Pharm. 59 (2001) 784.
[14] Y.H. Choi, H.S. Kang, M.A. Yoo, J. Biochem. Mol. Biol. 36 (2003) 223.
[15] C.W. Distelhorst, G. Dubyak, Blood 91 (1998) 731.
[16] M. Fang, H. Zhang, S. Xue, N. Li, L. Wang, Cancer Lett. 127 (1998) 113.

[17] A.R. Marks, Am. J. Phys. 272 (1997) H597.
[18] D.J. McConkey, P. Hartzell, J.F. Amador-Perez, S. Orrenius, M. Jondal, J. Immunol. 143 (1989) 1801.
[19] I.E. Wertz, V.M. Dixit, J. Biol. Chem. 275 (2000) 11470.
[20] A. Petersen, R.F. Castilho, O. Hansson, T. Wieloch, P. Brundin, Brain Res. 857 (2000) 20.
[21] J.J. Pink, S. Wuerzberger-Davis, C. Tagliarino, S.M. Planchon, X. Yang, C.J. Froelich, D.A. Boothman, Exp. Cell Res. 255 (2000) 144.
[22] C. Tagliarino, J.J. Pink, G.R. Dubyak, A.-L. Niemineni, D.A. Boothman, J. Biol. Chem. 276 (2001) 19150.
[23] D. Gupta, K. Podar, Y.T. Tai, B. Lin, T. Hideshima, M. Akiyama, R. LeBlanc, L. Catley, N. Mitsiades, C. Mitsiades, D. Chauhan, N.C. Munshi, K.C. Anderson, Exp. Hematol. 30 (2002) 711.
[24] H.R. Shah, R.M. Conway, K.R. Van Quill, M.C. Madigan, S.A. Howard, J. Qi, V. Weinberg, J.M. O'Brien, Eye 22 (2008) 454.
[25] H.-P. Tzeng, F.-M. Ho, K.-F. Chao, M.L. Kuo, S.-Y. Lin-Shiau, S.-H. Liu, Am. J. Respir. Crit. Care Med. 168 (2003) 85.
[26] K. Maruyama, Y. Naruta, Chem. Lett. 847 (1977).
[27] A.C.F. Amaral, R.A. Barnes, J. Heterocycl. Chem. 29 (1992) 1457.
[28] E. Perez-Sacau, A. Estevez-Braun, A. ravelo, E. Ferro, H. Tokuda, T. Mukainaka, H. Nishino, Bioorg. Med. Chem. 11 (2003) 483.
[29] X. Wang, Y. Chen, Y.R. Lee, Bull. Kor. Chem. Soc. 32 (2011) 153.
[30] O. Piccolo, L. Filippini, L. Tinucci, E. Valoti, A. Citterio, Tetrahedron 42 (1986) 885.
[31] H.W. Sandermann, M.H. Simatupang, Naturwissenschaften 54 (1968) 118.
[32] A.R. Burnett, R.H. Thomson, J. Chem. Soc. 854 (1968).
[33] B.K. Rohatgi, R.B. Gupta, R.N. Khanna, Indian J. Chem. 20 (1981) 501.

CHAPTER FIFTEEN

Bombykol

Abbreviations

Bu_4NF *tetra-n*-butylammonium fluoride
CH_2Cl_2 dichloromethane
$(COCl)_2$ oxalyl chloride
DIBAL-H diisobutylaluminum hydride
dppp 1,3-bis(diphenylphosphino)propane
rt room temperature
TBDMS *tetra*-butyldimethylsilyl
THF tetrahydrofuran

Systematic name: (10*E*,12*Z*)-10,12-Hexadecadien-1-ol
Compound class: Long-chain alkenyl alcohol
Structure:

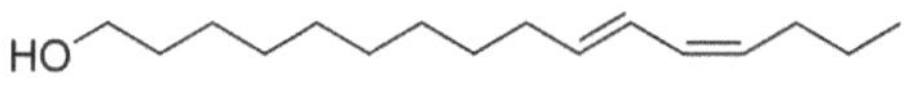

Bombykol (**1**)

Natural source: Silkworm moth *Bombyx mori* [1, 2]
Pharmaceutical potential: Sex-hormone pheromone [1, 2]
Synthetic route:

Bombykol (**1**) is famous as the first isolated insect sex pheromone. It is a component of the female sex pheromone of silkworm moth *Bombix mori*, and it was identified and synthesized by Butenant and collaborators in 1959 [1, 2]. Since its first synthesis, a number of reports on its total synthesis are available implementing a variety of techniques [3–7]. Bombykol possesses a consecutive *E*,*Z*-diene unit on C-16 carbon chain alcohol, and the most important part of this synthesis is the stereoselective preparation of this *E*,*Z*-diene moiety. In 2000, Uenishi and coworkers [8] reported a concise and efficient stereospecific total synthesis of geometrically pure bombykol (Scheme 1) using the Kumada-Tamao-Corriu coupling technique starting from the known aldehyde 10-((*tert*-butyldimethylsilyl)oxy)decanal (**2**) [9].

Total Synthesis of Bioactive Natural Products
https://doi.org/10.1016/B978-0-08-102822-3.00015-8

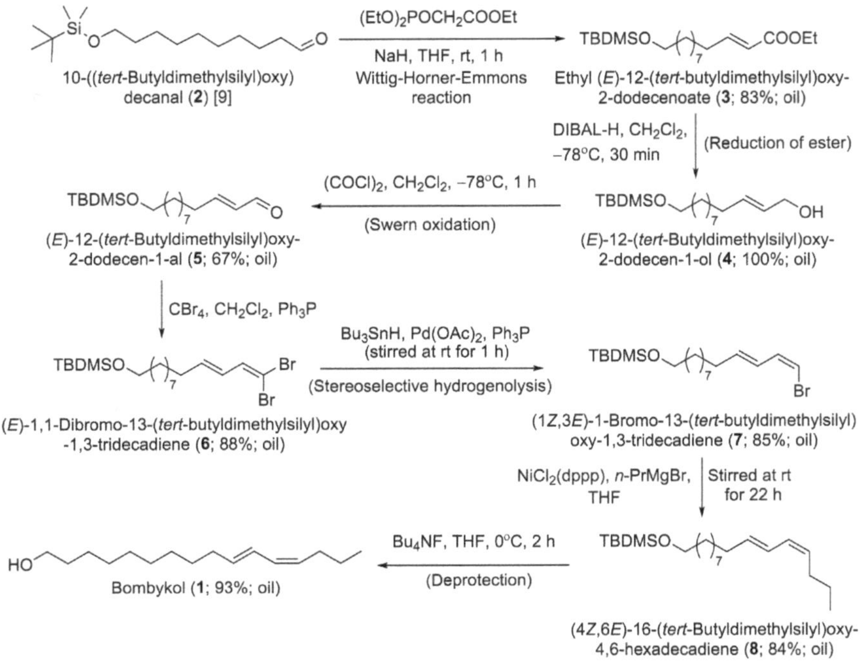

Scheme 1 Total synthesis of bombykol (**1**) [8].

Wittig-Horner-Emmons reaction of protected aldehyde **2** with triethyl phosphonoacetate produced α,β-unsaturated ester (**3**) in 83% yield. Reduction of the ester with DIBAL-H to alcohol followed by oxidation under Swern conditions afforded α,β-unsaturated aldehyde (**5**) in 67% yield in two steps. Dibromomethylenation of **5** with carbon tetrabromide and triphenylphosphine in dichloromethane gave (*E*)-1,1-dibromo-13-(*tert*-butyldimethylsilyl)oxy-1,3-tridecadiene (**6**; 88%) as oil. Stereoselective hydrogenolysis with Bu_3SnH in the presence of a Pd catalyst furnished the desired *Z,E*-bromodiene (**7**) exclusively in 85% yield. Cross-coupling reaction of **7** was carried out in the presence of $NiCl_2$(dppp) with propylmagnesium chloride in THF for 22 h at room temperature to obtain (4*Z*,6*E*)-16-(*tert*-butyldimethylsilyl)oxy-4,6-hexadecadiene (**8**; 84%; oil). Finally, deprotection of the TBDMS with Bu_4NF in THF afforded bombykol (**1**) in 93% yield as oil. Physical and spectral data of the synthetic bombykol are in accordance with those reported for the natural sample.

References

[1] A. Butenandt, R. Beckmann, D. Stamm, E.Z. Hecker, Naturforsch. 14B (1959) 283.
[2] A. Butenandt, E. Hecker, Angew. Chem. 73 (1961) 349.
[3] E. Negishi, G. Lew, T. Yoshida, J. Chem. Soc., Chem. Commun. (1973) 874 J.F. Normant, A. Commercon, J. Villieras, Tetrahedron Lett. (1975) 1465; D. Samain, C. Descoins, Bull. Soc. Chim. Fr. (1979) II-71; E. Negishi, T. Yoshida, A. Abramovitch, G. Lew, R.M. Williams, Tetrahedron Lett. 47 (1991) 343; J.A. Cabezas, A.C. Oehlschlager, Synthesis (1999) 107.
[4] H.J. Bestmann, O. Vostrowsky, H. Paulus, W. Billmann, W. Stransky, Tetrahedron Lett. (1977) 121 S. Ranganathan, V. Maniktala, R. Kumar, G.P. Singh, Indian J. Chem. 23B (1984) 1197; J.S. Yadav, K. Balakrishnan, L. Sivadasan, Indian J. Chem. 28B, 297; A. Alexakis, D. Jachiet, Tetrahedron 45 (1989) 381; L. Dasaradhi, P. Neelakantan, S.J. Rao, U.T. Bhalerao, Syn. Commun. 21 (1991) 183.
[5] J.F. Normant, A. Commercon, J. Villieras, Tetrahedron Lett. 18 (1975) 1465 N. Miyaura, H. Suginome, A. Suzuki, Tetrahedron 39 (1983) 3271; M. Gardette, N. Jabri, A. Alexakis, J.F. Normant, Tetrahedron 40 (1984) 2741; V. Fiandanese, G. Marchese, F. Naso, L. Ronzini, D. Rotunno, Tetrahedron Lett. 30 (1989) 243.
[6] S. Tsuboi, T. Masuda, H. Makino, A. Takeda, Tetrahedron Lett. 23 (1982) 209.
[7] B.M. Trost, J.M. Fortunak, J. Am. Chem. Soc. 102 (1980) 2841 A. Alexakis, D. Jachiet, Tetrahedron 45 (1989) 381; H. Miyake, K. Yamamura, Chem. Lett. (1994) 897.
[8] J. Uenishi, R. Kawahama, Y. Izaki, O. Yonemitsu, Tetrahedron 56 (2000) 3493.
[9] B.M. Trost, Synthesis (1991) 1235 P.E. Sonnet, S.F. Osman, H.C. Gerard, R.L. Dudley, Chem. Phys. Lipids 69 (1994) 121.

CHAPTER SIXTEEN

Bulbophylol-B

Abbreviations

Bn benzyl
CH_2Cl_2 dichloromethane
DMF *N,N*-dimethylformamide
***m*-CPBA** *meta*-chloroperbenzoic acid
$NaBH_4$ sodium borohydride
rt room temperature
PPh_3 triphenylphoshine
TBAF tetrabutylammonium fluoride
THF tetrahydrofuran
TIPSCl triisopropylsilyl chloride

Systematic name: 4-Methoxy-6,7-dihydro-[1,3]dioxolo[4′,5′:5,6]benzo[1,2-*b*]benzo[*f*]oxepin-9-ol

Compound class: Dihydrodibenz[*b,f*]oxepin

Structure:

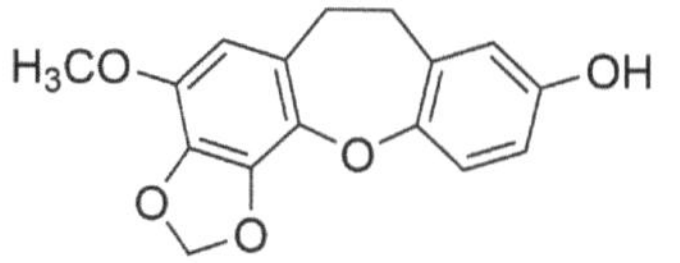

Bulbophylol-B (**1**)

Natural source: *Bulbophyllum kwangtungense* Schltr (leaves and stems; family: Orchidaceae) [1]; *Bulbophyllum odoratissimum* Lindl [2]

Pharmaceutical potential: Anticancer [1]; antioxidant [3]

Synthetic route:

The first total synthesis of bulbophylo-B (**1**) was performed by Yao and his group [2] in 12 steps with an overall yield of 17.9%. They designed a retrosynthetic route involving the Wittig reaction, selective reduction of a carbon-carbon double bond, and intramolecular Ullmann diaryl ether forming reaction as the key steps for construction of the dihydrodibenz[*b,f*]oxepin skeleton as depicted in Scheme 1.

Total Synthesis of Bioactive Natural Products
https://doi.org/10.1016/B978-0-08-102822-3.00016-X

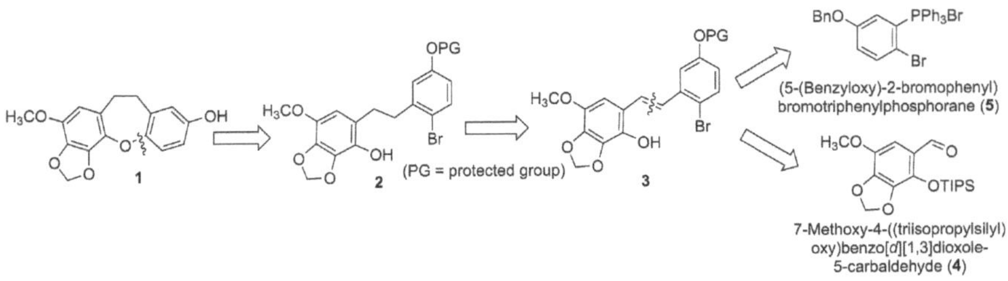

Scheme 1 Retrosynthesis route for bulbophylol-B (**1**).

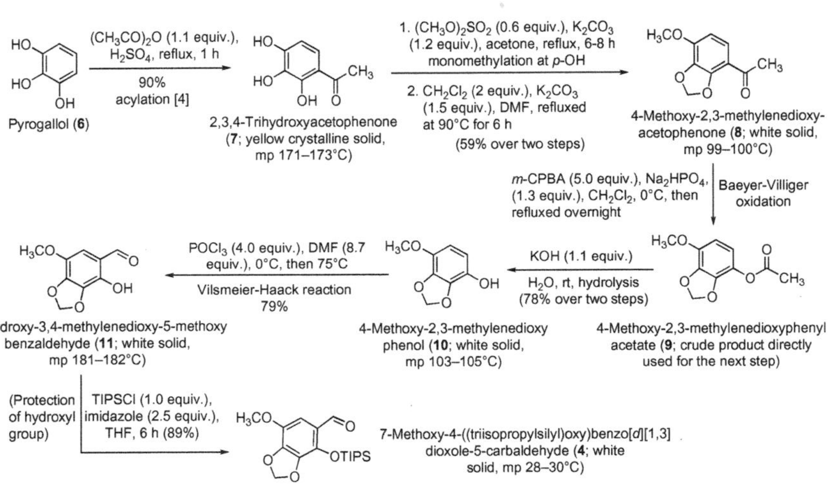

Scheme 2 Synthesis of substituted benzaldehyde **4**.

Based on their retrosynthetic strategy, the investigators first prepared substituted benzaldehyde **4** and benzyltriphenyl-phosphonium salt **5** starting from known compounds as shown in Schemes 2 and 3, respectively. Once having the starting compounds at hand, they then explored the total synthesis as described in Scheme 4.

Synthesis of substituted benzaldehyde 4:

Preparation of the synthon **4** was started with the commercially available pyrogallol (**6**; benzene-1,2,3-triol), which on acylation gave ketone **7** [4]. Monomethylation of *p*-OH was achieved by treating **7** with dimethyl sulfate under basic conditions, and the crude product was then treated with CH_2Cl_2 and K_2CO_3 in DMF to have 4-methoxy-2,3-methylenedioxyacetophenone (**8**) as white solid. Baeyer-Villiger oxidation of compound **7** with *m*-CPBA in the presence of Na_2HPO_4 yielded the corresponding ester **9**, which was hydrolyzed to furnish phenol **10**. Vilsmeier-Haack reaction of **10** by

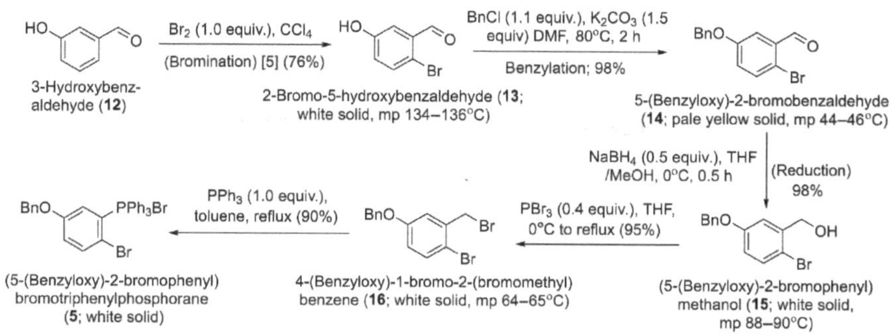

Scheme 3 Synthesis of benzyltriphenyl-phosphonium salt **5**.

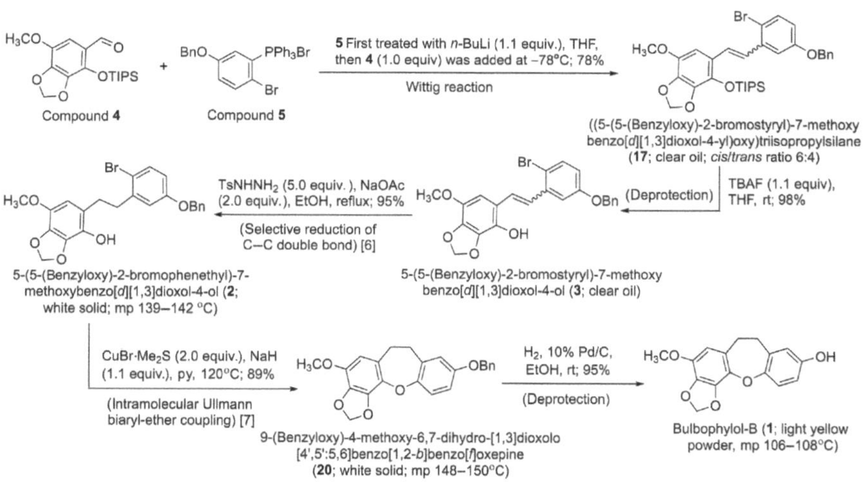

Scheme 4 Synthesis of bulbophylol-B (**1**) [2].

treatment with dimethylformamide (DMF) and phosphorus oxychloride gave the corresponding salicylaldehyde **11**, which was then protected with the triisopropyllsilyl (TIPS) group by the usual method to give the intermediate 7-methoxy-4-((triisopropylsilyl)oxy)benzo[*d*][1,3]dioxole-5-carbaldehyde (**4**) as white solid (Scheme 2).

Synthesis of benzyltriphenyl-phosphonium salt 5:

Another synthon, benzyltriphenylphosphonium salt (**5**), was prepared starting from 3-hydroxybenzaldehyde (**12**) in a five-step reaction sequence. Bromination of compound **12** afforded **13** [5], which was then protected with a benzyl group to give **14**. Reduction of **14** with sodium borohydride in THF/MeOH (1:1) yielded 4-(benzyloxy)-1-bromo-2-(bromomethyl) benzene (**15**), which was treated with phosphorus tribromide to furnish benzyl bromide **16**. Treatment of **16** with triphenylphosphine in toluene under reflux provided (5-(benzyloxy)-2-bromophenyl) bromotriphenylphosphorane (**5**) as white solid (Scheme 3).

Combination of synthons 4 and 5 to build up bulbophylol-B (1): With key synthones **4** and **5** at hand, the investigators then performed the key Wittig reaction using *n*-BuLi/THF to form the stilbene ((5-(5-(benzyloxy)-2-bromostyryl)-7-methoxybenzo[*d*][1,3]dioxol-4-yl)oxy) triisopropylsilane (**17**) with a *cis/trans* ratio of 6:4 as clear oil in 78% yield. Then deprotection of **17** with tetrabutylammonium fluoride gave phenol **3** in near quantitative yield. The olefinic double bond in stilbene **3** was selectively reduced upon refluxing with 4-methylbenzenesulfonohydrazide and anhydrous sodium acetate in ethanol under nitrogen atmosphere [6] for 2 h to obtain the desired dihydrostilbene **2** in 95% yield. On treatment with cuprous bromide/dimethylsulfide and NaH/THF, compound **2** underwent intramolecular Ullmann biaryl-ether coupling [7] to form 9-(benzyloxy)-4-methoxy-6,7-dihydro-[1,3]dioxolo[4′,5′:5,6]benzo[1,2-*b*]benzo[*f*]oxepine (**20**), which on deprotection with Pd-C/H_2 in ethanol afforded the target molecule bulbophylol-B (**1**) in excellent yield as light yellow powder. The overall reaction sequences are shown in Scheme 4. The ^{1}H NMR, ^{13}C-NNR, and MS spectral data of this synthetic compound were found to be identical to the reported data for natural bulbophylol-B [1].

References

[1] B. Wu, S. He, Y.J. Pan, Planta Med. 72 (2006) 1244.
[2] J. Lin, W. Zhang, N. Jiang, Z. Niu, K. Bao, L. Zhang, D. Liu, C. Pan, X. Yao, J. Nat. Prod. 71 (2008) 1938.
[3] J. Wang, L.Y. Wang, S. Kitanaka, J. Nat. Med. 61 (2007) 381.
[4] Y.K. Rao, C.V. Rao, P.H. Kishore, D. Gunasekar, J. Nat. Prod. 64 (2001) 368.
[5] W.A.L. Van Otterlo, J.P. Michael, M.A. Fernandes, C.B. De Koning, Tetrahedron Lett. 45 (2004) 5091.
[6] T.R. Kelly, Q. Li, V. Bhushan, Tetrahedron Lett. 31 (1990) 161.
[7] R. Olivera, R. SanMartin, F. Churruca, E. Dominguez, J. Org. Chem. 67 (2002) 7215.

CHAPTER SEVENTEEN

Caminoside A

Abbreviations

Ac acetyl
Azmb 2-(azidomethyl)benzoyl
Bn benzyl
Bt butyryl
Bz benzoyl
CH_2Cl_2 dichloromethane
DCC 1,3-dicyclohexylcarbodiimide
DDQ 2,3-dichloro-5,6-dicyano-1,4-benzoquinone
DMAP 4-dimethylaminopyridine
DMF *N,N*-dimethylformamide
DMSO dimethylsulfoxide
Lev levulinyl
MeCN acetonitrile
MeOH methanol
MP *p*-methoxyphenyl
NIS *N*-iodosuccinimide
PMB *p*-methoxybenzyl
TFA trifluoroacetic acid
THF tetrahydrofuran
TMS trimethylsilyl
TsOH *p*-toluenesulfonic acid

Systematic name: (2*R*,3*R*,4*S*,5*R*,6*S*)-3-Acetoxy-6-(((2*R*,3*R*,4*S*,5*S*,6*R*)-4,5-dihydroxy-6-(hydroxymethyl)-2-((2-oxononadecan-10-yl)oxy)tetrahydro-2*H*-pyran-3-yl)oxy)-5-(((2*S*,3*S*,4*R*,5*R*,6*S*)-3,4,5-trihydroxy-6-methyltetrahydro-2*H*-pyran-2-yl)oxy)-2-((((2*R*,3*S*,4*S*,5*R*,6*R*)-3,4,5-trihydroxy-6-methyltetrahydro-2*H*-pyran-2-yl)oxy)methyl)tetrahydro-2*H*-pyran-4-yl butyrate

Compound class: Glycolipid

Total Synthesis of Bioactive Natural Products
https://doi.org/10.1016/B978-0-08-102822-3.00017-1

Structure:

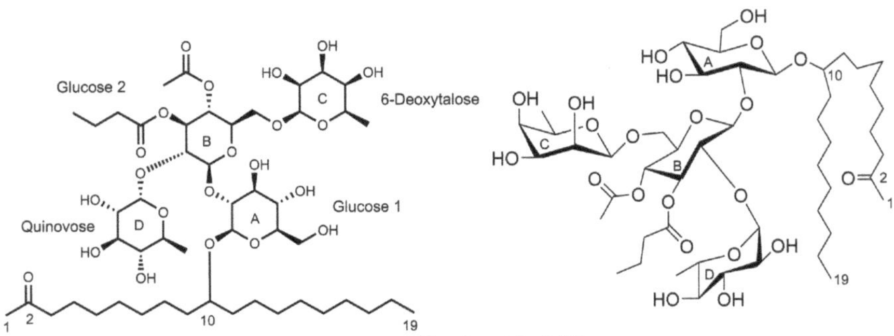

Natural source: *Caminus sphaeroconia* (marine sponge) [1]

Pharmaceutical potential: Antimicrobial [1]

Synthetic route:

The first total synthesis of caminoside A (**1**), a novel antimicrobial tetrasaccharide glycolipid from the marine sponge *Caminus sphaeroconia* acting as the first bacterial type III secretion inhibitor, was accomplished by Yu and his coworkers in a total of 57 steps starting from D-glucose, D-galactose, L-rhamnose, and 9-decenal [2]. The retrosynthetic approach for this molecule **1** as designed is shown in Scheme 1. Preparation of individual monosaccharide building blocks **2**–**5** was depicted, respectively, in Schemes 2–5.

Preparation of 2-*O*-acetyl-3,4,6-tri-*O*-benzyl-1-*O*-(2,2,2-trifluoro-1-(phenylimino)ethyl)-D-glucose (fragment 2):

The monosaccharide building block 2-*O*-acetyl-3,4,6-tri-*O*-benzyl-1-*O*-(2,2,2-trifluoro-1-(phenylimino)ethyl)-D-glucose (**2**) was prepared starting from D-glucose. 2-*O*-Acetyl-3,4,6-tri-*O*-benzyl-D-glucose (**8**) was first synthesized in six steps [3] with 42% overall yield from the sugar, which on treatment with 2,2,2-trifluoro-*N*-phenylacetimidoyl chloride in the presence of potassium carbonate in acetone at room temperature furnished fragment **2** in 90% yield [4] (Scheme 2).

Preparation of 4-*O*-acetyl-2-*O*-(2-azidomethylbenzoyl)-3-*O*-butyryl-1-*O*-*p*-methoxyphenyl-D-glucose (fragment 3):

4-*O*-Acetyl-2-*O*-azidomethylbenzoyl-3-*O*-butyryl-1-*O*-*p*-methoxyphenyl-β-D-glucose (**3**) was prepared starting from β-D-glucose (**7**); compound **7** was converted into 1,2,4,6-tetra-*O*-acetyl-3-*O*-allyl-D-glucose (**9**) in four steps with 70% yield as per reported procedure [5]. On glycosylation with *p*-methoxyphenol/trimethylsilyl

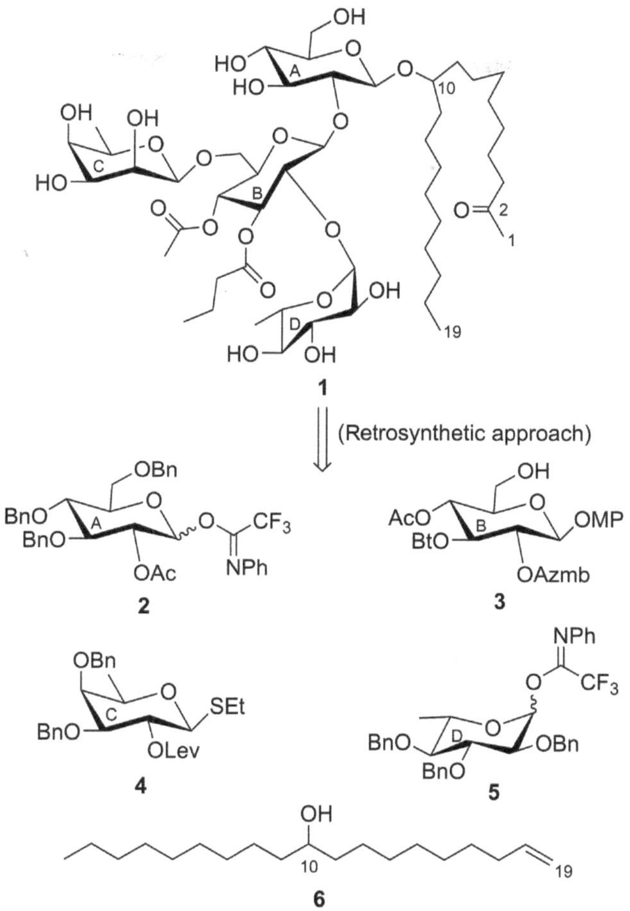

[Azmb = 2-(azidomethyl)benzoyl; Bn = benzyl; Bt = butyryl; Lev = levulinyl; MP = *p*-methoxyphenyl]

Scheme 1 Retrosynthetic approach for caminoside A (**1**) [2].

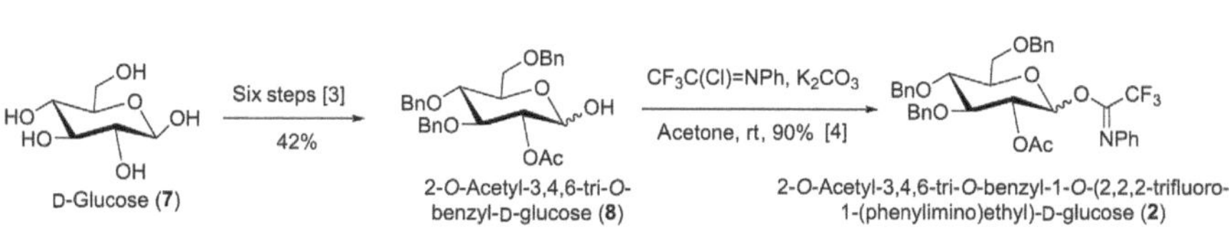

Scheme 2 Preparation of fragment **2** [2].

trifluoromethanesulfonate (TMSOTf) in dichloromethane [6], compound **9** gave β-D-glucoside **10**, which on removal of the 2,4,6-acetates followed by selective protection of the 4,6-OH groups with benzylidene furnished derivative **11**. The remaining 2-OH was then protected with 2-azidomethylbenzoyl (Azmb) group to give derivative **12**. Cleavage

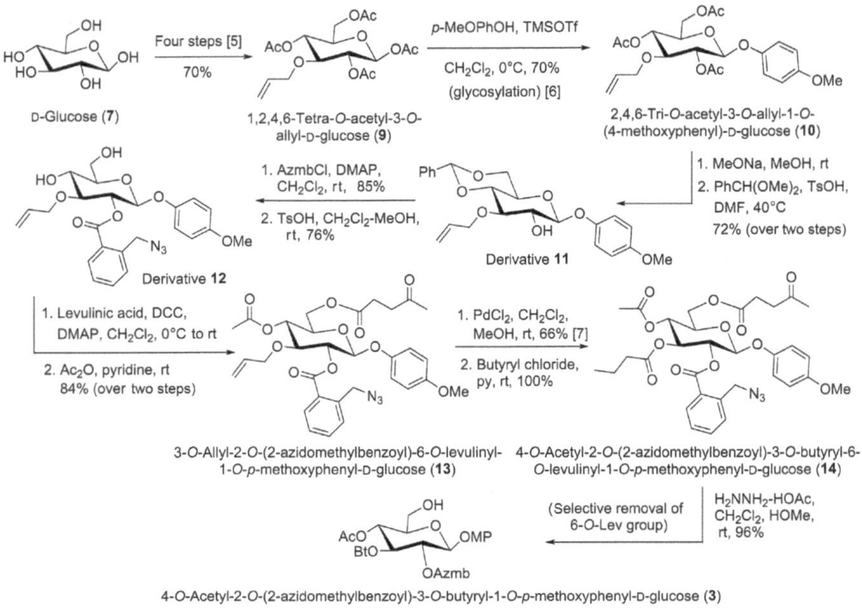

Scheme 3 Preparation of fragment **3** [2].

of the 4,6-benzylidene followed by selective protection of the resulting 6-OH with levulinyl (Lev) group and the remaining 4-OH with acetate produced the fully protected glucopyranoside 3-*O*-allyl-2-*O*-(2-azidomethylbenzoyl)-6-*O*-levulinyl-1-*O*-*p*-methoxyphenyl-D-glucose (**13**) in good yield. Selective cleavage of the 3-*O*-allyl group was then performed in the presence of $PdCl_2$ in methanol [7]. Substitution of the resulting 3-OH with Bt group formed compound **14** in excellent yield. Finally, selective removal of the 6-*O*-levulinyl moiety using hydrazine acetate afforded fragment **3**.

Preparation of ethyl 3,4-di-*O*-benzyl-2-*O*-levulinyl-1-thio-β-D-fucopyranoside (fragment 4):

Ethyl 3,4-di-*O*-benzyl-2-*O*-levulinyl-1-thio-β-D-fucopyranoside (fragment **4**) was synthesized starting from D-galactose (**15**) (Scheme 4). D-galactose (**15**) was converted into the thioglycoside derivative **20** employing reported procedures [8–11]. After temporary protection of the 2-OH with *p*-methoxybenzyl (PMB) group, the 3,4-*O*-isopropylidene within **21** was removed followed by blocking of the resulting 3,4-OH groups in **22** with the persistent benzyl (Bn) groups. Oxidative removal of the 2-*O*-PMB group with DDQ in CH_2Cl_2-H_2O yielded ethyl 3,4-di-*O*-benzyl-1-thio-β-D-fucopyranoside (**23**), which eventually furnished fragment **4** on protecting the 2-OH with levulinyl moiety in excellent yield.

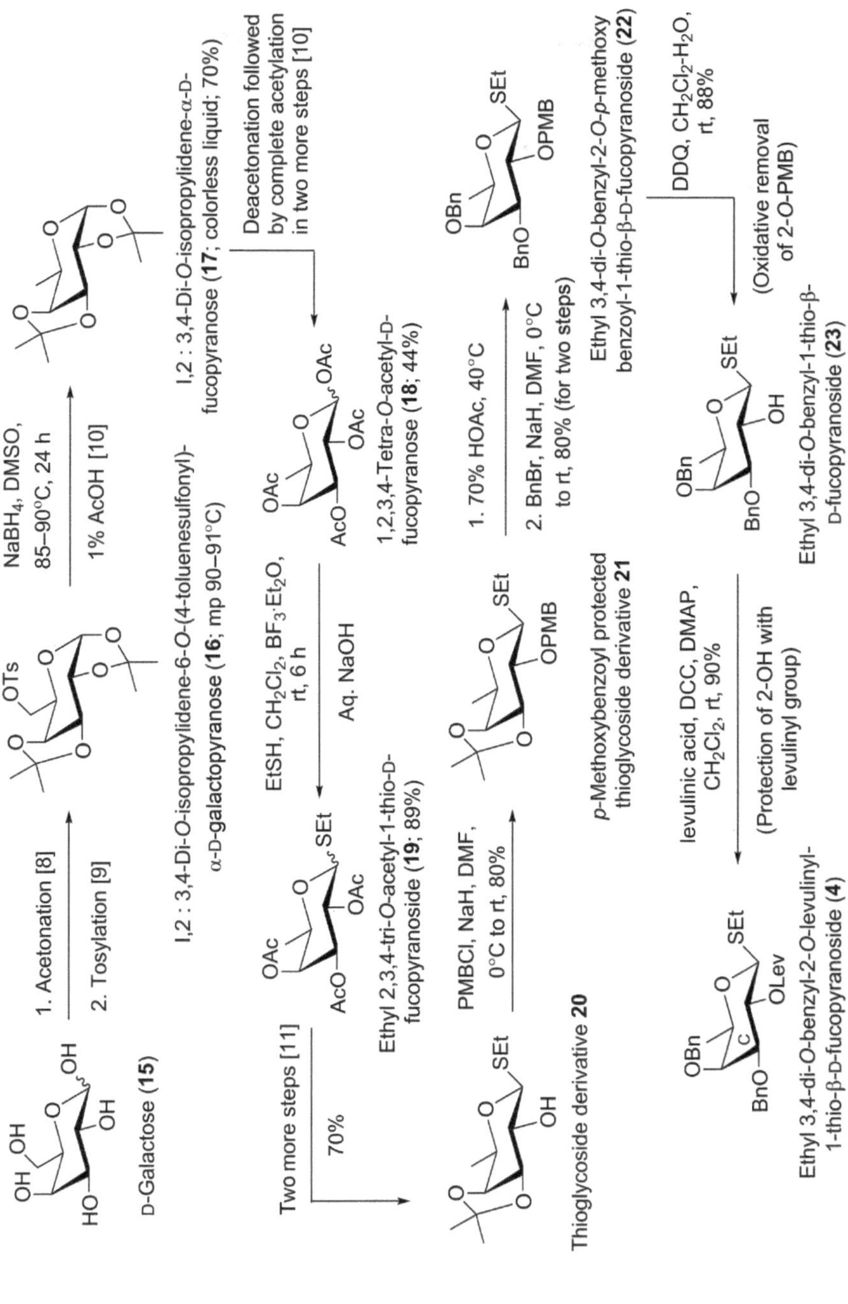

Scheme 4 Preparation of fragment **4** [2].

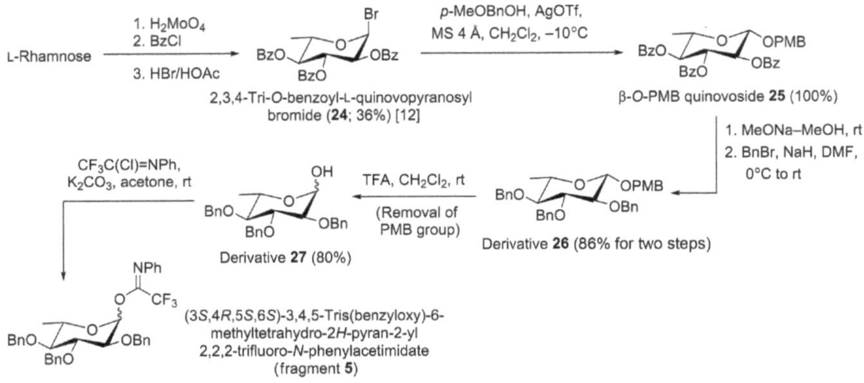

Scheme 5 Preparation of fragment **5** [2].

Preparation of 3,4,5-tris(benzyloxy)-6-methyltetrahydro-2*H*-pyran-2-yl 2,2,2-trifluoro-*N*-phenylacetimidate (fragment 5):

2,3,4-Tri-*O*-benzoyl-L-quinovopyranosyl bromide **24** was first prepared from L-rhamnose (Scheme 5) in three steps following reported methods [12]. Glycosylation of **24** with *p*-methoxybenzyl (PMB) alcohol in the presence of silver triflate yielded the β-*O*-PMB quinovoside **25** in quantitative yield, which eventually converted into fragment **5** through a series of reactions (Scheme 5).

Construction of C-B disaccharide fragment:

The C-B disaccharide fragment **31** was constructed first through coupling between fragments **3** and **4** (Scheme 6). Coupling of thioglycoside **4** with 6-OH-glucoside **3** took place in the presence of NIS/TMSOTf resulting in the expected (1→6)-β-disaccharide **28** in 76% yield [13]. Highly selective removal of the 2-*O*-Lev group then produced derivative **28** in good yield (90%). Conversion of the equatorial 2-OH in **29** into the axial one was then successfully achieved following an oxidation-reduction sequence [14] to give derivative **30** in a satisfactory 56% yield over the two steps. Then the investigators carried out an oxidative removal of the anomeric *p*-methoxyphenyl group to obtain a hemiacetal that was converted into the disaccharide trifluoroacetimidate **31** with an excellent yield (90% for two steps).

Final assembly for caminoside A (1):

Finally, caminoside A (**1**) was synthesized as depicted in Scheme 7. 1-Nonadecen-10-ol (**6**) was glycosylated with glucopyranosyl trifluoroacetimidate (**2**) under the catalysis of TMSOTf [15] to give the β-glucoside **32** in 92% yield. Selective hydrolysis of the 2-*O*-Ac group followed by Wacker oxidation [16] of the terminal carbon-carbon double bond yielded

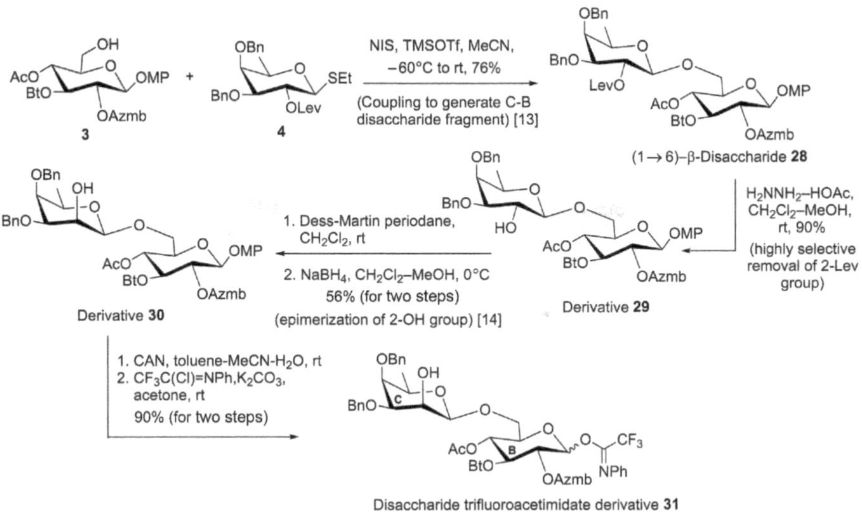

[Azmb = 2-(azidomethyl)benzoyl; Bn = benzyl; Bt = butyryl; Lev = levulinyl; MP = *p*-methoxyphenyl]

Scheme 6 Construction of C-B disaccharide fragment [2].

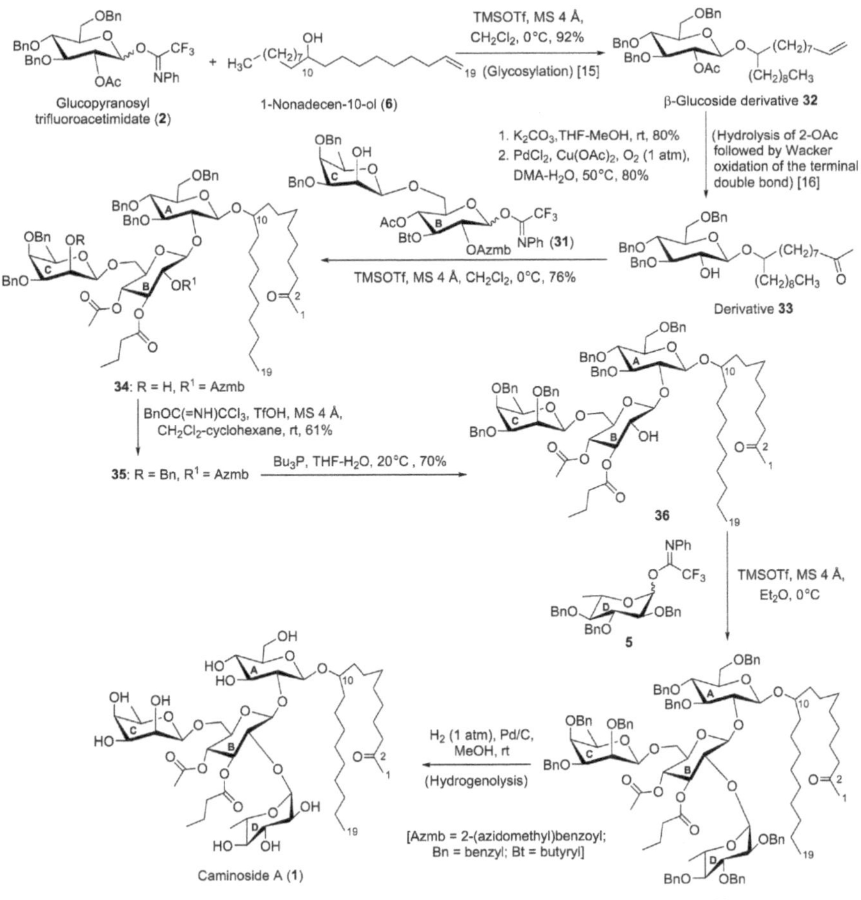

Scheme 7 Yu's total synthesis of caminoside A (**1**) [2].

derivative **33** that was then coupled with disaccharide trifluoroacetimidate **31** giving rise to the desired Talβ-(1 → 6)-Gluβ-(1 → 2)-Gluβ-trisaccharide **34** in 76% yield. The free axial 2-OH of the talose moiety was protected with Bn group on treatment with benzyl trichloroacetimidate in the presence of triflic acid to produce **35** in moderate yield (61%). Selective removal of the 2-*O*-Azmb group in **35**, in the presence of Ac and Bt groups, was achieved under the action of tributylphosphine, affording **36** in 70% yield. The resulting 2-OH was then glycosylated with the perbenzyl L-quinovopyranosyl trifluoroacetimidate **5** (fragment **5**) under TMSOTf catalysis to afford the desired tetrasaccharide **37** as the major product (53%). Hydrogenolysis of the Bn groups in the presence of Pd/C/MeOH eventually resulted in the target glycolipid caminoside A (**1**), the peracetate of which was directly compared with the reported analytical and spectral data of the peracetate derivative of the natural compound and found to be identical [1].

References

[1] R.G. Linington, M. Robertson, A. Gauthier, B. Brett Finlay, R. van Soest, R.J. Andersen, Org. Lett. 4 (2002) 4089.
[2] J. Sun, X. Han, B. Yu, Synlett (3) (2005) 437.
[3] A. Wotovic, J.-C. Jacquinet, P. Sinaÿ, Carbohydr. Res. 205 (1990) 235. M. Trumtel, P. Tavecchia, A. Veyrieres, P. Sinaÿ, Carbohydr. Res. 191 (1989) 29; J. Banoub, P. Boullanger, M. Potier, G. Descotes, Tetrahedron Lett. 27 (1986) 4145.
[4] B. Yu, H. Tao, J. Org. Chem. 67 (2002) 9099. B. Yu, H. Tao, Tetrahedron Lett. 42 (2001) 2405.
[5] K. Takeo, T. Nakaji, K. Shinmitsu, Carbohydr. Res. 133 (1984) 275.
[6] Y.S. Lee, E.S. Rho, Y.K. Min, B.T. Kim, K.H. Kim, J. Carbohydr. Chem. 20 (2001) 503.
[7] T. Ogawa, H. Yamamoto, Agric. Biol. Chem. 49 (1985) 475.
[8] O.T. Schmidt, Methods Carbohydr. Chem. 2 (1963) 318.
[9] O.T. Schmidt, Methods Carbohydr. Chem. 1 (1962) 191. A.B. Foster, W.G. Overend, M. Stacey, L.F. Wiggins, J. Chem. Sot. (1949) 2542.
[10] L.M. Lerner, Carbohydr. Res. 241 (1993) 291.
[11] K. Zegelaar-Jaarsveld, S.C. van der Plas, G.A. van der Marel, J.H. van Boom, J. Carbohydr. Chem. 15 (1996) 591. Z. Zhang, G. Magnusson, Carbohydr. Res. 262 (1994) 79.
[12] F.W. Lichtenthaler, T. Metz, Eur. J. Org. Chem. 2003 (2003) 3081. V. Bilik, W. Voelter, E. Bayer, Angew. Chem. Int. Ed. Engl. 10 (1971) 909.
[13] D.K. Hunt, P.H. Seeberger, Org. Lett. 4 (2002) 2751. P. Konradsson, U.E. Udodong, B. Fraser-Reid, Tetrahedron Lett. 31 (1990) 4313.
[14] F.W. Lichtenthaler, T. Schneider-Adams, J. Org. Chem. 59 (1994) 6728. K.K.-C. Liu, S.J. Danishefsky, J. Org. Chem. **59** (1994) 1892.
[15] M. Adinolfi, A. Iadonisi, A. Ravida, M. Schiattarella, Synlett (2004) 275. B. Yu, H. Tao, J. Org. Chem. 67 (2002) 9099; B. Yu, H. Tao, Tetrahedron Lett. 42 (2001) 2405.
[16] K.C. Nicolaou, D. Gray, J. Tea, Angew. Chem. Int. Ed. (2001) 3675.

CHAPTER EIGHTEEN

(+)-Chaetocin

Abbreviations

$BF_3 \cdot OEt_2$ boron trifluoride diethyl ether complex
Boc *tert*-butoxycarbonyl
Cbz carbobenzyloxy
$CoCl(PPh_3)_3$ *tris*(triphenylphosphine)cobalt chloride
DMAP 4-dimethylaminopyridine
DMF *N,N*-dimethylformamide
EDC·HCl 1-ethyl-3-(3-dimethylaminopropyl)carbodiimide hydrogen chloride
HOBt hydroxybenzotriazole
NBS *N*-bromosuccinamide
TBDMSCl *tert*-butyldimethylsilylchloride
V-70 (2,2′-azobis(4-methoxy-2,4-dimethylvaleronitrile)

Systematic name: (3*S*,3′*S*,6*R*,6′*R*,14′*R*,16*S*,16′*S*)-3,3′-bis(hydroxymethyl)-2,2′-dimethyl-2,2′,3,3′,6,6′,7,7′-octahydro-1*H*,1′*H*-[14,14′-bi(3,11a-epidithiopyrazino[1′,2′:1,5]pyrrolo[2,3-*b*]indole)]-1,1′,4,4′(15*H*,15′*H*)-tetraone

Compound class: Epidithiodiketopiperazine alkaloid (mycotoxin class)

Structure:

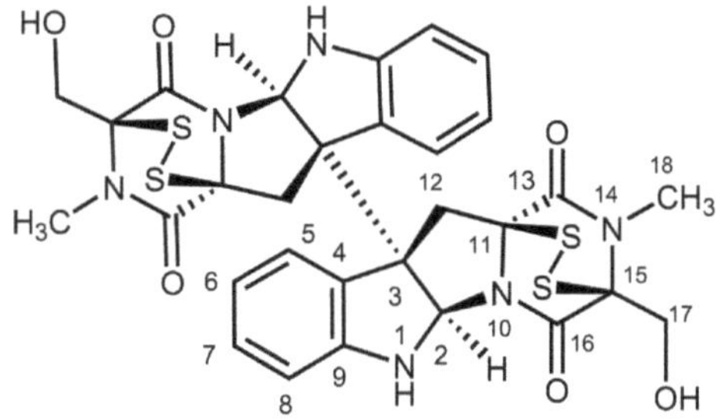

(+)-Chaetocin (**1**)

Natural source: *Chaetomium minutum* (fungus) [1]

Total Synthesis of Bioactive Natural Products
https://doi.org/10.1016/B978-0-08-102822-3.00018-3

Pharmaceutical potentials: Antibacterial and cytostatic [1, 2]; anticancer [3–5]; histone methyltransferases inhibitor [6]; thioredoxin reductase-1 inhibitor [7]; anti-HIV [8, 9]; antileukemia [10]; antihepatoma [11]

Synthetic route:

(+)-Chaetocin (**1**), a potentially bioactive fungal metabolite belonging to the 3,6-epidithiodioxopiperazine class, is a molecular dimer of two *cis* fused five-membered rings [1, 12]. Iwasa et al. [5] first reported the total synthesis of this alkaloid **1** (Scheme 1) from the (+)-diketopiperazine intermediate **7** prepared in five steps starting from known *N*-Cbz-protected *N*-methyl-D-serine (**1**) [13] and commercially available D-tryptophan methyl ester hydrochloride (**2**) with an overall yield of 69%. The obtained (+)-diketopiperazine intermediate was then subjected to the stereoselective bromocyclization reaction with *N*-bromosuccinimide in acetonitrile at −30°C to afford (+)-tetracyclic diketopiperazine bromide (**8**) in 88% yield. Compound **8** underwent radical bromination with NBS in the presence of radical initiator V-70 at room temperature in a stereoselective manner to give the tribromide derivative **9** followed by its hydrolysis in phosphate buffer at pH 7 to furnish (+)-tetracyclic diketopiperazine bromo-diol (**10**) as a major stereoisomer in 47% yield. This unprotected bromo-diol **10** was then directly converted into the desired dimeric octacyclic diol **11** through a reductive coupling reaction using $CoCl(PPh_3)_3$ complex in acetone as a single isomer in 55% yield. For constructing the disulfide bridges, at this stage, the investigators carried out the reaction of **11** with condensed H_2S at −78°C in the presence of $BF_3{\cdot}OEt_2$ in a sealed glass tube, when a dimeric tetrathiol intermediate was formed presumably via iminium ions followed by nucleophilic attack by the sulfur nucleophile in a stereoselective manner. After aqueous work-up of this dimeric tetrathiol intermediate, the crude mixture in ethyl acetate was treated with I_2, and pure (+)-chaetocin (**1**) was isolated as white solid ($[\alpha]_D^{26}$ +537° ($CHCl_3$, *c* 0.20)) in 44% yield. The synthetic compound was found to be spectroscopically identical to a natural sample of (+)-chaetocin ($[\alpha]_D^{26}$ +530° ($CHCl_3$, *c* 0.04)) [1].

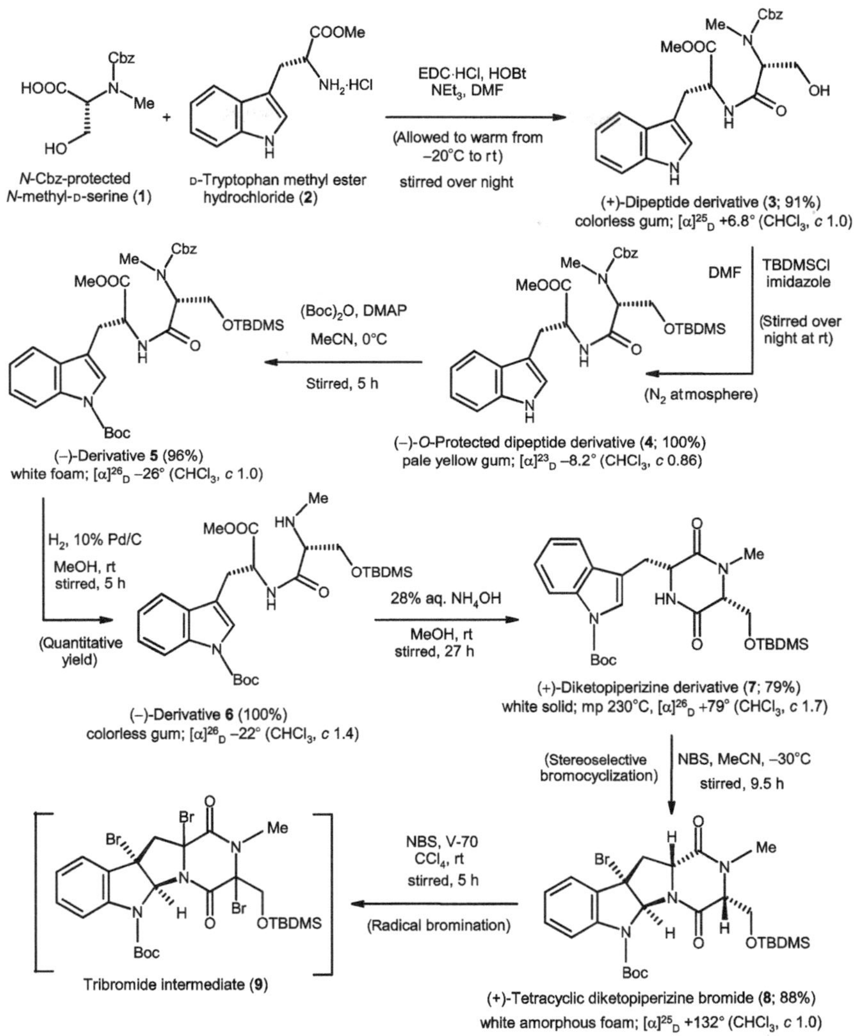

Scheme 1—Cont'd

(Continued)

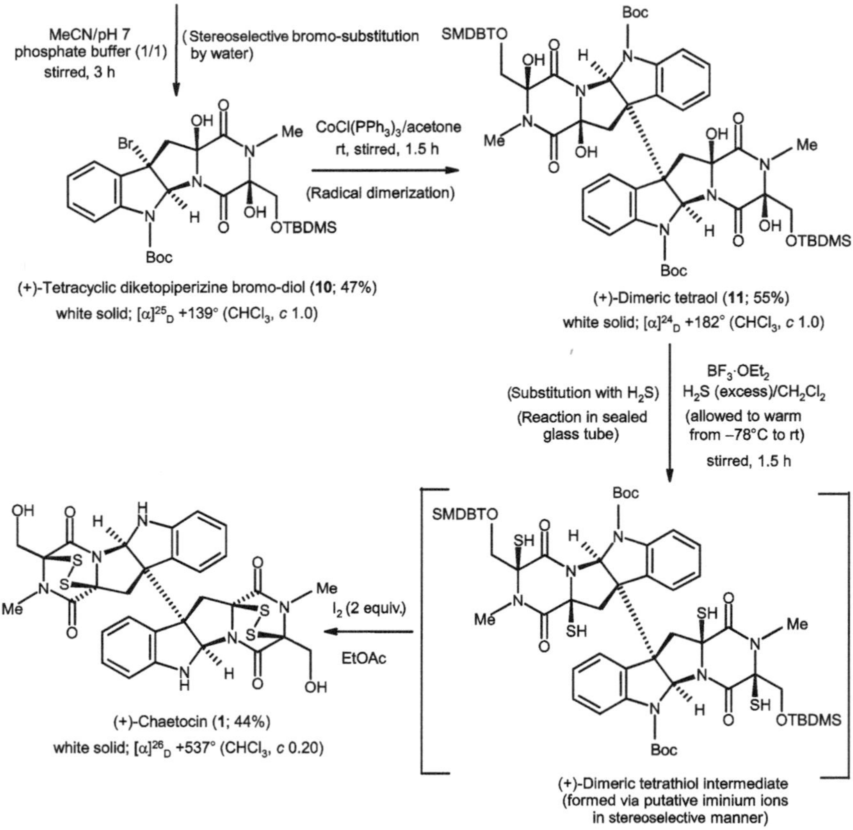

Scheme 1 Total synthesis of (+)-chaetocin (**1**).

References

[1] D. Hauser, H.P. Weber, H.P. Sigg, Helv. Chim. Acta 53 (1970) 1061.

[2] S. Utagawa, T. Muroi, H. Kurata, Can. J. Microbiol. 25 (1979) 170.

[3] C.R. Isham, J.D. Tibodeau, W. Jin, R. Xu, M.M. Timm, K.C. Bible, Blood 109 (2007) 2579.

[4] C.R. Isham, J.D. Tibodeau, A.R. Bossou, J.R. Merchan, K.C. Bible, Br. J. Cancer 17 (2012) 314.

[5] E. Iwasa, Y. Hamashima, S. Fujishiro, E. Higuchi, A. Ito, M. Yoshida, M. Sodeoka, J. Am. Chem. Soc. 132 (2010) 4078.

[6] D. Greiner, T. Bonaldi, R. Eskeland, E. Roemer, A. Imhof, Nat. Chem. Biol. 1 (2005) 143.

[7] J.D. Tibodeau, L.M. Benson, C.R. Isham, W.G. Owen, K.C. Bible, Antioxid. Redox Signal. 11 (2009) 1097.

[8] S. Bouchat, J.S. Gatot, K. Kabeya, C. Cardona, L. Colin, G. Herbein, S. de Wit, N. Clumeck, O. Lambotte, C. Rouzioux, O. Rohr, C. van Lint, AIDS 26 (2012) 1473, https://doi.org/10.1097/QAD.0b013e32835535f5.

[9] W. Bernhard, K. Barreto, A. Saunders, M.S. Dahabieh, P. Johnson, I. Sadowski, FEBS Lett. 585 (2011) 3549.
[10] H. Chaib, A. Nebbioso, T. Prebet, R. Castellano, S. Garbit, A. Restouin, N. Vey, L. Altucci, Y. Collette, Leukemia 26 (2012) 662.
[11] Y.M. Lee, J.H. Lim, H. Yoon, Y.S. Chun, J.W. Park, Hepatology 53 (2011) 171.
[12] H.P. Weber, Acta Crystallogr. B28 (1972) 1945.
[13] L. Aurelio, J.S. Box, R.T.C. Brownlee, A.B. Hughes, M.M. Sleebs, J. Org. Chem. 68 (2003) 2652.

CHAPTER NINETEEN

Ciliatamides A and B

Abbreviations

DCE dichloroethane
DCM dichloromethane
DIPEA *N,N*-diisopropylethylamine
DMAP 4-dimethylaminopyridine
DMF *N,N*-dimethylformamide
EDC 1-ethyl-3-(3-dimethylaminopropyl)carbodiimide hydrochloride
HOBt 1-hydroxybenzotriazole
MeOH methanol
MW microwave irradiation
NMM *N*-methylmorpholine
PS-DCC *N*-cyclohexylcarbodiimide-*N′*-propyloxymethyl polystyrene
rt room temperature

Systematic names:

N-Methyl-*N*-((*S*)-1-oxo-1-(((*S*)-2-oxoazepan-3-yl)amino)-3-phenylpropan-2-yl)dec-9-enamide (Ciliatamide A); *N*-methyl-*N*-((*S*)-1-oxo-1-(((*S*)-2-oxoazepan-3-yl)amino)-3-phenylpropan-2-yl) octanamide (Ciliatamide B)

Compound class: Lipopeptides

Structures:

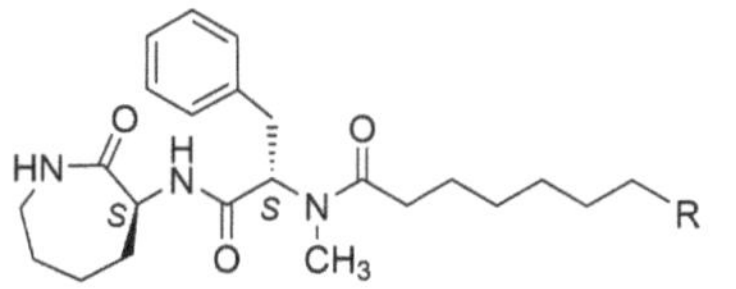

Ciliatamide A (**1**): R = allyl ($CH_2CH{=}CH_2$)
Ciliatamide A (**2**): R = methyl (CH_3)
[natural isolates were reported as (*S*,*S*)]

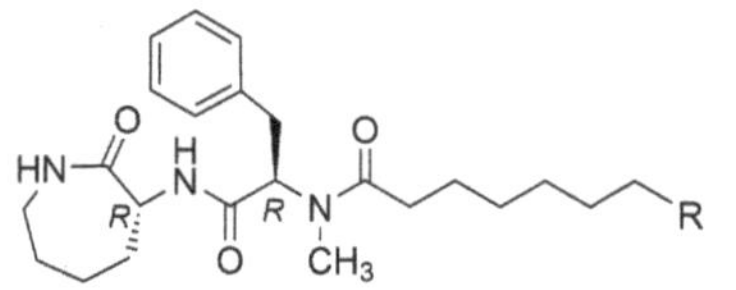

Ciliatamide A (**1**): R = allyl ($CH_2CH{=}CH_2$)
Ciliatamide A (**2**): R = methyl (CH_3)
[upon synthesis revised as (*R*,*R*)]

Natural source: *Aaptos ciliata* (family: Suberitidae; deep-sea sponge) [1]
Pharmaceutical potential: Antileishmanial [1]

Total Synthesis of Bioactive Natural Products
https://doi.org/10.1016/B978-0-08-102822-3.00019-5

Synthetic routes:

The first total syntheses of antileishmanial lipopeptides Ciliatamides A (**1**) and B (**2**) were demonstrated by Lindsley and his group [2] in just three steps (Scheme 1) from the commercially available starting materials, and they also revised the reported stereochemistry for the naturally isolated molecules from (*S*,*S*)-Ciliatamides to (*R*,*R*)-Ciliatamides.

The synthetic scheme started with the first amide bond formation between Boc-*N*-methyl-L-phenylalanine (**3**) and (*S*)-3-aminoazepan-2-one (**4**) in the presence of PS-DCC and HOBt (1-hydroxybenzotriazole) in DMF/DCM upon stirring at room temperature for 16 h to form *tert*-butyl methyl((*S*)-1-oxo-1-(((*S*)-2-oxoazepan-3-yl)amino)-3-phenylpropan-2-yl)carbamate (**5**), which in turn underwent deprotection on treatment with 4 M HCl in dioxane to produce the free base (*S*)-2-(methylamino)-*N*-((*S*)-2-oxoazepan-3-yl)-3-phenylpropanamide (**6**) as colorless oil with 59% yield over the two steps. The free base **6** on stirring with 9-decenoic acid (**7**) in the presence of EDC/HOBt/DIPEA in *N*,*N*-dimethylformamide at room temperature for 16 h afforded (*S*,*S*)-Ciliatamide A (**1**; pale yellow oil; $[\alpha]_D^{20}$ −35° (CH_2Cl_2, *c* 0.05)) in 56% yield. Intermediate **6** when reacted with octanoyl chloride in the presence of NMM/DMAP/DMF under the influence of MW at 160°C for 15 min furnished (*S*,*S*)-Ciliatamide B (**2**; pale yellow oil; $[\alpha]_D^{20}$ −44° (CH_2Cl_2, *c* 0.1)) in 58% yield (Scheme 1). The ^1H and ^{13}C NMR spectra of these synthetic compounds were found to be identical to those reported for the natural products; however, the optical rotations were of comparable magnitude, but opposite in sign (namely, $[\alpha]_D^{20}$ +40° (MeOH, *c* 0.05) for natural Ciliatamide A and $[\alpha]_D^{20}$ +55° (MeOH, *c* 0.1) for natural Ciliatamide B). Based on these results, the investigators synthesized the four possible stereoisomers ((*S*,*S*), (*S*,*R*), (*R*,*S*), and (*R*,*R*)) of Ciliatamide A and Ciliatamide B, following the same the route depicted in Scheme 1, compared NMR spectra, and obtained optical rotations. From these experimental outcomes, it was then revealed that the respective NMR spectra as well as optical rotations for natural Ciliatamides matched the respective (*R*,*R*)-stereoisomer of the synthetic compounds. Hence, stereochemistries of the natural products were revised. Later on (2016), Avula and Mahapatra also accomplished total synthesis of Ciliatamides A and B based on ring-closing methathesis with further confirmation of the revised stereochemistry [3].

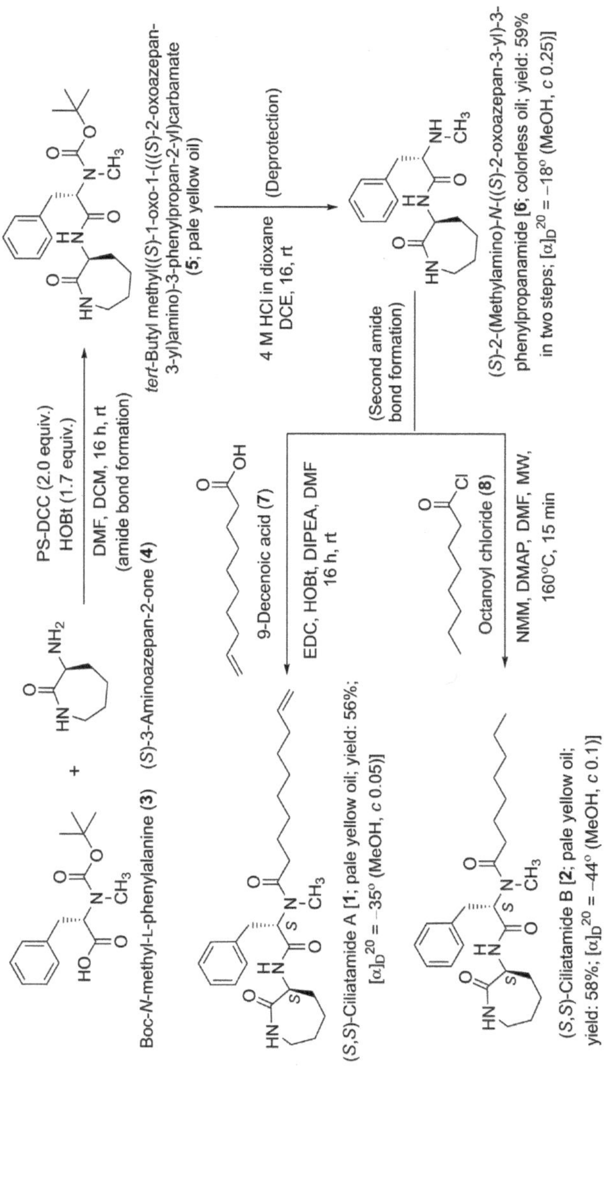

Scheme 1 Synthesis of Ciliatamides A (**1**) and B (**2**).

References

[1] Y. Nakao, S. Kawatsu, C. Okamoto, M. Okamoto, Y. Matsumoto, S. Matsunaga, R.W.M. van Soest, N. Fusetani, J. Nat. Prod. 71 (2008) 469.
[2] J.A. Lewis, R.N. Daniels, C.W. Lindsley, Org. Lett. 10 (2008) 4545.
[3] K. Avula, D.K. Mohapatra, Tetrahedron Lett. 57 (2016) 1715.

CHAPTER TWENTY

Cylindol A

Abbreviations

$AlCl_3$ aluminum trichloride
BBr_3 boron tribromide
(BINAP)$PdCl_2$ bis(diphenylphosphine)-1,1′-binaphthyl palladium (II) dichloride; 1,1′-binaphthalene-2,2′-diylbis(diphenylphosphine)-dichloropalladium
CH_2Cl_2 dichloromethane
DMF *N,N*-dimethylformamide
DMSO dimethylsulfoxide
Et_3N triethylamine
$KMnO_4$ potassium permanganate
Me_2SO_4 dimethyl sulfide
NBS *N*-bromosuccinimide
rt room temperature

Systematic name: Dimethyl 3,3′-oxybis(4-hydroxybenzoate)
Compound class: Biphenyl ether derivative
Structure:

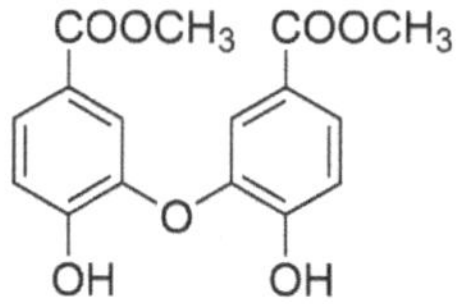

Cylindol A (**1**)

Natural source: *Imperata cylindrica* Beauvois (rhizomes; family: Gramineae) [1]
Pharmaceutical potential: 5-Lipoxygenase inhibitor [1]
Synthetic routes:
Ohizumi and his group first isolated the symmetrical biphenyl ether cylindol A (**1**) from the rhizomes of *Imperata cylindrica*, a Japanese medicinal plant (locally known as "Chigaya"), and deduced its structure both on the basis of detailed spectral studies as well as total synthesis (Scheme 1) [1]. The investigators carried out an Ullmann-type coupling reaction between

Total Synthesis of Bioactive Natural Products
https://doi.org/10.1016/B978-0-08-102822-3.00020-1

the phenol derivatives **3** and **4** to obtain the biphenyl ether 2,2′-dimethoxy-5,5′-dimethyl-diphenyl ether (**5**) in 31% yield. Compound **5** was then oxidized to the corresponding dicarboxylic acid derivative **6** (27%) on refluxing with alkaline potassium permanganate, which ultimately furnished cylindol A (**1**; colorless amorphous powder, mp 217–219°C) on demethylation with BBr_3 in dichloromethane followed by esterification with methanol. The spectral data of the synthetic compound were found to be completely identical with those of natural cylindol A [1]; however, the overall yield of the product was only 4.5%.

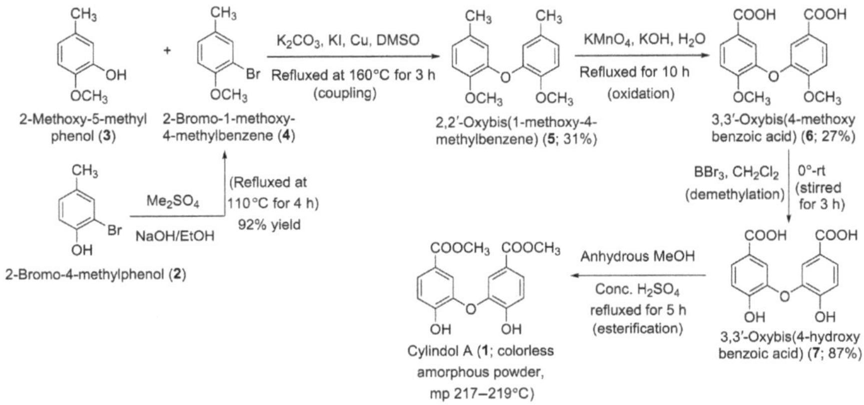

Scheme 1 Ohizumi's total synthesis of cylindol A (**1**) [1].

Since the overall yield of cylindol A in the above method is too low (4.5%), later on a modified total synthesis of the target compound **1** was reported by Shin and coworkers starting from the symmetrical 2,2′-dihydroxy diphenyl ether (**8**) (Scheme 2) [2]. Compound **8** on bromination with *N*-bromosuccinimide produced the *para*-substituted dibromide **9** in a regioselective manner [3], which on methylation formed 2,2′-dimethoxy-5,5′-dibromo-diphenyl ether (**10**). The diemthyl derivative **10** underwent smooth methoxycarbonylation in the presence of (BINAP)$PdCl_2$ catalyst in methanol under 50 psi of carbon monoxide [4] to furnish the diester 2,2′-dimethoxy-5,5′-dimethoxycarbonyl-diphenyl ether (**11**) as a yellow solid (92%), which was then demethylated selectively by aluminum chloride [5] to afford cylindol A (**1**) as a beige powder in an excellent overall yield of 39%. Spectroscopic data of the synthetic compound were found to be consistent with that of the reported value [1].

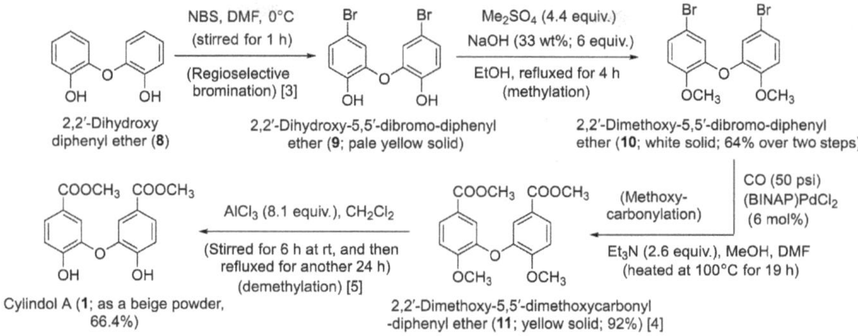

Scheme 2 Shin's total synthesis of cylindol A (**1**) [2].

References

[1] K. Matsunaga, M. Ikeda, M. Shibuya, Y. Ohizumi, J. Nat. Prod. 57 (1994) 1290.
[2] J.H. Chang, D.H. Nam, H. Shin, Bull. Kor. Chem. Soc. 29 (2008) 1003.
[3] N. Fujikawa, T. Ohta, T. Yamaguchi, T. Fukuda, F. Ishibashi, M. Iwao, Tetrahedron 62 (2006) 594.
[4] J. Albaneze-Walker, C. Bazaral, T. Leavey, P.G. Dormer, J.A. Murry, Org. Lett. 6 (2004) 2097.
[5] M. Mondal, V.G. Puranik, N.P. Argade, J. Org. Chem. 72 (2007) 2068.

CHAPTER TWENTY ONE

Daedalin A

Abbreviations

Ac acetyl
Ag(DPAH)$_2$ silver(II) dipicolinate
CH_2Cl_2 dichloromethane
DDQ 2,3-dichloro-5,6-dicyanobenzoquinone
DMAP 4-dimethylaminopyridine
DMF *N*,*N*-dimethylformamide
DMP Dess-Martin periodinane
L-(+)-DET L-(+)-diethyl L-tartarate
$LiAlH_4$ lithium aluminum hydride
MeOH methanol
MOMCl methoxymethyl chloride
***p*-TsOH** *para*-toluenesulfonic acid
TBAF tetrabutylammonium fluoride
TBHP *tert*-butyl hydroperoxide
TBSCl *tert*-butyldimethylsilyl chloride
THF tetrahydrofuran

Systematic name: (2*R*)-6-Hydroxy-2-hydroxymethyl-2-methyl-2*H*-chromene

Compound class: Chromene-derivative

Structure:

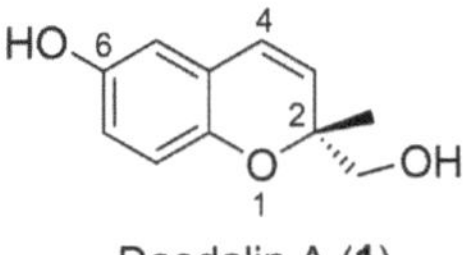

Daedalin A (**1**)

Natural source: *Daedalea dickinsii* (mycelial culture) [1, 2]

Pharmaceutical potential: Tyrosinase inhibitor [1]; antioxidant [1]; inhibitor to melanin synthesis [2]

Total Synthesis of Bioactive Natural Products
https://doi.org/10.1016/B978-0-08-102822-3.00021-3

Synthetic route:

Makabe and his group revealed the asymmetric total synthetic route for daedalin A (**1**) (Scheme 1) [3] starting from commercially available 4-methoxyphenol (**2**) and 2-methyl-2-vinyloxirane (**3**). Compounds **2** and **3** underwent a modified Kirschleger's reaction [4] to give 2-(4-methoxyphenoxy)-2-methylbut-3-en-1-ol (**4**), which on acetylation followed by Claisen rearrangement afforded (*E*)-4-(2-hydroxy-5-methoxyphenyl)-2-methylbut-2-en-1-yl acetate (**6**) in 99% yield. Through a sequence of reactions, compound **6** was converted into ((2*R*,3*R*)-3-(5-methoxy-2-(methoxymethoxy)benzyl)-2-methyloxiran-2-yl)methanol (**9**) using Sharpless asymmetric epoxidation as a key step. On reduction with $LiAlH_4$ followed by protecting 1,2-diol with 2,2-dimethoxypropane, compound **9** produced (*R*)-4-(5-methoxy-2-(methoxymethoxy)phenethyl)-2,2,4-trimethyl-1,3-dioxolane (**11**) that underwent facile oxidative demethylation with silver(II) dipicolinate $[Ag(DPAH)_2]$ in the presence of sodium acetate and aqueous MeCN to give compound **12** in 96% yield [5]. Transformation from **12** to (*R*)-2-(hydroxymethyl)-2-methylchroman-6-ol (**14**) was then achieved using Kirschleger's procedure [6]. Protection of the hydroxy groups of **14** with TBSCl and imidazole afforded **15** in quantitative yield. On treatment with DDQ **15** underwent dehydrogenation reaction to furnish (*R*)-*tert*-butyl((6-((*tert*-butyldimethylsilyl)oxy)-2-methyl-2*H*-chromen-2-yl)methoxy)dimethylsilane (**16**; 81%) [7]. Finally, deprotection of the TBS ether of **16** with TBAF resulted in the formation of daedalin A (**1**) in good yield. Recrystallization from chloroform gave colorless solid, mp 136–138°C. All the spectral data of synthetic compound **1** were found to be in good agreement with those of natural daedalin A [1, 2].

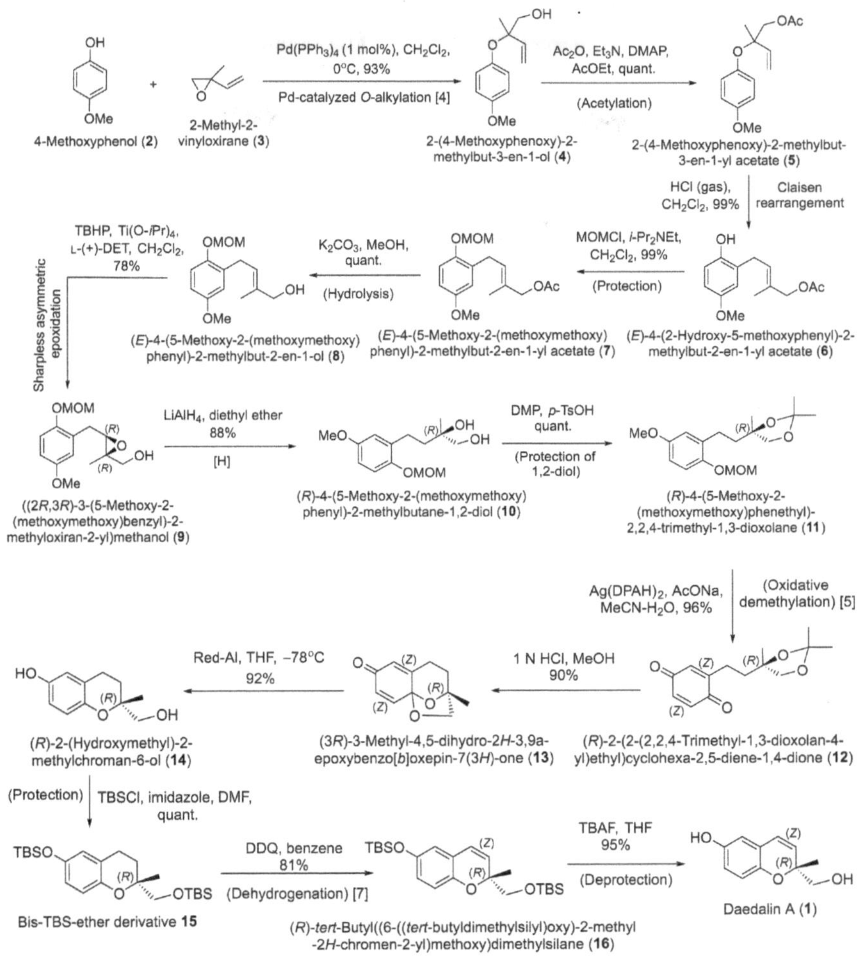

Scheme 1 Makabe's total synthesis of daedalin A (**1**) [3].

References

[1] K. Morimura, C. Yamazaki, Y. Hattori, H. Makabe, T. Kamo, M. Hirota, Biosci. Biotechnol. Biochem. 71 (2007) 2837.

[2] K. Morimura, K. Hiramatsu, C. Yamazaki, Y. Hattori, H. Makabe, M. Hirota, Biosci. Biotechnol. Biochem. 73 (2009) 627.

[3] M. Sekimoto, Y. Hattori, K. Morimura, M. Hirota, H. Makabe, Bioorg. Med. Chem. Lett. 20 (2010) 1063.

[4] J.-Y. Goujon, A. Duval, B. Kirschleger, J. Chem. Soc. Perkin Trans. 1 (2002) 496.

[5] K. Kloc, J. Mlochowski, L. Syper, Chem. Lett. (1980) 725.

[6] S. Bouzbouz, J.-Y. Goujon, J. Deplanne, B. Kirschleger, Eur. J. Org. Chem. (2000) 3223.

[7] B.M. Trost, H.C. Shen, J.-P. Surivet, Angew. Chem. Int. Ed. 42 (2003) 3943.

CHAPTER TWENTY TWO

6-Deoxypladienolide D

Abbreviations

Ac_2O acetic anhydride
CDI 1,1′-carbonyldiimidazole
CH_2Cl_2 dichloromethane
DMAP 4-methylaminopyridine
DMF *N,N*-dimethylformamide
EDC·HCl N^1-((ethylimino)methylene)-N^3, N^3-dimethylpropane-1,3-diamine hydrochloride
Et_3N triethylamine
Hoveyda Grubbs second-generation catalyst (1,3-dimesitylimidazolidin-2-ylidene)(2-isopropoxybenzylidene)ruthenium(VI) chloride
(−)-$(Ipc)_2$BOMe (−)-B-methoxydiisopinocampheylborane
rt room temperature
TBAF tetrabutylammonium fluoride
TBME *tert*-butyl methyl ether
TBSCl *tert*-butyldimethylsilyl chloride
THF tetrahydrofuran

Systematic name: (2*S*,3*S*,6*R*,7*S*,10*R*,*E*)-10-hydroxy-2-((*R*,2*E*,4*E*)-6-hydroxy-7-((2*R*,3*R*)-3-((2*R*,3*S*)-3-hydroxypentan-2-yl)oxiran-2-yl)-6-methylhepta-2,4-dien-2-yl)-3,7-dimethyl-12-oxooxacyclododec-4-en-6-yl acetate

Compound class: Macrocyclic polyketide

Structure:

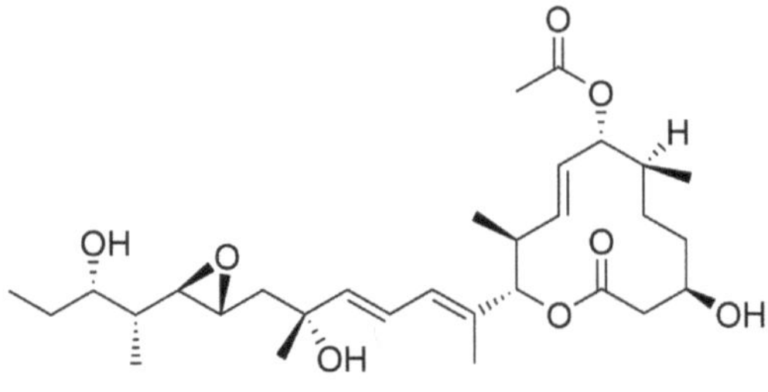

6-Deoxypladienolide D (**1**)

Total Synthesis of Bioactive Natural Products
https://doi.org/10.1016/B978-0-08-102822-3.00022-5

Natural source: *Streptomyces* strain [1]

Pharmaceutical potential: Anticancer (exhibited potent growth inhibitory activity in a mutant SF3B1 cancer cell line, high binding affinity to the SF3b complex, and inhibition of pre-mRNA splicing) [1]

Synthetic route:

Keaney and his group reported the first total synthesis of 6-deoxypladienolide D (**1**), a promising anticancer natural microbial polyketide using inexpensive and commercially available starting materials; they envisioned a retrosynthetic approach for the target compound (Scheme 1) [1]. Synthesis of individual fragments followed by their assembly affording **1** is depicted herein.

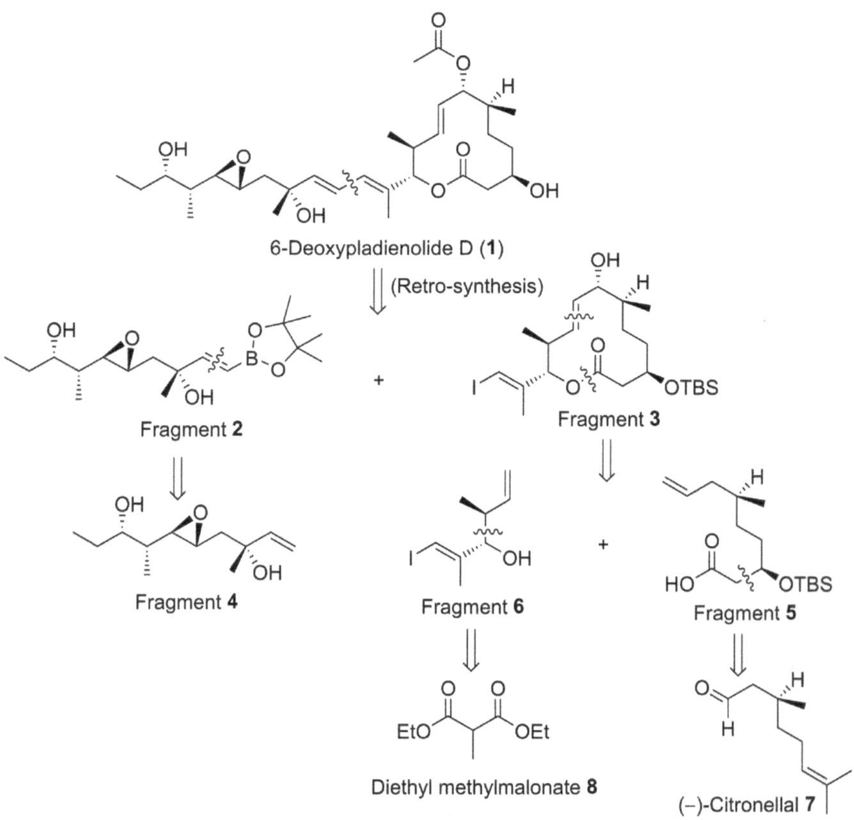

Scheme 1 Retrosynthetic approach for 6-deoxypladienolide D (**1**) [1].

Synthesis of fragment 5:

Synthesis of (3*R*,6*S*)-3-((*tert*-butyldimethylsilyl)oxy)-6-methylnon-8-enoic acid (fragment **5**) was accomplished starting from an inexpensive and commercially available (−)-citronellal (**7**) in 12 steps (Scheme 2). It was first methylenated using Peterson olefination conditions to give (*S*)-4,8-dimethylnona-1,7-diene (**9**; colorless oil), which on chemoselective dihydroxylation of the monosubstituted olefin followed by oxidative cleavage with sodium periodate provided the (*S*)-4-methylhept-6-enal (**10**) in 44% yield over four steps. The carboxylic acid derivative obtained on Pinnick oxidation [2] of **10** was then activated with 1,1′-carbonyldiimidazole (CDI) to undergo a Claisen condensation and decarboxylation to afford ester **11**. Reagent-controlled reduction of **11** with $NaBH_4$/L-tartaric acid afforded ester alcohol **12** (as a 4:1 ratio of diastereomers favoring the desired β-epimer) in 66% yield [3], which was then hydrolyzed into carboxylic acid derivative **13**. A chiral resolution of the corresponding carboxylic acid in the presence of (*R*)-(+)-α-methylbenzylamine provided crystalline amine salt **14** with >98% diastereopurity (C-3 stereochemistry was supported by X-ray studies). A three-step sequence (acidification, silylation, and regeneration of the free carboxylic acid) ultimately furnished the fragment **5** in moderate yield.

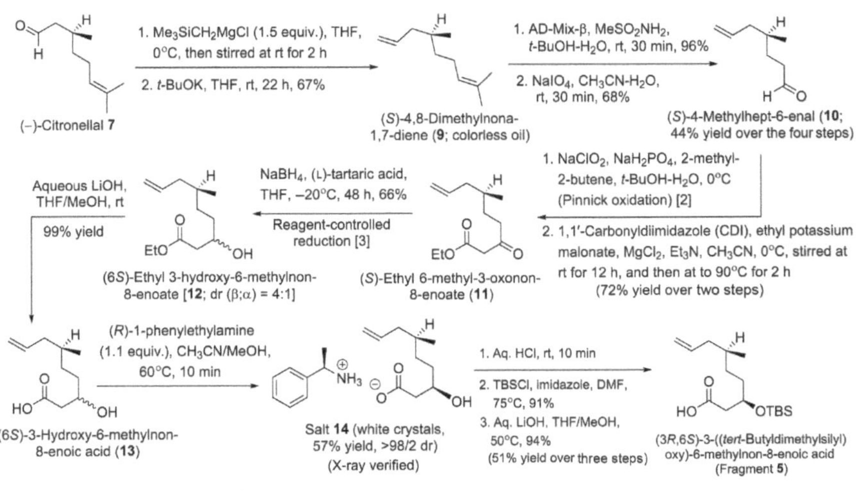

Scheme 2 Synthesis of fragment **5** [1].

Synthesis of fragment 6:

Preparation of fragment **6** is described in Scheme 3. Diethyl 2-methylmalonate (**8**) was first alkylated with iodoform, and a base-mediated hydrolysis/decarboxylation/elimination of the resulting compound produced carboxylic acid **15** as a single isomer with 71% overall yield. On lithium aluminum hydride reduction the acid gave alcohol **16**, which on oxidation with manganese dioxide resulted in (*E*)-3-iodo-2-methylacrylaldehyde (**17**; 63% over the two steps). Fragment **6** [(3*S*,4*S*,*E*)-1-iodo-2,4-dimethylhexa-1,5-dien-3-ol] was finally obtained from the Brown crotylation of **17** in 56% yield and 94% *ee*.

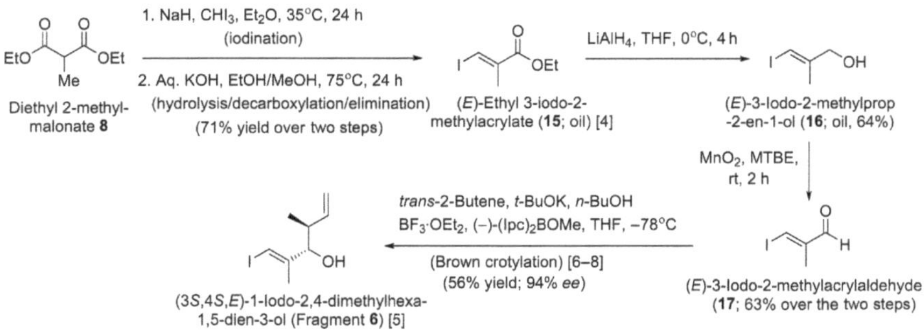

Scheme 3 Synthesis of fragment **6** [1].

Synthesis of fragment 2:

Fragment **2** (vinyl pinacol boronate) was prepared from (2*R*,3*S*)-2-((2*R*,3*R*)-3-((*R*)-2-hydroxy-2-methylbut-3-en-1-yl)oxiran-2-yl)pentan-3-ol (fragment **4**; green oil) obtained following the method of Kanada et al. [9] in a single step by the reaction with 4,4,5,5-tetramethyl-2-vinyl-1,3,2-dioxaborolane (3.0 equiv.), and 1,3-bis(2,4,6-trimethylphenyl)-4,5-dihydroimidazol-2-ylidene[2-(*i*-propoxy)-5-(*N*,*N*-dimethylaminosulfonyl)phenyl]methyleneruthenium(II) dichloride (Zhan catalyst-1B, 0.2 equiv) (Scheme 4).

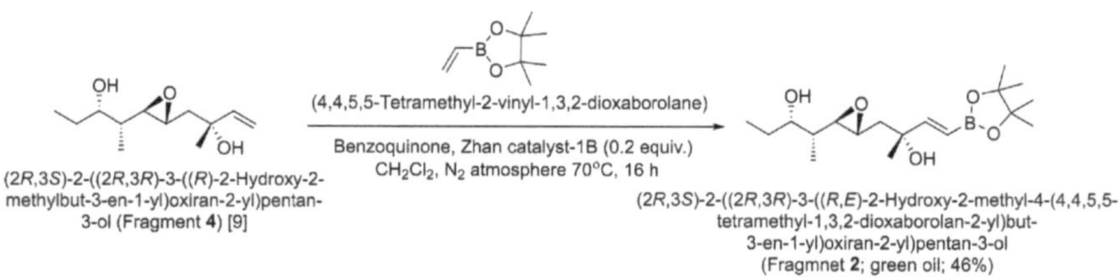

Scheme 4 Synthesis of fragment **2** [1].

Synthesis of macrocyclic ring and final assembly leading to 6-deoxypladienolide D (1):

Next, the investigators marched forward for macrocyclization with fragments **5** and **6** in hand; the fragments underwent EDC-mediated coupling to give derivative **18**, which on ring-closing metathesis under the catalysis of

Hoveyda-Grubbs second-generation catalyst [10] in toluene at 50°C in the presence of 1,4-benzoquinone gave macrocycle **19** as a green oil. Very interestingly, on exposure to selenium dioxide the macrocycle yielded the desired alcohol (4*R*,7*S*,8*R*,11*S*,12*S*,*E*)-4-((*tert*-butyldimethylsilyl)oxy)-8-hydroxy-12-((*E*)-1-iodoprop-1-en-2-yl)-7,11-dimethyloxacyclododec-9-en-2-one (**20**) in 77% yield with complete chemo-, regio-, and diastereoselectivity. The acetyl derivative **21** was then coupled with vinyl pinacol boronate (fragment **2**) under room temperature Suzuki conditions, followed by desilylation with TBAF, which afforded the natural product 6-deoxypladienolide D (**1**) (Scheme 5).

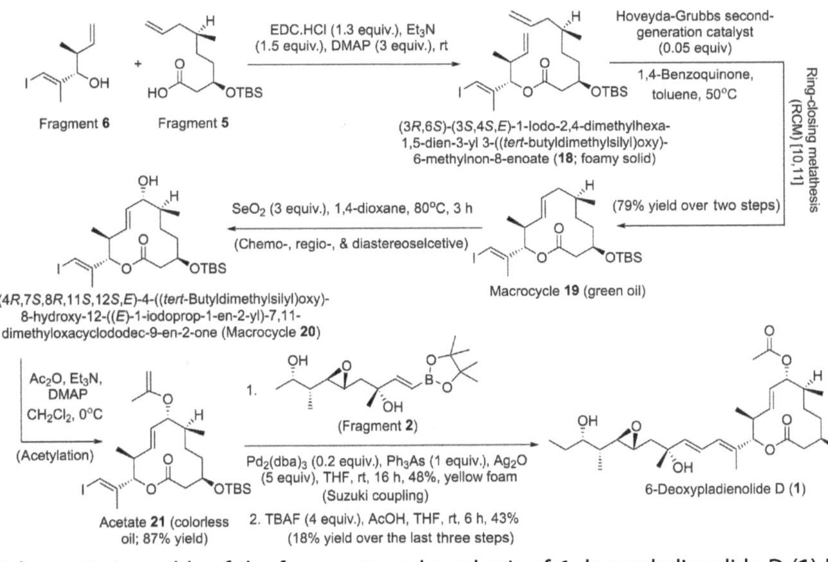

Scheme 5 Assembly of the fragments and synthesis of 6-deoxypladienolide D (**1**) [1].

References

[1] K. Arai, S. Buonamici, B. Chan, L. Corson, A. Endo, B. Gerard, M.-H. Hao, C. Karr, K. Kira, L. Lee, X. Liu, J.T. Lowe, T. Luo, L.A. Marcaurelle, Y. Mizui, M. Nevalainen, M.W. O'Shea, E.S. Park, S.A. Perino, S. Prajapati, M. Shan, P.G. Smith, P. Tivitmahaisoon, J.Y. Wang, M. Warmuth, K.-M. Wu, L. Yu, H. Zhang, G.-Z. Zheng, G.F. Keaney, Org. Lett. 16 (2014) 5560.

[2] B.S. Bal, W.E. Childers, H.W. Pinnick, Tetrahedron 37 (1981) 2091.

[3] A.K. Ghosh, D.D. Anderson, Org. Lett. 14 (2012) 4730.

[4] B.M. Trost, J. Waser, A. Meyer, J. Am. Chem. Soc. 129 (2007) 14556.

[5] A.L. Mandel, B.D. Jones, J.J. La Clair, M.D. Burkart, Bioorg. Med. Chem. Lett. 17 (2007) 5159.

[6] H.C. Brown, P.K. Jadhav, J. Am. Chem. Soc. 105 (1983) 2092.

[7] H.C. Brown, K.S. Bhat, J. Am. Chem. Soc. 108 (1986) 293.

[8] H. Brown, K. Bhat, J. Org. Chem. 108 (1986) 293.
[9] R.M. Kanada, D. Itoh, M. Nagai, J. Niijima, N. Asai, Y. Mizui, S. Abe, Y. Kotake, Angew. Chem. Int. Ed. 46 (2007) 4350.
[10] S.B. Garber, J.S. Kingsbury, B.L. Gray, A.H. Hoveyda, J. Am. Chem. Soc. 122 (2000) 8168.
[11] S. Hong, D. Sanders, C. Lee, R. Grubbs, J. Am. Chem. Soc. 127 (2005) 17160.

CHAPTER TWENTY THREE

7-Desmethoxyfusarentin and Its Methyl Ether

Abbreviations

Bn benzyl
CH_2Cl_2 dichloromethane
DMAP 4-dimethylaminopyridine
rt room temperature
TBAF tetrabutylammonium fluoride
TBS *tert*-butyldimethylsilyl
TFA trifluoroacetic acid
THF tetrahydrofuran
TPP triphenylphosphine

Systematic name: (*S*)-8-Hydroxy-3-((*S*)-2-hydroxypentyl)-6-methoxyisochroman-1-one (7-desmethoxyfusarentin); (*S*)-3-((*S*)-2-hydroxypentyl)-6,8-dimethoxyisochroman-1-one (methyl ether derivative)

Compound class: Dihydroisocoumarins

Structures:

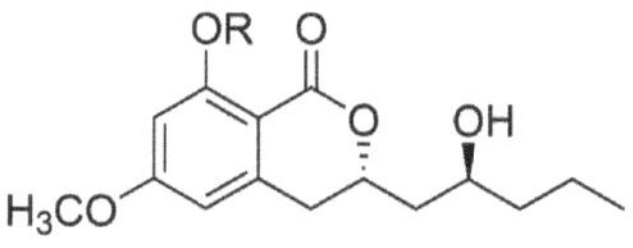

7-Desmethoxyfusarentin (**1**): R = H
Methyl ether (**2**): R = CH_3

Natural source: *Ophiocordyceps communis* BCC 16475 (an insect pathogenic fungus) [1]

Pharmaceutical potential: Cytotoxic against MCF-7 cells [1, 2]

Synthetic route:

Reddy et al. [2] reported a concise total synthesis of 7-desmethoxyfusarentin (**1**) and its methyl ether (**2**) employing a sequence of reactions such as Prins cyclization, ring opening of tetrahydropyran ring,

Total Synthesis of Bioactive Natural Products
https://doi.org/10.1016/B978-0-08-102822-3.00023-7

and Alder-Rickerts reaction as key steps. Their approach for the total synthesis of **1** and **2** is summarized in Scheme 1.

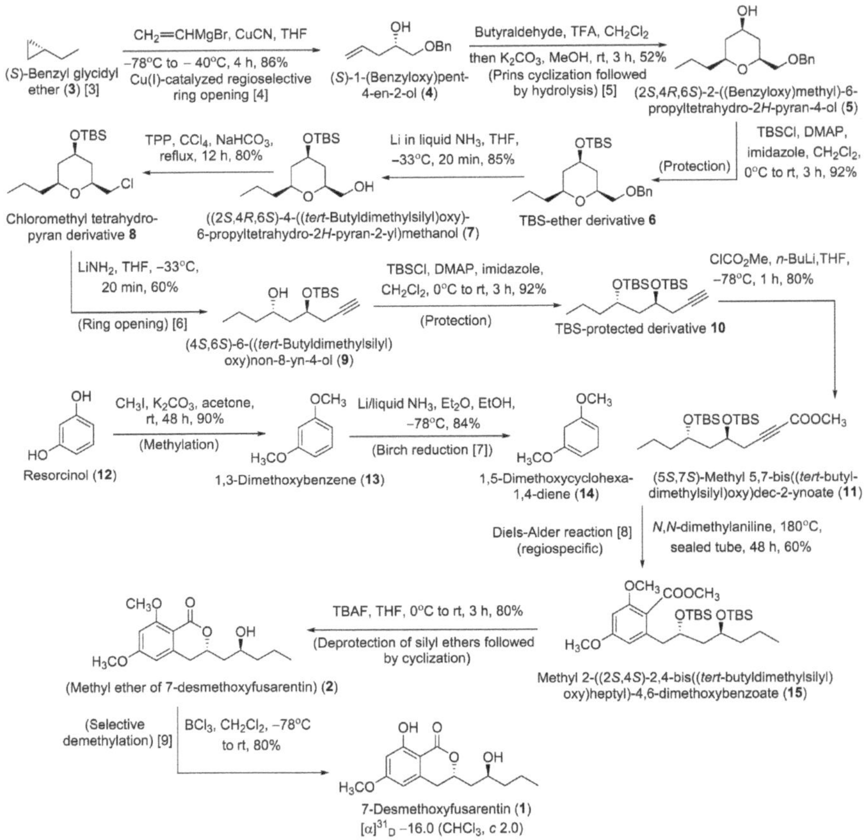

Scheme 1 Total synthesis of 7-desmethoxyfusarentin (**1**) and its methyl ether (**2**) [2].

They started the synthesis with (*S*)-benzyl glycidyl ether (98% *ee*; **3**) [3], which on regioselective Cu(I)-catalyzed ring opening with vinyl magnesium bromide resulted in the homoallyl alcohol (*S*)-1-(benzyloxy)pent-4-en-2-ol (**4**) in 86% yield [4]. Treatment of **4** with butyraldehyde in the presence of TFA (under Prins cyclization conditions), followed by hydrolysis of the resulting trifluoroacetate with potassium carbonate in methanol, furnished the trisubstituted tetrahydropyran **5** in 52% yield [5]. Cleavage of the primary benzyl ether moiety of TBS-ether derivative **6** by treating with lithium in liquid ammonia afforded ((2*S*,4*R*,6*S*)-4-((*tert*-butyldimethylsilyl)oxy)-6-propyltetrahydro-2*H*-pyran-2-yl)methanol (**7**) in 78% yield (over two steps). Further treatment of this pyranyl methanol **7** with TPP in CCl_4 in the presence of a catalytic amount of $NaHCO_3$ under reflux conditions produced the chloromethyl tetrahydropyran **8** in 80% yield, which in the next step underwent ring opening on treatment with lithium amide giving rise to the partially protected *anti*-4,6-diol of alkyne (4*S*,6*S*)-6-((*tert*-butyldimethylsilyl)oxy)non-8-yn-4-ol (**9**) in good yield [6]. The fully protected disilyl ether **10** underwent addition reaction with methyl chloroformate in the presence of *n*-BuLi/THF to generate the acetylenic ester **11** in 80% yield. The other key component, 1,5-dimethoxycyclohexa-1,4-diene (**14**), was prepared in 84% yield starting from resorcinol (**12**) through methylation followed by Birch reduction [7].

In the next phase, a regiospecific Diels-Alder reaction between these two key components **11** and **14** was performed in a sealed tube at 180°C in the presence of a catalytic amount of *N*,*N*-dimethylaniline to obtain the aromatic precursor methyl 2-((2*S*,4*S*)-2,4-bis((*tert*-butyldimethylsilyl)oxy) heptyl)-4,6-dimethoxybenzoate (**15**) [8]. Deprotection of this silyl ether **15** with TBAF in THF afforded the dihydroisocoumarin **2** that underwent selective demethylation of its 8-methoxyl group on treatment with boron trichloride in dichloromethane at −78°C to afford the target molecule 7-desmethoxyfusarentin (**1**) in good yield. The optical rotation and spectral data of synthetic compounds **1** and **2** were found to be in agreement with those of natural products [1].

References

[1] R. Haritakun, M. Sappan, R. Suvannakad, K. Tasanathai, M. Isaka, J. Nat. Prod. 73 (2010) 75.

[2] P.J. Reddy, A.S. Reddy, J.S. Yadav, B.V.S. Reddy, Tetrahedron Lett. 53 (2012) 4051.

[3] M.E. Furrow, S.E. Schaus, E.N. Jacobsen, J. Org. Chem. 68 (1998) 6776.

[4] C. Bonini, L. Chiummiento, M.T. Lopardo, M. Pullex, F. Colobert, G. Solladie, Tetrahedron Lett. 44 (2003) 2695.

[5] C.S.J. Barry, S.R. Crosby, J.R. Harding, R.A. Hughes, C.D. King, G.D. Parker, C.L. Willis, Org. Lett. 5 (2003) 2429.

[6] J.S. Yadav, M.C. Chander, B.V. Joshi, Tetrahedron Lett. 29 (1988) 2737 J.S. Yadav, P. Deshpande, G.V.M. Sharma, Tetrahedron 46 (1990) 7033.

[7] A.J. Birch, J. Chem. Soc. (1944) 430 A.J. Birch, G.S.R. Subba Rao, Adv. Org. Chem. 8 (1972) 1.

[8] A.J. Birch, N.S. Mani, G.S.R. Subba Rao, J. Chem. Soc., Perkin Trans. 1 (1990) 1423 C.C. Kanakum, N.S. Mani, H. Ramanathan, G.S.R. Subba Rao, J. Chem. Soc. Perkin Trans. 1 (1907), 1989.

[9] F.M. Dean, J. Goodchild, L.E. Houghton, J.A. Martin, R.B. Morton, B. Parton, A.W. Price, N. Somvichien, Tetrahedron Lett. 35 (1966) 4153 G. Jenner, M. Papadopoulos, High Pressure Res. 1 (1988) 67.

CHAPTER TWENTY FOUR

Etnangien

Abbreviations

Bz benzoyl
DDQ 2,3-dichloro-5,6-dicyano-1,4-benzoquinone
CSA (±)-camphor sulfonic acid
DIBAL-H diisobutylaluminum hydride
DMAP 4-dimethylaminopyridine
DMP dimethoxypropane
DMSO dimethylsulfoxide
ee enantiomeric excess
HOAc acetic acid
Ipc_2BCl diisopinocampheylborane
KHMDS potassium hexamethyldisilazane
$[Ph_3PCH_2I]^+I^-$ (iodomethyl)triphenylphosphonium iodide
PMB 4-methoxy benzyl
PPh_3 triphenylphosphine
Py pyridine
***p*-TsCl** *p*-toluenesulfonyl chloride
TBAF tetra-*n*-butylammonium fluoride
TBS *tert*-butyl-dimethyl-silanyl
TBSCl *tert*-butyldimethylsilyl chloride
TBSOTf *tert*-butyldimethylsilyl triflate
TEA triethylamine
THF tetrahydrofuran

Systematic name: (4*E*,6*S*,8*E*,10*E*,12*E*,14*E*,16*E*,18*E*,20*R*,21*R*)-6,20-dihydroxy-4,18-dimethyl-21-((2*S*,4*R*,8*S*,10*Z*,12*E*,15*S*,16*R*,17*S*,18*S*,19*R*,20*R*)-4,16,18,20-tetrahydroxy-8-methoxy-15,17,19-trimethyl-22-oxooxacyclodocosa-10,12-dien-2-yl)docosa-4,8,10,12,14,16,18-heptaenoic acid

Compound class: Polyketide macrolide (macrolactone antibiotic)

Total Synthesis of Bioactive Natural Products
https://doi.org/10.1016/B978-0-08-102822-3.00024-9

Structure:

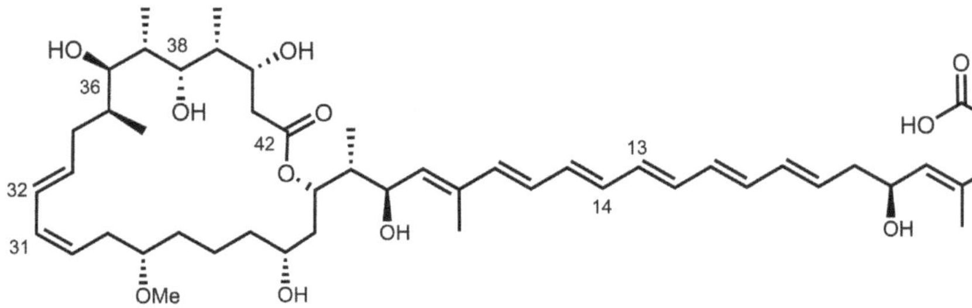

Etnangien (**1**)

Natural source: Culture broth of the myxobacterium *Sorangium cellulosum* (strains So ce750 and So ce1045) [1, 2]

Pharmaceutical potential: Antibiotic (antibacterial against Gram-positive bacteria) [1, 3, 4]

Synthetic route:

In 2009, Menche and coworkers [5] accomplished the first total synthesis of the huge polyketide macrolide etnangien (**1**), and thereby established its relative and absolute configuration as well. Their retrosynthetic approach is shown in Scheme 1. Thus they initially prepared the subunits **2**, **3**, and **4** separately, and ultimately assembled them to obtain the target compound **1**. The overall synthetic approach is depicted sequentially.

Synthesis of C_{32}–C_{42} subunit **2** (Scheme 2) involves a tin-mediated aldol condensation of ketone **6** with Roche ester-derived aldehyde **5**, to give the expected (1,4)-*syn*-product **7** with high levels of stereocontrol (dr = 7:1) and 85% yield. Subsequent (1,3)-*syn* reduction of the hindered β-hydroxyketone **7** was efficiently carried out after chelation [(*c*Hex)$_2$BCl/LiBH$_4$] (dr > 20:1) followed by the acetonide protection of the derived diol. On selective removal of the primary TBS group in compound **8**, the desired C-2 homologation to **9** was performed by nucleophilic substitution of the derived tosylate with sodium acetylide. Completion of the synthesis of **2** proceeded smoothly by PMB deprotection, oxidation to the acid, and Lindlar reduction of the alkyne as colorless oil with an overall yield of 63% over the last four steps.

Construction of the C_{15}–C_{31} subunit **4** (Scheme 3) commenced with the starting material (2*E*,4*E*)-ethyl 3-methyl-6-oxohexa-2,4-dienoate (**12**), prepared from ethyl 3-methyl-4-oxobut-2-enoate (**10**) and $Ph_3PCHCHO$ (**11**) [8]. Compound **15** was produced with high diastereoselectivity

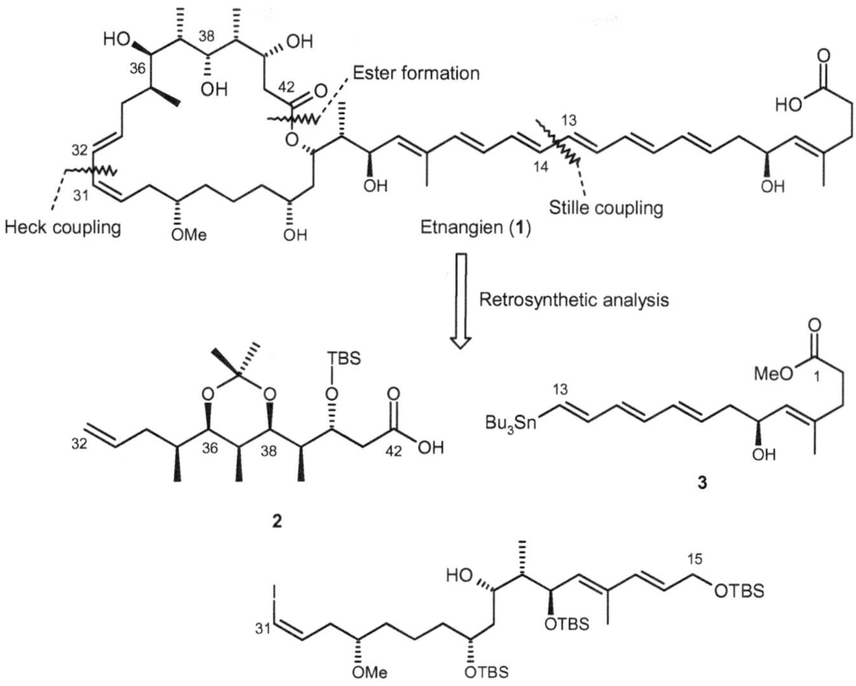

Scheme 1 Retrosynthetic analysis of etnangien (**1**) [5].

(dr > 20:1) and yield (97%) from the derived aldehyde **13** by its homologation with lactate-derived ethyl-ketone **14** by means of a boron-mediated Paterson aldol reaction [9]. It was then converted into (3*R*,4*R*,5*E*,7*E*)-4,9-*bis*(*tert*-butyldimethylsilyloxy)-3,6-dimethylnona-5,7-dien-2-one (**16**) in five consecutive steps with an overall yield of 74%. The aldehyde coupling partner (*S*,*Z*)-8-iodo-5-methoxyoct-7-enal (**19**) was realized in six steps (69%) from epoxide **17**, readily available by Jacobsen hydrolytic kinetic resolution methodology [10]. Terminal *Z*-vinyl iodide was introduced with excellent selectivity (dr = 27:1) utilizing an optimized Wittig-Stork-Zhao olefination protocol [11]. Finally, the pivotal aldol coupling of methyl ketone **16** with aldehyde **19** to develop the desired chiral C-24 center was accomplished with high diastereoselectivity (dr = 14:1) and yield (77%) by an Ipc-boron-mediated aldol condensation. Subsequent 1,3-*anti* reduction of the derived hydroxyl ketone **20** with the Evans-Carreira protocol [12] and protection of the less hindered C_{24}–OH offered the building block **4** in good yields (71%).

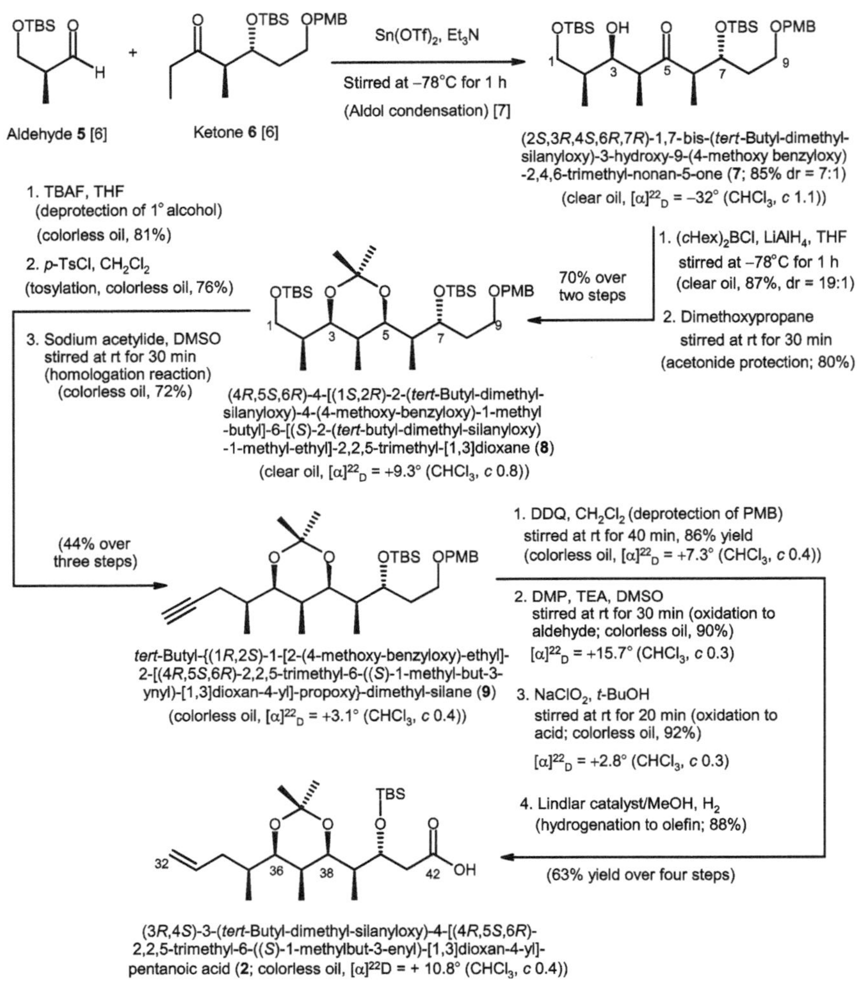

Scheme 2 Synthesis of the C_{32}–C_{42} subunit **2** [5].

The investigators carried out the synthesis of side chain subunit **3** utilizing a Brown allylation of readily available aldehyde **21** to afford homoallylether **22** on subsequent TBS protection (Scheme 4). Homologation of the methylester **23** to enal **24** proceeded smoothly by cross-metathesis in the presence of Grubbs(II) catalyst; finally, the required stannane was introduced by Horner-Wadsworth-Emmons reaction with phosphonate derivative **25** [13] to furnish the desired building block **3** after deprotection of the C_6–OH function.

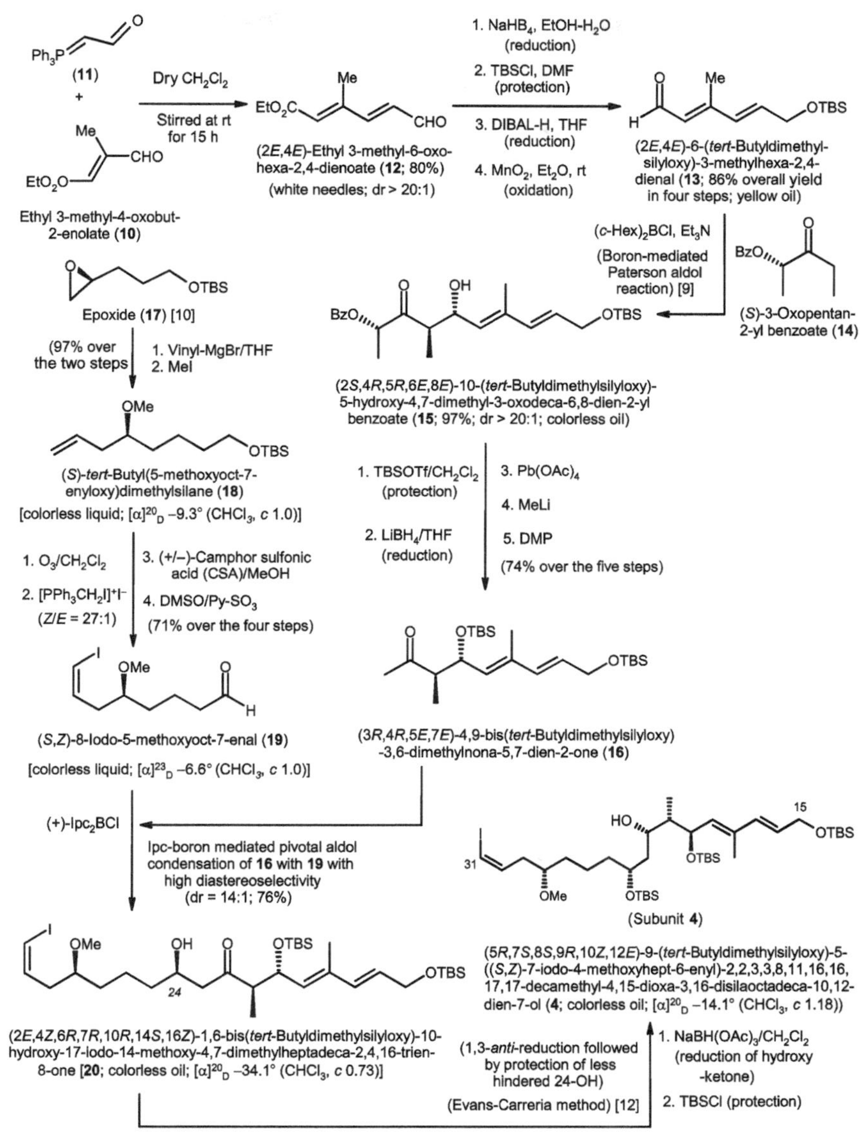

Scheme 3 Synthesis of the C_{15}–C_{31} subunit **4** [5].

After synthesizing the subunits **2**, **3**, and **4**, the investigators then assembled them (Scheme 5); the macrocyclic core was first developed from the esterification of subunits **2** and **4** by means of the Yamaguchi protocol (97%) followed by Heck macrocyclization with excellent yield and diastereoselectivity (70%, $E/Z > 20{:}1$) to give **26** as a colorless oil with an

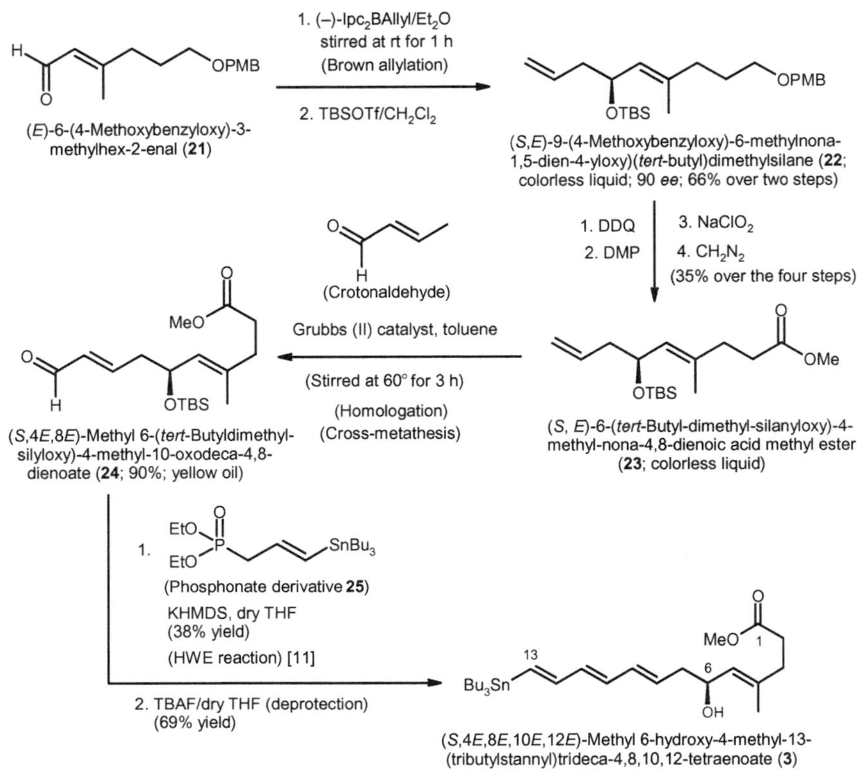

Scheme 4 Synthesis of the C_1–C_{13} subunit **3** [5].

overall yield of 68%. After selective removal of the primary TBS group (deprotection with acetic acid) and allylic oxidation, the required *E*-vinyl iodide was introduced by a Takai reaction (92%, *E*/*Z*=4:1). All the TBS groups in **26** were then removed (deprotection with TBAF), and the resulting compound was made to undergo Stille coupling with subunit **3** for attachment of the side chain, followed by subsequent acetonide cleavage under only mildly acidic conditions (65% HOAc) to yield the acetonide-protected etnangien methyl ester **28**. Finally, enzyme-catalyzed ester cleavage [14] afforded etnangien (**1**) as a colorless oil, $[\alpha]_D^{20}$ +16.2° (MeOH, *c* 0.65 mg/mL), which was identical to an authentic sample of etnangien [1]. This first total synthesis of etnangien proceeds in 23 steps with an overall yield of 0.25%, and establishes unequivocally the relative and absolute configuration [5].

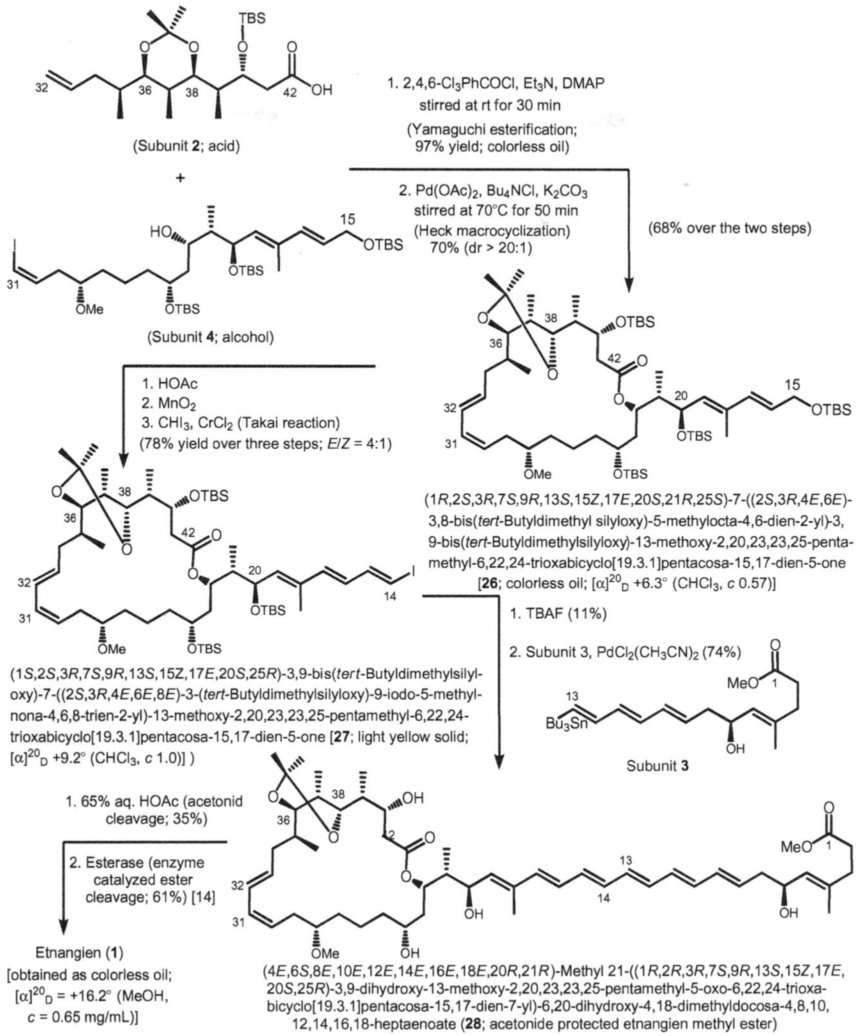

Scheme 5 Assembly of the subunits **2**, **3**, and **4** [5].

References

[1] Höfle, G., Reichenbach, H., Irschik, H., and Schummer, D. (1998) German Patent DE 196 30 980 A1: 1–7 (5.2.1998); H. Irschik, D. Schummer, G. Höfle, H. Reichenbach, H. Steinmetz, R. Jansen, J. Nat. Prod. 70 (2007) 1060.

[2] K. Carsten, G. Klaus, M. Rolf, J. Biotechnol. 121 (2006) 201.

[3] D.V. Haebich, F. Nussbaum, Angew. Chem. Int. Ed. 48 (2009) 3397.

[4] D. Menche, F. Arikan, O. Perlova, N. Horstmann, W. Ahlbrecht, S.C. Wenzel, R. Jansen, H. Irschik, R. Müller, J. Am. Chem. Soc. 130 (2008) 14234.

[5] P. Li, J. Li, F. Arikan, W. Ahlbrecht, M. Dieckmann, D. Menche, J. Am. Chem. Soc. 131 (2009) 11678.
[6] F. Arikan, J. Li, D. Menche, Org. Lett. 10 (2008) 3521.
[7] I. Paterson, R.D. Tillyer, Tetrahedron Lett. 33 (1992) 4233.
[8] L.J. Famer, K.S. Marron, S.S.C. Koch, C.K. Hwang, E.A. Kallel, L. Zhi, A.M. Nadzan, D.W. Robertson, Y.L. Bennani, Bioorg. Med. Chem. Lett. 16 (2006) 2352.
[9] C.J. Cowden, I. Paterson, Org. React. 51 (1997) 1.
[10] A.G. Myers, B.A. Lanman, J. Am. Chem. Soc. 124 (2002) 12969.
[11] G. Stork, K. Zhao, Tetrahedron Lett. 30 (1989) 2173.
[12] D.A. Evans, K.T. Chapman, E.M. Carreira, J. Am. Chem. Soc. 110 (1988) 3560.
[13] A.B. Smith III, Z. Wan, J. Org. Chem. 65 (2000) 3738.
[14] D. Menche, F. Arikan, O. Perlova, N. Horstmann, W. Ahlbrecht, S.C. Wenzel, R. Jansen, H. Irschik, R. Müller, J. Am. Chem. Soc. 130 (2008) 14234.

CHAPTER TWENTY FIVE

(+)-Gliocladin B

Abbreviations

Boc *tert*-butoxycarbonyl
DTBMP 2,6-di-*tert*-butyl-4-methylpyridine
EDC·HCl 1-ethyl-3-(3-dimethylaminopropyl)carbodiimide hydrogen chloride
HOBt hydroxybenzotriazole
LHMDS lithium *bis*(trimethylsilyl)amide
MS molecular sieves
NEt_3 triethylamine
Py pyridine
TFA trifluoroacetic acid
THF tetrahydrofuran
TIPS triisopropylsilyl

Systematic name: (3*S*,5a*R*,10b*S*,11a*S*)-10b-(1*H*-Indol-3-yl)-2-methyl-3,11a-bis(methylthio)-2,3,5a,6,11,11a-hexahydro-1*H*-pyrazino[1′,2′:1,5]pyrrolo[2,3-*b*]indole-1,4(10b*H*)-dione

Compound class: Epidithiodiketopiperazine alkaloid (mycotoxin class)

Structure:

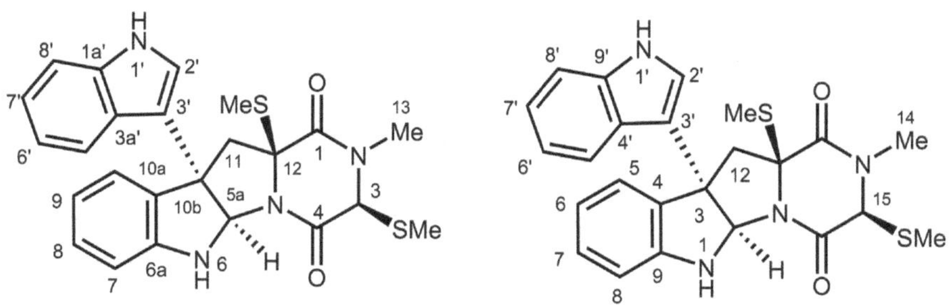

(Usami's numbering report) [1] (Movassaghi's numbering report) [3]

(+)-Gliocladin B (**1**)

Natural source: *Gliocladium roseum* (syn. *Clonostachys rosea f. rosea*; family Bionectriaceae; fungal strain OUPS-N132) [1]

Pharmaceutical potential: Cytotoxic [1, 2]

Total Synthesis of Bioactive Natural Products
https://doi.org/10.1016/B978-0-08-102822-3.00025-0

Synthetic routes:

The first total synthesis of (+)-glioclandin B has recently been reported by Boyer and Movassaghi [3]. Their first-generation (Scheme 1) and second-generation (Scheme 2) total syntheses of this mycotoxin alkaloid are depicted herein.

First-generation total synthesis of (+)-glioclandin B (1):

The investigators [3] prepared the starting material (-)-diketopiperazine (**6**) from commercially available (−)-*N*-Boc-L-tryptophan (**2**) in three steps. Compound **2** was sulfonylated with phenylsulfonyl chloride in the presence of lithium *bis*(trimethylsilyl)amide (LHMDS) in THF to give (−)-*N'*-sulfonylated *N*-Boc-L-tryptophan (**3**; 71%). This derivative **3** reacted with L-sarcosin methyl ester (**4**) in the presence of EDC·HCl/HOBt/NEt_3 yielding (+)-(*S*)-methyl 2-(2-((*tert*-butoxycarbonyl)amino)-*N*-methyl-3-(1-(phenylsulfonyl)-1*H*-indol-3-yl)propanamido) acetate (**5**; 82%), which in turn afforded the (−)-starting diketopiperazine intermediate (*S*)-1-methyl-3-((1-(phenylsulfonyl)-1*H*-indol-3-yl)methyl)piperazine-2,5-dione (**6**; 99%) in the next step. (−)-Diketopiperazine **6** underwent cyclization and bromination reaction with molecular bromine in dichloromethane to form (+)-*endo*-(2*S*,3*S*)-tetracyclic *N*-methyl diketopiperazine bromide (**7**) with a high level of diastereoselection (*endo*:*exo* ~97:3) in 75% yield. This *endo*-tetracyclic bromide (**7**) took part in Friedel-Crafts-type coupling with 5-bromo-1-triisopropylsilylindole (**8**) in a regio- and stereoselective manner via intermediate **9** promoted by $AgBF_4$/DTBPM in nitroethane to afford the desired indole adduct (+)-C3-(5-bromo-1-TIPS-indol-3-yl)-pyrrolidinoindoline (**10**) in 83.6% yield. This indole adduct (**7**) was then deprotected quantitatively to (+)-C3-(indol-3-yl)-pyrrolidinoindoline (**11**; 100%) followed by its dihydroxylation at C-11 and C-15 with tetra-*n*-butylammonium permanganate (*n*-Bu_4NMnO_4, 3.79 equiv.) in dichloromethane to obtain the (−)-hexacyclic diol (**12**; 41%) as a single diastereomer.

On reaction with trifluoroacetic acid in hydrogen sulfide-saturated dichloromethane solution at 0°C, the diol derivative **12** produced the corresponding hemithioaminal **14** in a highly diastereoselective fashion (>10:1 dr) via trapping of iminium ion **13**, which in turn yielded (+)-hexacyclic thioisobutyrate (**12**; 82%) on subsequent reaction with isobutyryl chloride and pyridine in dichloromethane. Desulfonylation of this intermediate **15** to the desired aminothioisobutyrate **16** in 57% yield was carried out by using an aqueous sodium ascorbate–ascorbic acid mixture in combination with UV irradiation at 350 nm; the desulfonylated product then underwent hydrazinolysis of both of its thioester and ester groups followed by chemoselective *S*-sulfenylation with triphenylmethanesulfenyl chloride to give (+)-hexacyclic triphenylmethanedisulfide (**17**) in 81% yield over two steps. The investigators tactically utilized the efficacy of hafnium trifluoromethanesulfonate [$Hf(OTf)_4$] in acetonitrile for smooth cyclization of (+)-triphenylmethanedisulfide **17** to obtain the corresponding epidisulfide (+)-12-deoxybionectin (**19**) via the putative C15-iminium ion **18** in overall 80% yield. Ultimately, reduction of the bridgehead disulfide with $NaBH_4$ followed by in situ *S*-methylation with methyl iodide afforded (+)-gliocladin B (**1**; $[\alpha]_D^{24}$ +200.4° ($CHCl_3$, *c* 0.062)) in 80% yield. The physical and spectral data of this synthetic compound were found to be matched with those reported for natural (+)-glioclandin B [1]; in addition, this structural representation **1** with all relative and absolute configurations was further verified by X-ray crystallographic analysis [3].

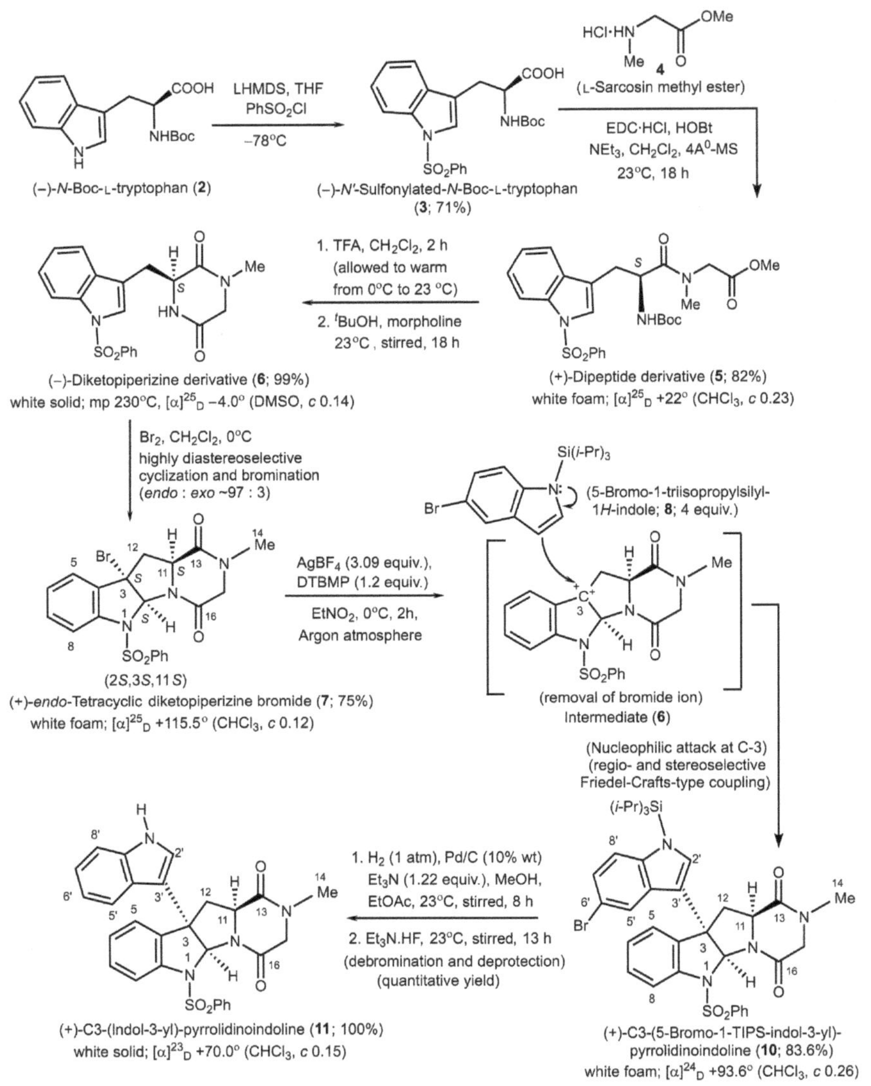

Scheme 1—Cont'd

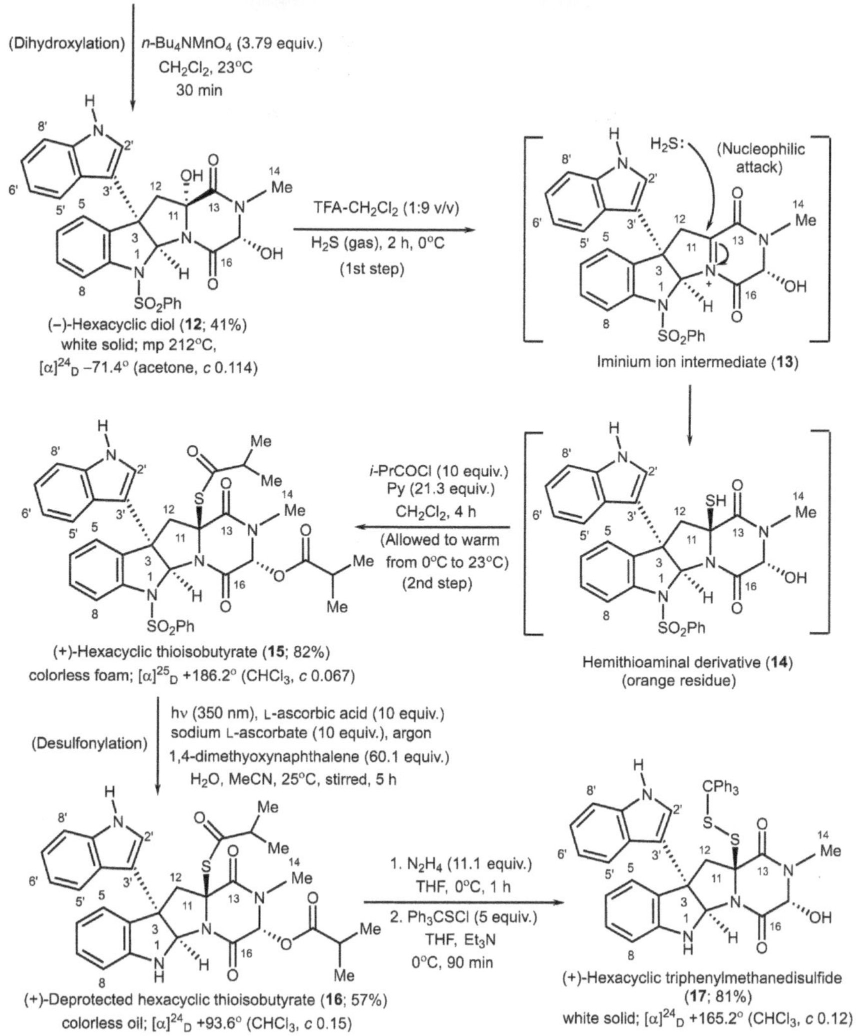

Scheme 1—Cont'd

(Continued)

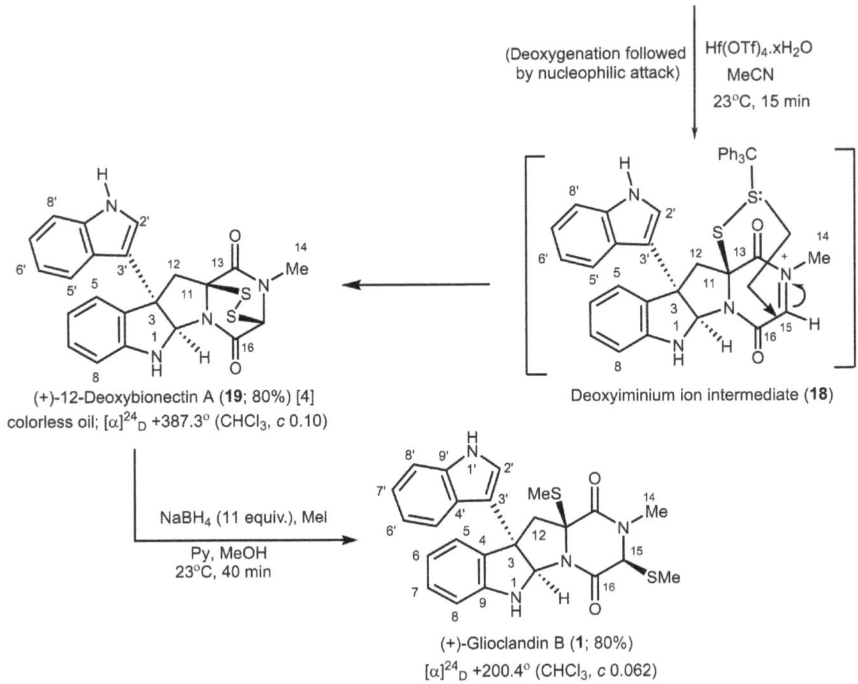

Scheme 1 First-generation total synthesis of (+)-glioclandin B (**1**).

Second-generation total synthesis of (+)-glioclandin B (1):

In the second-generation total synthesis strategy, the investigators [3] developed a more streamlined and expedient route to (+)-glioclandin B (**1**) from the (−)-hexacyclic diol (**12**) with fewer steps and higher yield (~10%) [3]. The (−)-diol derivative **12** on exposure to sodium thiomethoxide and trifluoroacetic acid in nitromethane resulted in the formation of (+)-hexacyclic bis(methylthioether) **20** with quite a good level of diastereoselection (C-15β:C-15α ~7:1) in 77% yield (single diastereomer). Benzenesulfonyl photodeprotection at N-1 in **20** eventually afforded (+)-gliocladin B (**1**; 88%) in 68% yield over the last two steps.

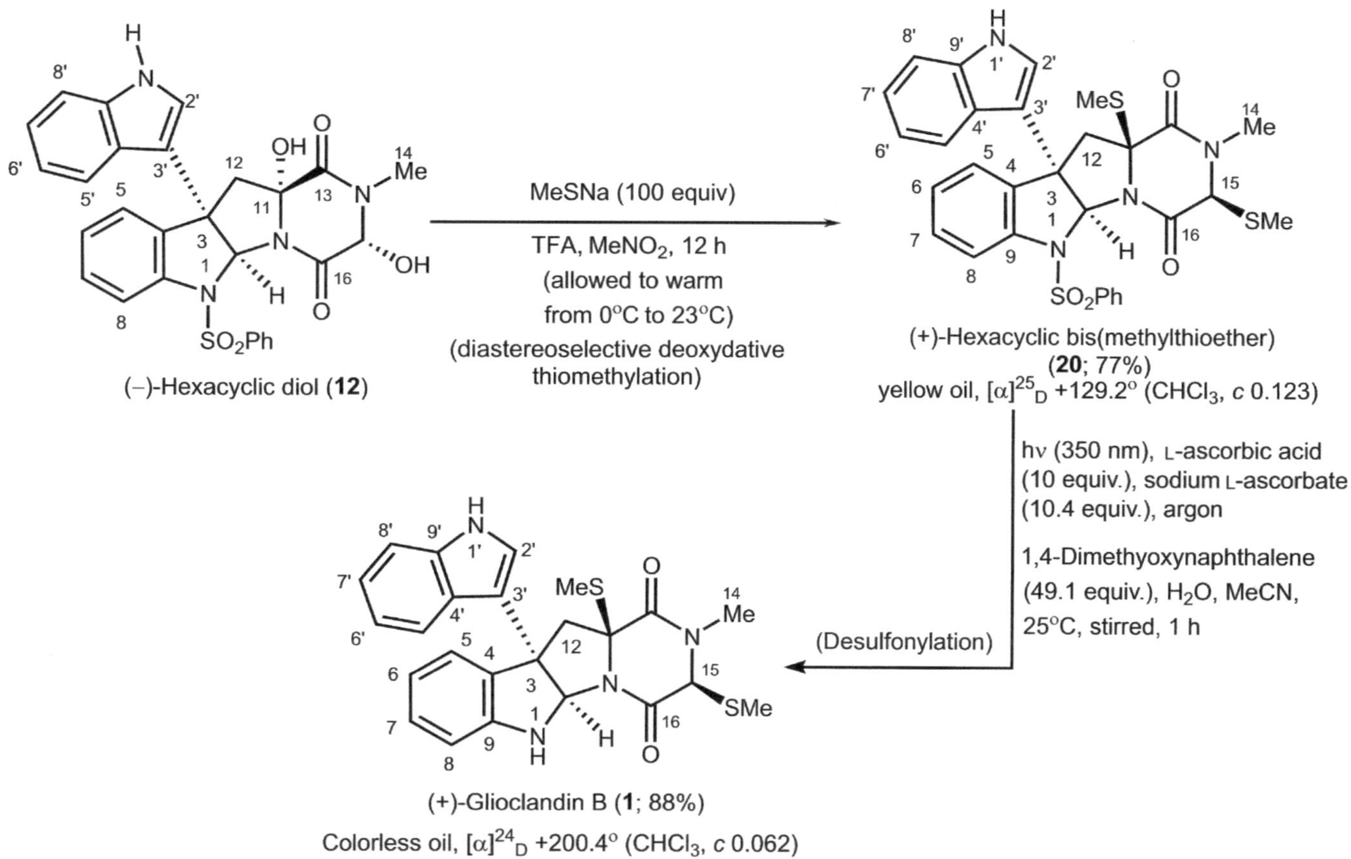

Scheme 2 Second-generation total synthesis of (+)-glioclandin B (**1**).

References

[1] Y. Usami, J. Yamaguchi, A. Numata, Heterocycles 63 (2004) 1123.
[2] S. Bräse, A. Encinas, J. Keck, C.F. Nising, Chem. Rev. 109 (2009) 3903.
[3] N. Boyer, M. Movassaghi, Chem. Sci. 3 (2012) 1798.
[4] C.-J. Zheng, C.-J. Kim, K.S. Bae, Y.-H. Kim, W.-G. Kim, J. Nat. Prod. 69 (2006) 1816.

CHAPTER TWENTY SIX

(+)-Harziphilone

Abbreviations

$BH_3 \cdot SMe_2$ borane-methylsulfide complex
DABCO 1,4-diazobicyclo[2.2.2]octane
DMAP 4-*N*,*N*-dimethylaminopyridine
4-DMSO dimethylsulfoxide
Et_3N triethylamine
HIV human immunodeficiency virus
REV regulation of virion expression
TBDMSCl *tert*-butyldimethylsilylchloride
TESCl triethylsilyl chloride
THF tetrahydrofuran

Systematic name: (6*R*,7*R*)-6,7-Dihydroxy-7-methyl-3-((1*E*,3*E*)-penta-1,3-dien-1-yl)-6,7-dihydro-1*H*-isochromen-8(5*H*)-one

Compound class: Antibiotic

Structure:

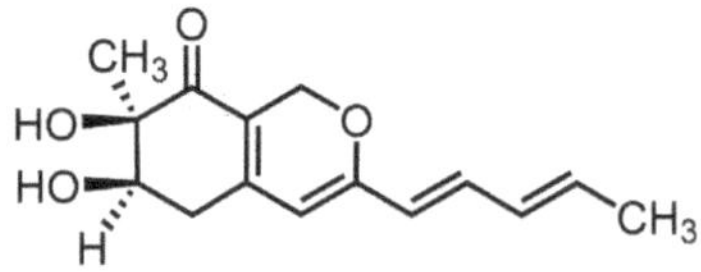

(+)-Harziphilone (**1**)

Natural source: *Trichoderma harzianum* (fungal fermentation broth; family: Hypocreaceae) [1]

Pharmaceutical potential: Anti-HIV (HIV-1 REV/REV-responsive element inhibitor) [1]

Synthetic route:

Sorensen and his group [2] accomplished an enantioselective total synthesis (Scheme 1) of (+)-harziphilone (**1**), a naturally occurring fungal metabolite with significant anti-HIV potential through a nucleophile-catalyzed cycloisomerization of an acyclic polyunsaturated diketone **18** prepared from commercially available 2-methyl-2-cyclopenten-1-one (**2**). A silylated derivative **4** was obtained on asymmetric reduction of ketone **2** by the powerful method of Corey, Bakshi, and Shibata [3] followed by

https://doi.org/10.1016/B978-0-08-102822-3.00026-2

silylation of the resulting secondary alcohol **3**. On treatment with osmium tetraoxide, compound **4** underwent an efficient and highly diastereoselective dihydroxylation of the ring alkene to afford the vicinal diol **5** in 86% yield. After protection of the vicinal diol system in the form of an acetonide, the silyl ether **6** was then cleaved with fluoride ion and the resulting alcohol **7** was oxidized with SO_3-pyridine complex to optically active cyclopentanone **8** (71% yield).

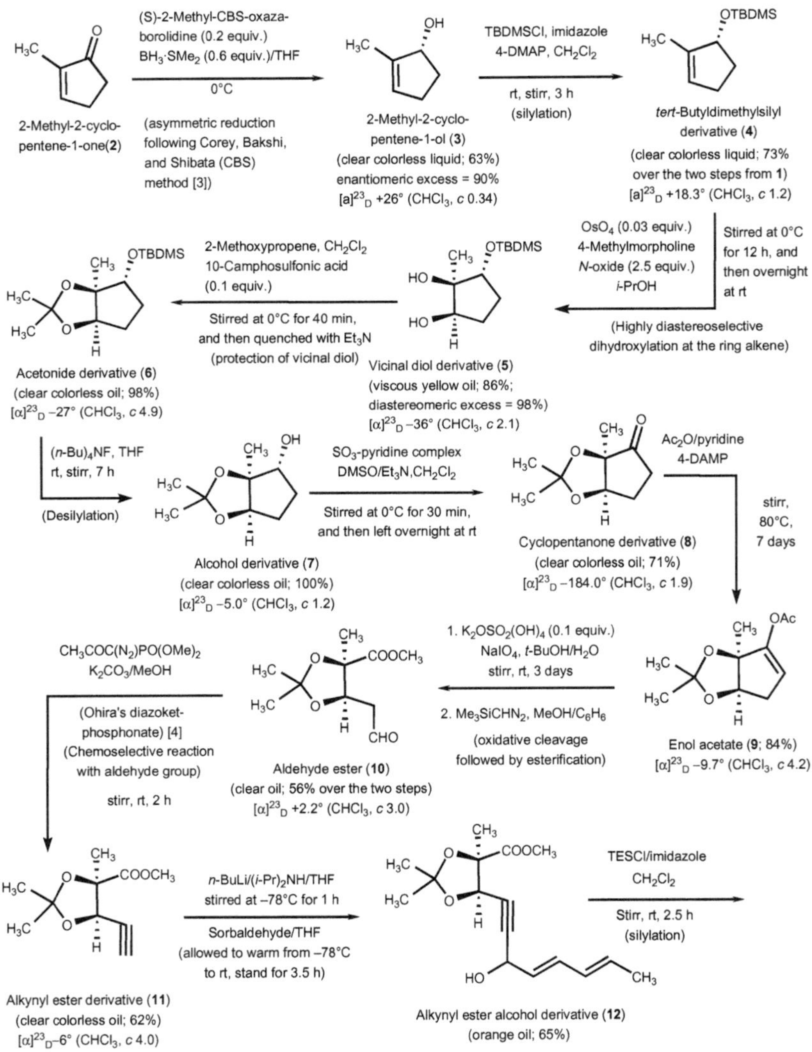

Scheme 1—Cont'd

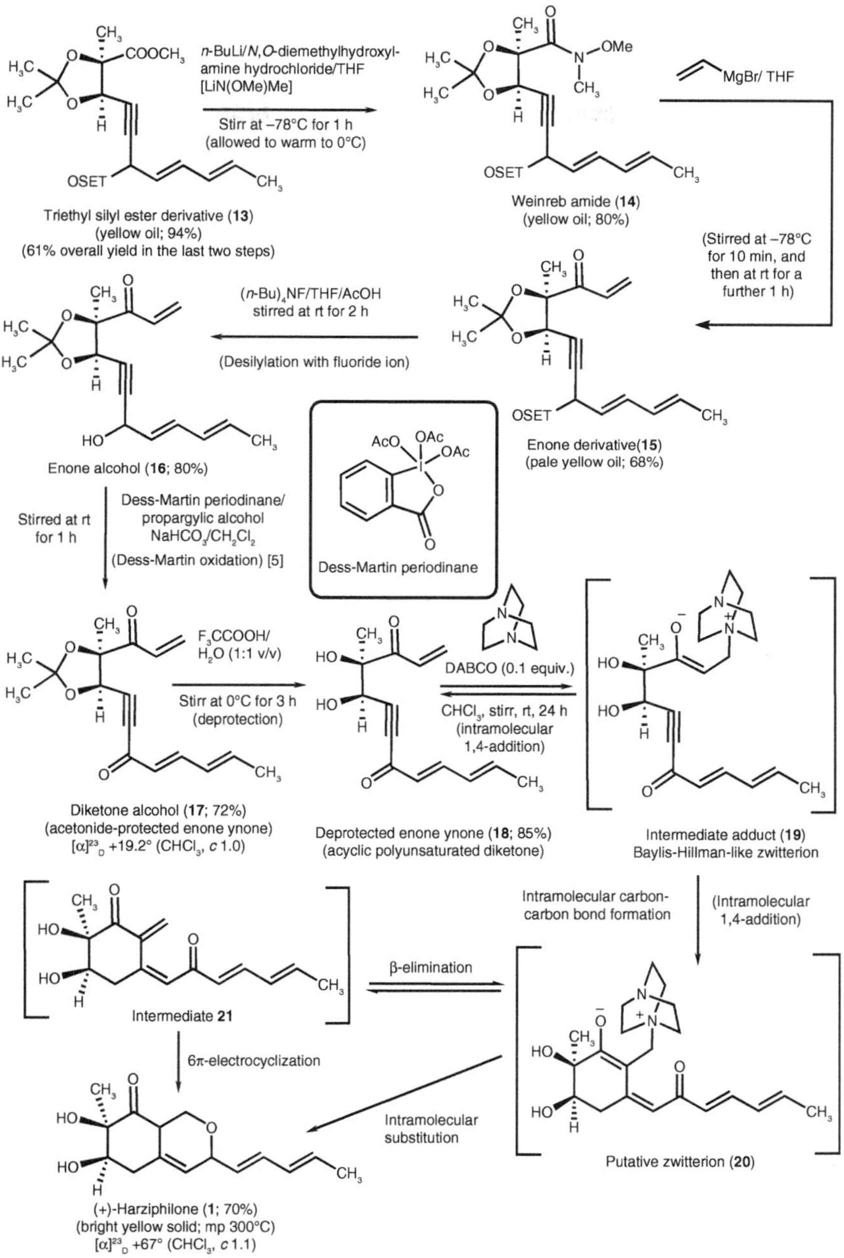

Scheme 1 Total synthesis of (+)-harziphilone (**1**).

On treatment with acetic anhydride/pyridine, compound **8** was enolized and acetylated on an oxygen atom; oxidative cleavage of this intermediate enol acetate **9** afforded an aldehyde carboxylic acid, which was esterified with (trimethylsilyl)diazomethane to furnish the corresponding aldehyde methyl ester **10**. The desired alkynyl ester **11** was obtained from derivative **10** through a chemoselective reaction of the aldehyde group with Ohira's diazoketophosphonate [4] in basic methanol. The investigators then prepared silyl ester derivative **13** in 61% overall yield from alkynyl ester **11** on reaction of its lithium acetylide (formed by a low-temperature deprotonation on treatment with lithium diisopropylamide in THF) with commercially available sorbaldehyde. The resulting Weinreb amide **14** underwent a smooth reaction with vinylmagnesium bromide in THF to give the corresponding enone **15**. Desilylation of the enone derivative **15** followed by Dess-Martin oxidation [5] of the resulting enone alcohol **16** afforded the desired diketone **17** (72%). Finally, an acidic hydrolysis of the acetonide protecting group afforded the polyunsaturated diketone **18** in 85% yield, bicycloisomerization of which eventually furnished (+)-harziphilone (**1**) in the following steps.

The investigators [2] carried out nucleophile-catalyzed bicycloisomerization of the polyunsaturated diketone **18** in the presence of DABCO (0.1 equiv.) under mild reactions conditions (room temperature stirring in chloroform solution for 24 h). They suggested the formation of a Baylis-Hillman-like zwitterion **19** arising out of a reversible and intramolecular 1,4-addition of DABCO with the unsubstituted and potentially activated enone system of **18**; an intramolecular carbon-carbon bond formation could then generate an allenolate ion and subsequently the putative zwitterion **20** through a proton transfer. The nucleophilic catalyst (DABCO) could then be detached by means of a simple β-elimination reaction to give intermediate **21**, thereby facilitating the 6π-electrocyclization to (+)-harziphilone (**1**). The investigators also suggested that the oxacyclic ring of (+)-harziphilone (**1**) could also alternatively be accomplished from zwitterion **20** by an intramolecular displacement of a neutral molecule of DABCO [2]. (+)-Harziphilone (**1**) was obtained as a bright yellow solid in 70% yield (starting from diketone **18**), mp 300°C, $[\alpha]_D^{23}$ +67° ($CHCl_3$, *c* 1.1); the spectroscopic data obtained for the synthesized (+)-harziphilone (**1**) were found to be matched with data reported for natural harziphilone [1].

References

[1] J. Qian-Cutrone, S. Huang, L.-P. Chang, D.M. Pirnik, S.E. Klohr, R.A. Dalterio, R. Hugill, S. Lowe, M. Alam, K.F. Kadow, J. Antibiot. 49 (1996) 990.
[2] L.M. Stark, K. Pekari, E.J. Sorensen, Proc. Natl. Acad. Sci. U. S. A. 101 (2004) 12064.
[3] E.J. Corey, C.J. Helal, Angew. Chem. Int. Ed. 37 (1998) 1986.
[4] S. Ohira, Synth. Commun. 19 (1989) 561.
[5] D.B. Dess, J.C. Martin, J. Org. Chem. 48 (1983) 4155.

CHAPTER TWENTY SEVEN

Hibarimicinone

Abbreviations

AIBN azobisisobutyronitrile
Bz benzoyl
CH_2Cl_2 dichloromethane
DBTC dibutyltin dichloride
DBU 1,8-diazabicyclo[5.4.0]undec-7-ene
DDQ 2,3-dichloro-5,6-dicyano-1,4-benzoquinone
DMSO dimethylsulfoxide
Et_3N triethylamine
IBX 2-iodoxybenzoic acid
MeCN acetonitrile
MOM methoxymethyl
NaHMDS sodium hexamethyldisilazide
PivCl pivaloyl chloride
Py pyridine
rt room temperature
TBS *tert*-butyldimethylsilyl
THF tetrahydrofuran

Systematic name: (1*S*,2*S*,3*R*,4*S*,4a*S*,5*S*,12a*S*)-9-((6a*R*,7*S*,8*R*,9*S*,10*S*,10a*S*)-1,7,8,9,10,10a,12-Heptahydroxy-3,4-dimethoxy-11-oxo-10-propyl-6,6a,7,8,9,10,10a,11-octahydrotetracen-2-yl)-2,3,4,6,11,12a-hexahydroxy-8-methoxy-1-propyl-1,3,4,4a,5,12a-hexahydro-1,5-epoxytetracene-7,10,12(2*H*)-trione

Compound class: Antibiotic

Structure:

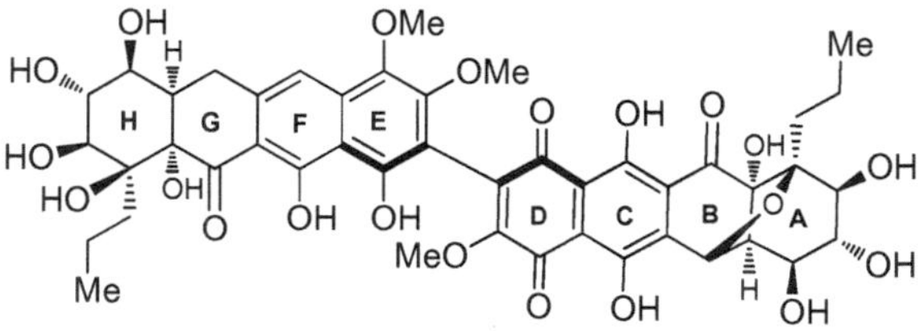

Hibarimicinone (**1**)

Total Synthesis of Bioactive Natural Products
https://doi.org/10.1016/B978-0-08-102822-3.00027-4

Natural source: *Microbispora rosea* subsp. *hibaria* (Actinomycetes; culture broth) [1, 2]

Pharmaceutical potential: *v-Src* tyrosine kinase inhibitor [1]

Synthetic route:

The first total synthesis of hibarimicinone (**1**) was achieved by Tatsuta et al. [3] featuring the chemistry of a chiral biaryl thiolactone and involvement of double Michael-Dieckmann-type cyclization and aromatization. They envisioned the retrosynthetic species for the target molecule as the chiral polyhydroxydecalin **2** and the chiral biaryl thiolactone **3** (Scheme 1).

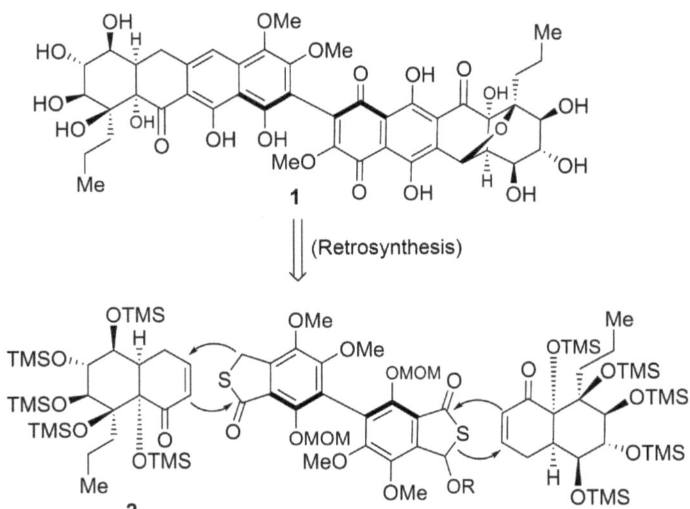

Scheme 1 Retrosynthetic approach for hibarimicinone (**1**) [3].

Synthesis of polyhydroxydecalin subunit 2:

Initially, the investigators prepared the polyhydroxydecalin subunit **2** starting from (4*S*,5*R*,6*S*)-4,5,6-tris((*tert*-butyldimethylsilyl)oxy)-2-(phenylsulfonyl)cyclohex-2-enone (**4**); on stereoselective Diels-Alder reaction with (1-trimethylsiloxy-1,3-butadiene) compound **4** formed adduct **5** in 90% yield. Jones oxidation of **5** and the subsequent reduction of the enone intermediate formed with SmI_2 to remove the sulfone moiety [4] afforded the tertiary alcohol **6**. It was then subjected to regioselective desilylation at C-10 position and the resulting alcohol was oxidized with IBX to produce triketone **7** in good yield. The regio- and stereoselective reduction of triketone **7** was accomplished in excellent yield by treatment with $NaBH(OAc)_3$ in ethanol resulting in the formation of (2*S*,3*R*,4*S*,4a*R*,8a*R*)-2,3-bis((*tert*-butyldimethylsilyl)oxy)-4,8a-dihydroxy-3,4,4a,5-tetrahydronaphthalene-1,8(2*H*,8a*H*)-dione (**8**). TMS-protected derivative of compound **8** was allowed to react under Grignard conditions for introducing an allyl group in a stereoselective manner to produce the tertiary alcohol **9**. After silylation of this tertiary alcohol **9**, it was then treated with DBU in hot toluene in the presence of *i*-PrOH resulting in a regioselective desilylation to afford the enone derivative **10**. Hydrogenation of the *exo*-olefin of **10** ultimately afforded the polyhydroxydecalin subunit **2** (Scheme 2).

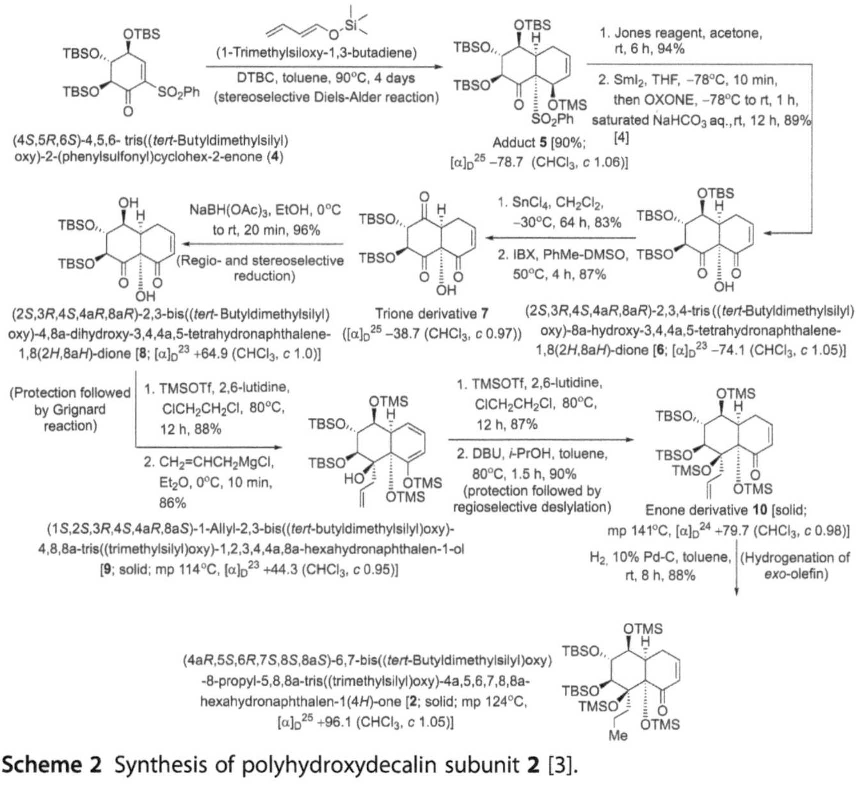

Scheme 2 Synthesis of polyhydroxydecalin subunit **2** [3].

Synthesis of polyhydroxydecalin subunit 3:

Preparation of the chiral biaryl thiolactone **3** is depicted in Scheme 3; synthesis started with the commercially available 2,4,5-trimethoxybenzoic acid (**11**). It was converted into amide, which on *ortho*-lithiation-methylation afforded *N*,*N*-diethyl-3,4,6-trimethoxy-2-methylbenzamide

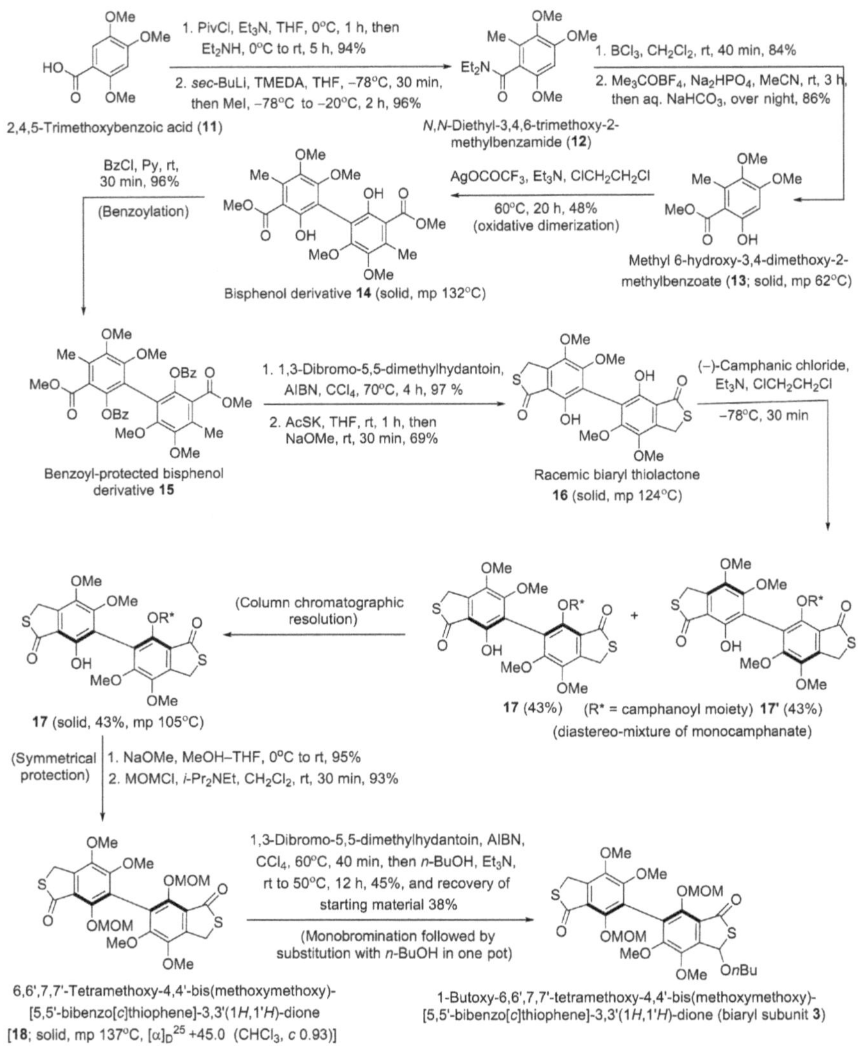

Scheme 3 Synthesis of biaryl thiolactone subunit **3** [3].

(**12**). Regioselective de-*O*-methylation was followed by the transformation of amide into ester to yield methyl 6-hydroxy-3,4-dimethoxy-2-methylbenzoate (**13**). Compound **13** was then subjected to oxidative dimerization in the presence of silver trifluoroacetate, and the resulting bisphenol **14** was protected as benzoate derivative **15**. Bromination at the benzylic positions of **15** followed by sequential substitution with thioacetate ion and methanolysis eventually furnished racemic biaryl thiolactone **16** in good yield. This racemic biaryl **16** was then converted into the diastereomixture of monocamphanates **17** and **17′**, which was resolved by column chromatography and subsequent crystallization to give optically pure **17** and **17′** in good yields. The free hydroxyl functions of the desired isomer **17** were protected as MOM ether **18**; monobromination of **18** and substitution with *n*-butanol were performed in one pot to afford the chiral biaryl subunit **3** (Scheme 3) [3].

Combination of subunits 2 and 3: total synthesis of hibarimicinone (1):

On getting these two key segments **2** and **3** on hand, the investigators next proceeded to connect them. They accomplished the combination of the subunits lucidly using certain conventional techniques in somewhat modified forms as illustrated in Scheme 4. Eventually, the huge molecule hibarimicinone (**1**) was obtained in appreciable yield.

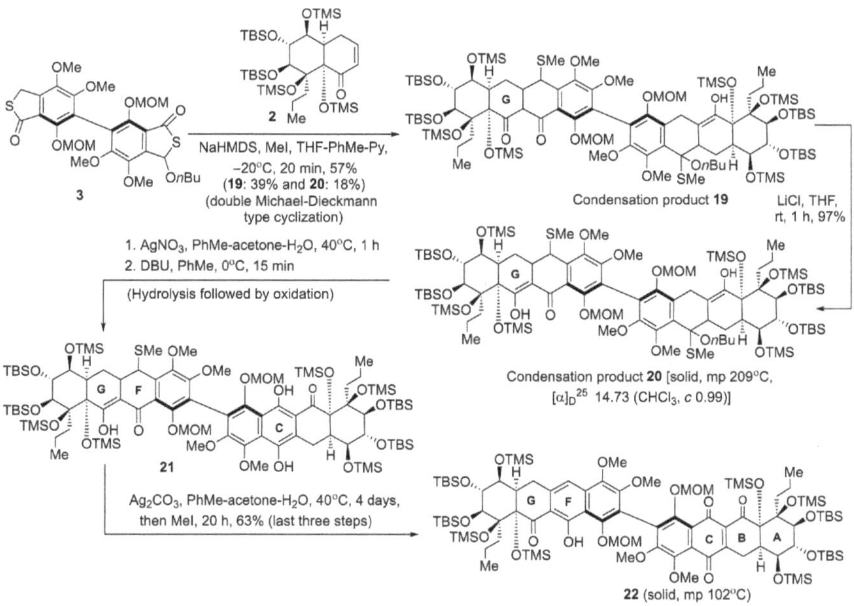

Scheme 4—Cont'd

(Continued)

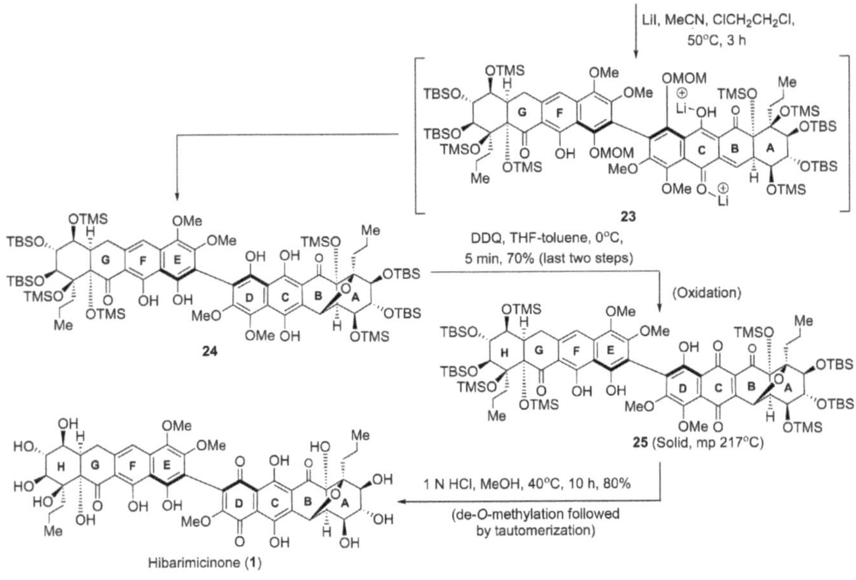

Scheme 4 Total synthesis of hibarimicinone (**1**) [3].

References

[1] S.I. Cho, H. Fukazawa, Y. Honma, T. Kajiura, H. Hori, Y. Igarashi, T. Furumai, T. Oki, Y. Uehara, *J. Antibiot.* 55 (2002) 270.
[2] H. Hori, Y. Igarashi, T. Kajiura, S. Sato, T. Furumai, K. Higashi, T. Ishiyama, Y. Uehara, T. Oki, Tennen Yuki Kagobutsu Toronkai Koen Yoshishu 46 (2004) 49.
[3] K. Tatsuta, T. Fukuda, T. Ishimori, R. Yachi, S. Yoshida, H. Hashimoto, S. Hosokawa, *Tetrahedron Lett.* 53 (2012) 422.
[4] G.A. Molander, G. Hahn, *J. Org. Chem.* 51 (1986) 1135.

CHAPTER TWENTY EIGHT

(±)-3-Hydroxy-β-Ionone

Abbreviations

CH_2Cl_2 dichloromethane
DBU 1,8-diazabicyclo[5.4.0]undec-7-ene
DMAP 4-dimethylaminopyridine
Et_3N triethylamine
HPLC high performance liquid chromatography
TBS *tert*-butylsilyl
THF tetrahydrofuran

Systematic name: (*E*)-4-(4-Hydroxy-2,6,6-trimethylcyclohex-1-en-1-yl) but-3-en-2-one

Compound class: Bisnorsesquiterpene

Structure:

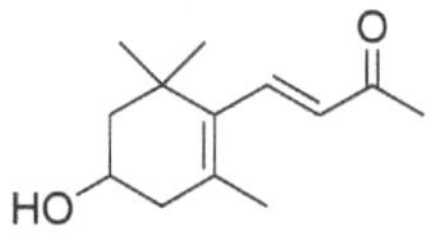

(±)-3-Hydroxy-β-ionone (**1**)

Natural source: Various tobacco leaves as neutral aroma constituents and Kudzu oil (Pueraria lobata Ohwi; family: Fabaceae) [1]; dwarf bean shoots (Phaseolus vulgaris cv Morocco; Leguminosae) [2]; Vulpia myuros (family: Poaceae; rattail fescue; aqueous extract) [3]

Pharmaceutical potential: Allelopathic [1–5]

Synthetic route:

Shishido and his group successfully completed the total synthesis of (±)-3-hydroxy-β-ionone (**1**), a bisnorsesquiterpene having considerable allelopathic activity, in eight steps starting from 3,3-dimethylpent-4-ynal (**2**) by employing a ring-closing enyne metathesis for the construction of the C_1–C_8 segment and a two-carbon elongation via nitrile oxide-alkene [3 + 2] cycloaddition as the key steps with an overall yield of 13% (Scheme 1) [6]. Reaction of **2** with (2-methylallyl)magnesium chloride (**3**),

https://doi.org/10.1016/B978-0-08-102822-3.00028-6

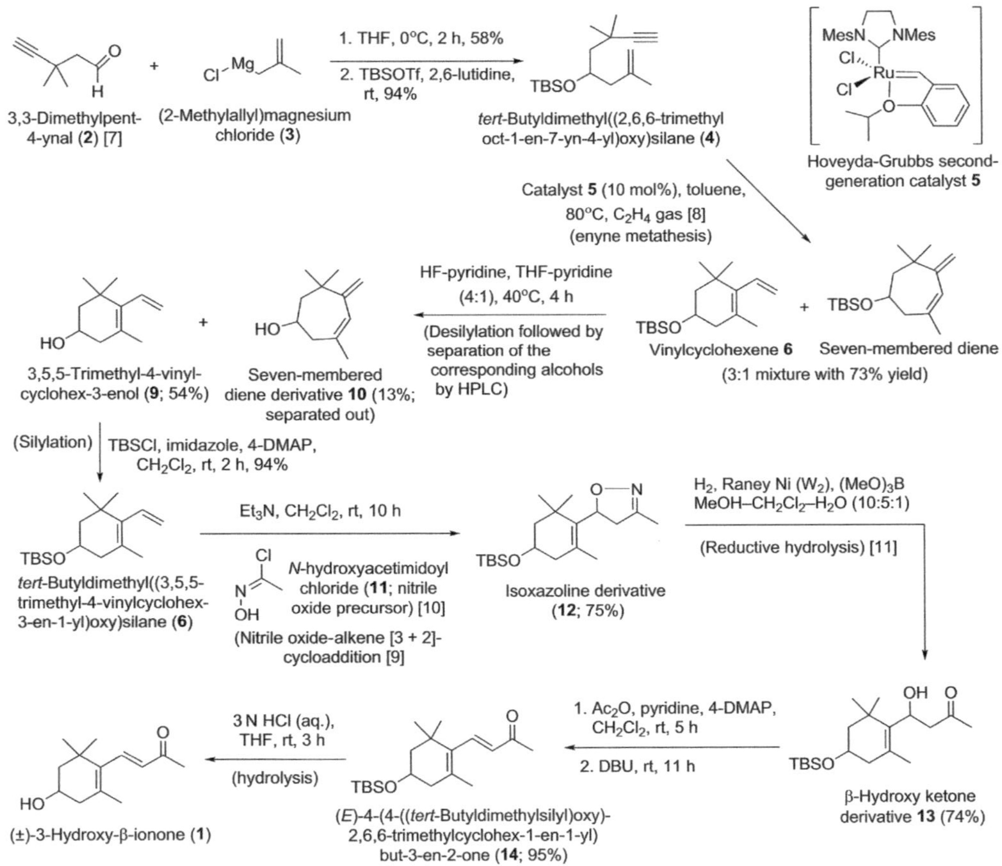

Scheme 1 Total synthesis of (±)-3-hydroxy-β-ionone (**1**) [6].

followed by protection of the alcohol formed initially as a TBS ether, produced *tert*-butyldimethyl((2,6,6-trimethyloct-1-en-7-yn-4-yl)oxy)silane (**4**) in excellent yield. The ring-closing enyne metathesis of **4** was then performed with 10 mol% of the Hoveyda-Grubbs second-generation catalyst **5** and ethylene gas [8] in toluene solution at 80°C when a 3:1 mixture of the desired vinylcyclohexene **6** and the seven-membered diene **7** was obtained in 73% yield. Desilylation of this mixture with HF/THF/pyridine provided a mixture of the corresponding alcohols **9** and **10** from which the desired alcohol **9** was separated by HPLC in 54% yield. Compound **9** was then protected at the alcoholic group to afford the desired vinylcyclohexene **6**.

The investigators carried out a nitrile oxide-alkene [3 + 2] cycloaddition for the carbon chain elongation [9]. For this purpose, vinylcyclohexene **6** was treated with a solution of *N*-hydroxyacetimidoyl chloride (**11**) used as a nitrile oxide precursor [10] in dichloromethane in the presence of triethylamine at room temperature for 10 h, when an isoxazoline derivative (**12**) was formed in 75% yield. The isoxazoline **12** thus obtained was then subjected to reductive hydrolysis [11] under an atmosphere of hydrogen in the presence of Raney Ni (W_2) and trimethyl borate in a methanol: dichloromethane:water (10:5:1) solvent mixture at room temperature for 4 h as a result of which the β-hydroxy ketone derivative **13** was produced in good yield (74%). Dehydration of compound **13** to obtain the α,β-unsaturated ketone **14** was achieved by acetylation followed by DBU treatment. On hydrolysis with 3 N aq. HCl in THF, the α,β-unsaturated ketone **14** furnished (±)-3-hydroxy-β-ionone (**1**) in 87% yield. All spectral data of the synthetic material were found to be identical to those reported in the literature.

References

[1] T. Fujimori, R. Kasuga, M. Noguchi, H. Kaneko, Agric. Biol. Chem. 38 (1974) 891. T. Fujimori, R. Kasuga, H. Matsushita, H. Kaneko, M. Noguchi, Agric. Biol. Chem. 40 (1976) 303; S. Shibata, A. Katsuyama, M. Noguchi, Agric. Biol. Chem. 42 (1978) 195.
[2] H. Kato-noguchi, S. Kosemura, S. Yamamura, K. Hasegawa, Phytochemistry 33 (1993) 553.
[3] H. Kato-Noguchi, M. Yamamoto, K. Tamura, T. Teruya, K. Suenaga, Y. Fujii, Plant Growth Regul. 60 (2010) 127.
[4] H. Kato-Noguchi, T. Seki, H. Shigemori, J. Plant Physiol. 167 (2010) 468.
[5] H. Kato-Noguchi, T. Seki, Plant Signal. Behav. 5 (2010) 702.
[6] D. Kikuchi, M. Yoshida, K. Shishido, Synlett 23 (2012) 577.

[7] J.E. McMurry, J.R. Mats, K.L. Kees, Tetrahedron 43 (1987) 5489.
[8] M. Mori, N. Sakakibara, A. Kinoshita, J. Org. Chem. 63 (1998) 6082.
[9] A.P. Kozikowski, Acc. Chem. Res. 17 (1984) 410.
[10] Z. Zhang, D.P. Curran, J. Chem. Soc. Perkin Trans. 1 (1991) 2627.
[11] D.P. Curran, J. Am. Chem. Soc. 104 (1982) 4042.

CHAPTER TWENTY NINE

Hyperforin

Abbreviations

Ac_2O acetic anhydride
CH_2Cl_2 dichloromethane
(+)-DET L-(+)-diethyltartarate
$(DHQD)_2PHAL$ hydroquinidine 1,4-phthalazinediyl diether
DIBAL-H diisobutylaluminum hydride
DMAP 4-methylaminopyridine
DMF *N,N*-dimethylformamide
DMP 2,2-dimethoxypropane
DMSO dimethylsulfoxide
Et_3N triethylamine
HMPA hexamethylphosphoramide
KHMDS potassium hexamethyldisilazide
LDA lithium diisopropylamide
LTMP/LiTMP lithium 2,2,6,6-tetramethylpiperidide
MsCl methanesulfonyl chloride
PTSA *para*-toluenesulfonic acid
py pyridine
rt room temperature
Sia_2BH disiamylborane
TBAF tetrabutylammonium bromide
TBHP *tert*-butyl hydroperoxide
TBS *tert*-butyldimethylsilyl
TESCl triethylsilyl chloride
THF tetrahydrofuran
TIPSOTf triisopropylsilyl trifluoromethanesulfonate
TMSOTf trimethylsilyl trifluoromethanesulfonate
TsOH *p*-toluenesulfonyl chloride

Systematic name: (1*R*,5*S*,6*R*,7*S*)-4-Hydroxy-5-isobutyryl-6-methyl-1,3,7-tris(3-methylbut-2-en-1-yl)-6-(4-methylpent-3-en-1-yl)bicyclo[3.3.1]non-3-ene-2,9-dione

Compound class: Polycyclic polyprenylated acylphloroglucinol (PPAP)

Total Synthesis of Bioactive Natural Products
https://doi.org/10.1016/B978-0-08-102822-3.00029-8

Structure:

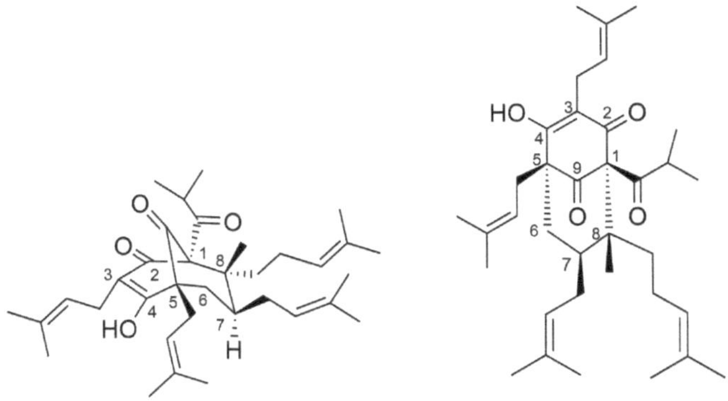

1 1

Hyperforin (**1**)

Natural source: *Hypericum perforatum* L. (St. John's wort; family: Hypericaceae) [1–6]

Pharmaceutical potentials: Antidepressant (hyperforin is reported to block the reuptake of a variety of neurotransmitters through a unique mechanism of action, possibly by selectively activating TRPC6, a classical transient receptor potential protein) [7–14]; accelerates hepatic drug metabolism through activation of the pregnane X receptor [15]; antitumor [16]; antibacterial [17].

Synthetic routes:

Uwamori and Nakada [18], for the first time, achieved the total synthesis of (±)-hyperforin (**1**) starting from commercially available methyl 2,6-dimethoxybenzoate in 35 steps (Scheme 1) involving the preparation of a bicyclo[3.3.1]nonane derivative via a three-step sequence: intramolecular cyclopropanation, construction of the C8 all-carbon quaternary stereogenic center, and subsequent regioselective ring opening of the cyclopropane. The total synthetic route as adopted by these investigators is depicted in Scheme 1.

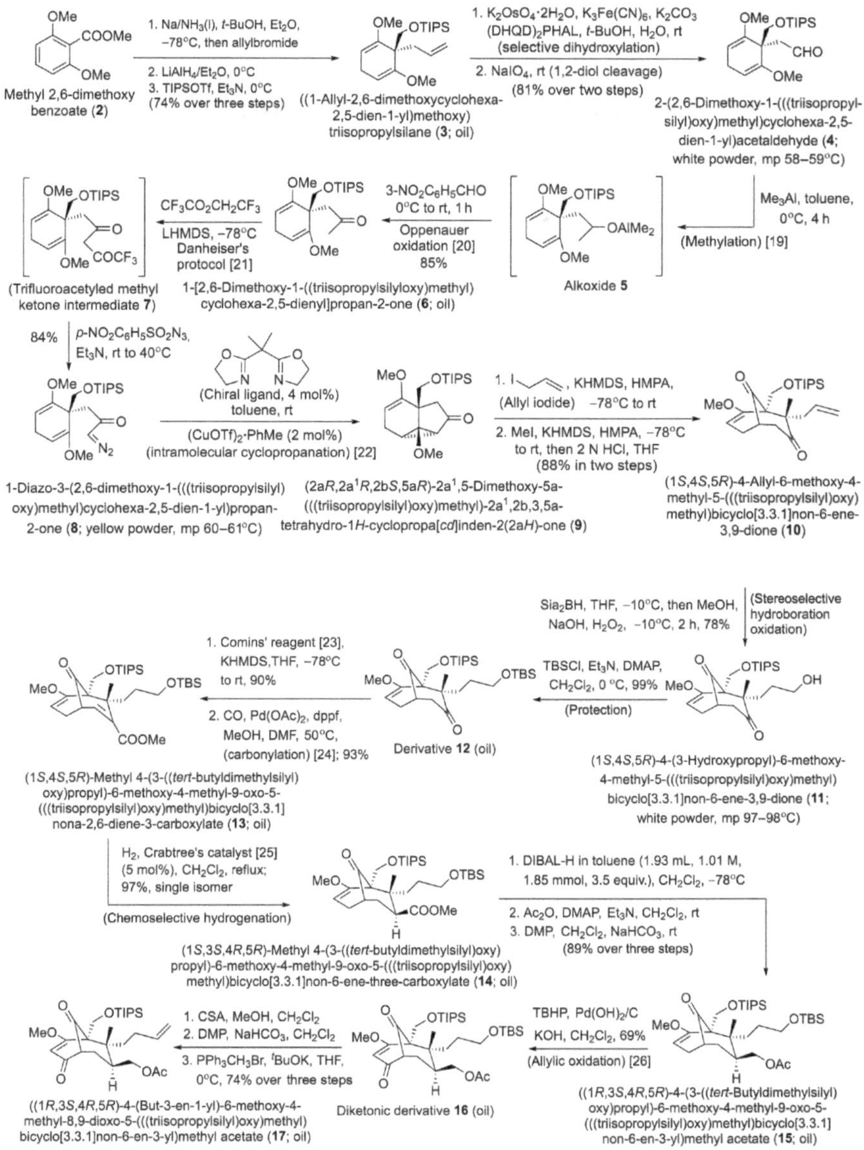

Scheme 1—Cont'd

(Continued)

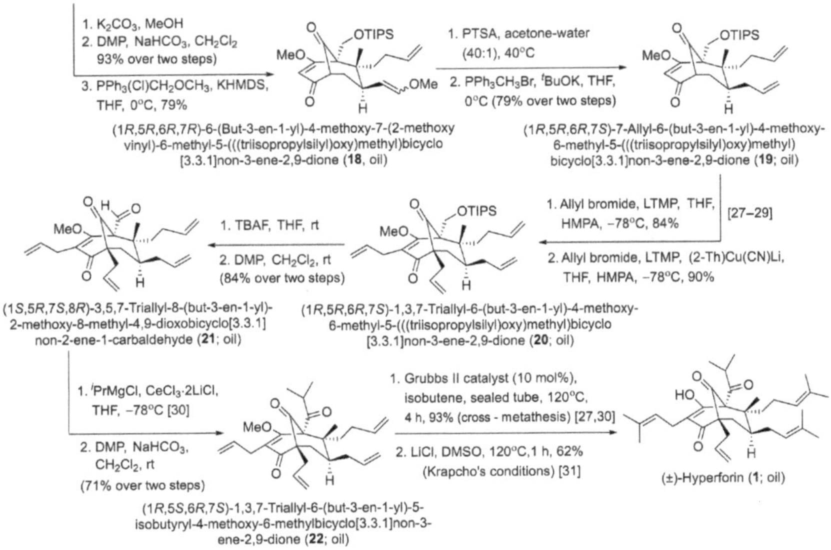

Scheme 1 Total synthesis of (±)-hyperforin (**1**) [18].

Methyl 2,6-dimethoxybenzoate (**2**) was subjected to Birch reduction, followed by a one-pot reaction with allylbromide, reduction of the methyl ester with lithium aluminum hydride, and protection of the primary hydroxyl group as a TIPS ether to afford ((1-allyl-2,6-dimethoxycyclohexa-2,5-dien-1-yl)methoxy)triisopropylsilane (**3**) as an oil; selective dihydroxylation of **3** followed by 1,2-diol cleavage afforded aldehyde **4**, which was then converted to 1-[2,6-dimethoxy-1-((triisopropylsilyloxy)methyl)cyclohexa-2,5-dienyl]propan-2-one (**6**; oil) using one-pot methylation—Oppenauer oxidation protocol in good yield [19, 20]. In the following steps, 1-diazo-3-(2,6-dimethoxy-1-(((triisopropylsilyl)oxy)methyl)cyclohexa-2,5-dien-1-yl)propan-2-one (**8**; yellow powder) was successfully prepared from the methyl ketone **6** using Danheiser's protocol [21]. Intramolecular cyclopropanation (IMCP) of **6** was carried out in the presence of a chiral ligand to obtain bicyclic ketone **9**, which in turn was converted into 4-allyl-6-methoxy-4-methyl-5-(((triisopropylsilyl)oxy)methyl)bicyclo[3.3.1] non-6-ene-3,9-dione (**10**) [22].

Chemo- and stereoselective hydroboration of **10** with disiamylborane followed by protection of the resultant hydroxyl as a TIPS ether furnished derivative **12**, the enol triflate of which was then reacted with Comins' reagent [23] following palladium-catalyzed carbonylation to afford compound **13** in good yield [24]. Chemo- and stereoselective reduction of the olefinic double bond within **13** was performed by hydrogenation using Crabtree's catalyst [25] under refluxing in dichloroethane to produce **14** as the sole product. In the next steps, the oxo and ester functionalities within **14** were reduced with DIBAL-H to obtain the corresponding diol, the primary hydroxyl was selectively acetylated, and Dess-Martin oxidation of the secondary hydroxyl afforded compound **15**. Allylic oxidation of **15** was achieved using alladium-catalyzed oxidation as per the procedure reported by Corey and Yu [26] to give diketonic derivative **16**, which on sequential removal of its TBS group, Dess-Martin oxidation, and Wittig reaction furnished compound **17**.

Compound **17** on saponification with potassium carbonate in methanol followed by Dess-Martin oxidation and Wittig reaction resulted in compound **18**. It was then subjected to acid hydrolysis and the hydrolysis product on subsequent Wittig reaction produced compound **19**. Successful allylations at desired positions using suitable reagents such as lithium 2,2,6,6-tetramethylpiperidide (LTMP) and thienylcuprate were achieved to obtain 1,3,7-triallyl-6-(but-3-en-1-yl)-4-methoxy-6-methyl-5-(((triisopropylsilyl)oxy)methyl)bicyclo[3.3.1]non-3-ene-2,9-dione (**20**) [27–29]. To construct the isopropyl ketone moiety, first an aldehyde functionality was imposed to have compound **21** by reacting **20** with TBAF followed by Dess-Martin oxidation of the alcohol formed. On reaction with the isopropylcerium reagent, prepared in situ from isopropylmagnesium chloride and $CeCl_3 \cdot 2LiCl$ [30], compound **21** successfully yielded the desired intermediate product, which on subsequent Dess-Martin oxidation yielded **22**. Finally, compound **22** was subjected to cross-metathesis on reaction with isobutene in the presence of Grubbs II reagent [27, 30] and the methyl ether as formed was cleaved under Krapcho's conditions [31] to furnish (±)-hyperforin (**1**); all the relevant spectroscopic properties of the synthetic compound were found to be matched in all respects with those for naturally occurring hyperforin [1].

Later on, Shair and coworkers reported an 18-step enantioselective total synthesis of (+)-hyperforin (**1**) in the following year (Scheme 2) [32]. Epoxygeranyl bromide **26** was prepared from geraniol (**23**) following previously reported methods [33, 34]. Oxymercuration followed by reductive work-up of **26** and subsequent TESCl-mediated silylation produced triethylsilyl derivative **28**, which was then coupled with 1,5-dimethoxy-6-(3-methylbut-2-en-1-yl)cyclohexa-1,4-diene (**31**, prepared from 1,3-dimethoxybenzene **29**; Scheme 2) in the presence of sec-BuLi to furnish the cyclization precursor **32** in 85% yield. Intramolecular Lewis acid-mediated opening of the epoxide ring within **32** in the presence of TMSOTf

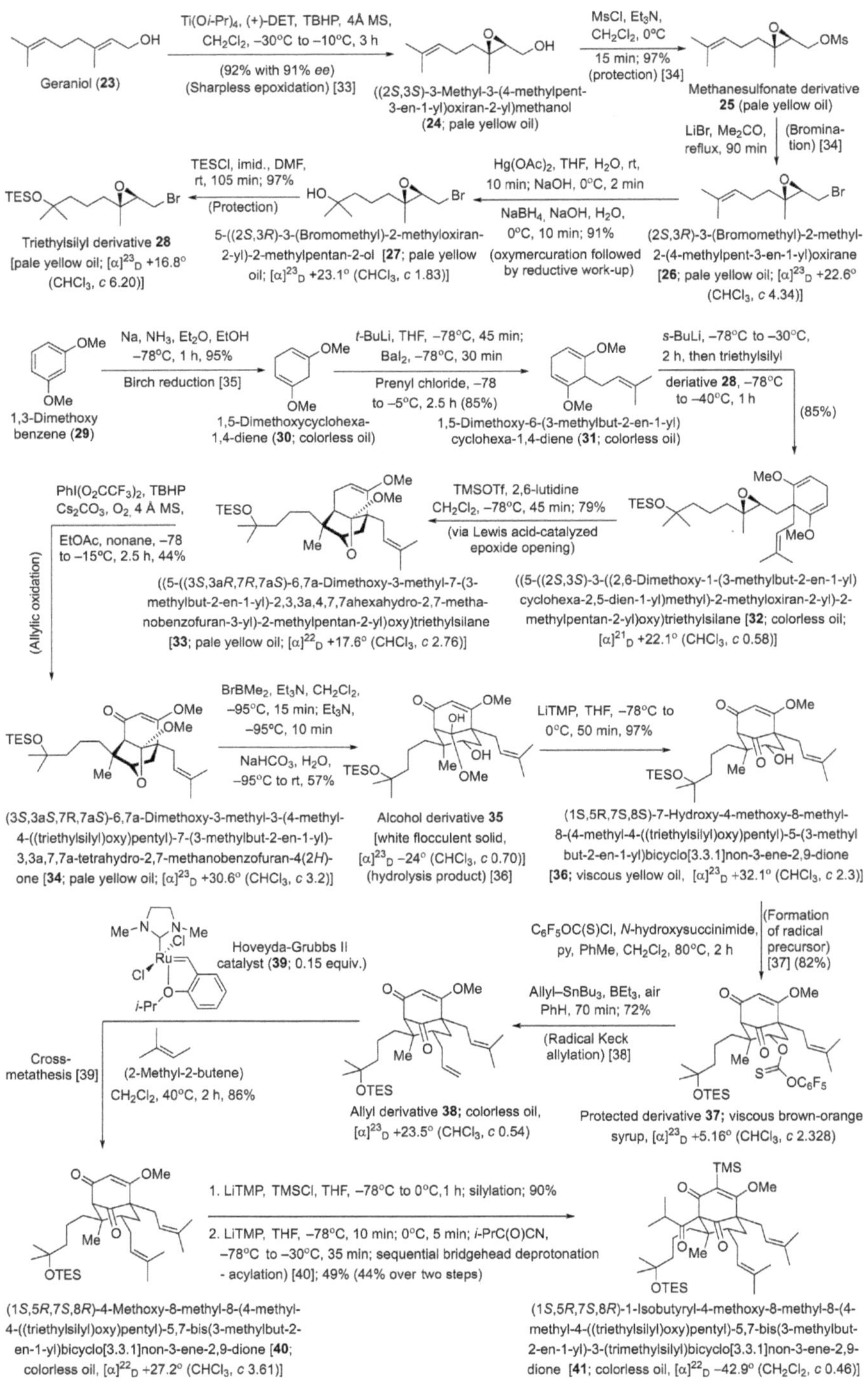

Scheme 2—Cont'd

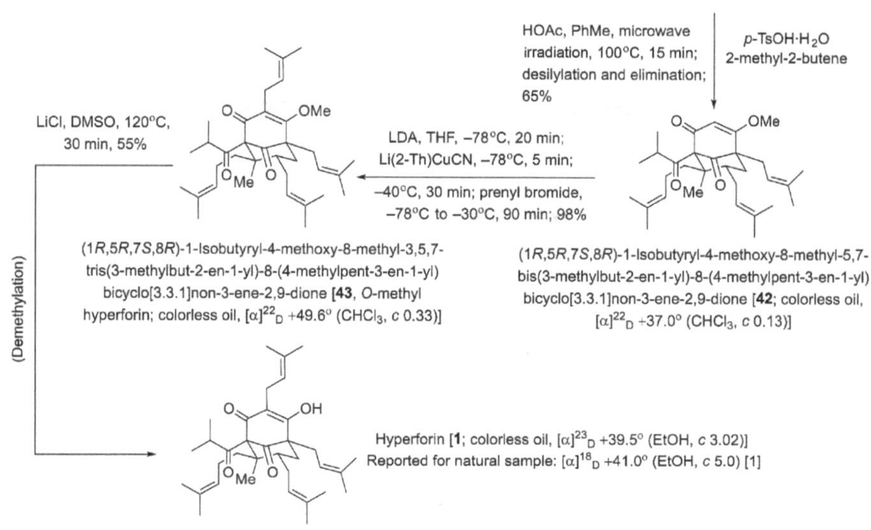

Scheme 2 Enantioselective total synthesis of (+)-hyperforin (**1**) [32].

and 2,6-lutidine afforded the ketal **33** with the key bicyclo[3.3.1]nonane core as a single product in 79% yield; in the next step, compound **33** underwent smooth allylic oxidation with TBHP, $PhI(O_2CCF_3)_2$, and oxygen in a highly chemoselective fashion to produce vinylogous ester **34**. Hydrolysis of the ketal present in **34** to yield alcohol **35** was accomplished in two steps: treatment of **14** with $BrBMe_2$ (below −90°C) followed by LiTMP-mediated methanol extrusion from the intermediate hemiketal **35**.

To install a prenyl group at C-7 position, the investigators successfully applied a radical Keck allylation approach [38]. For this purpose, they prepared thionocarbonate **37**, a radical precursor, from the reaction of alcohol **36** with $ClC(S)OC_6F_5$ [37]. The radical initiator **37**, in turn, underwent radical allylation with allyl-$SnBu_3$ in the presence of BEt_3/air as an initiator, affording allyl derivative **38** containing a C-7 allyl group as a single diastereomer. Compound **38** then proceeded with cross-metathesis [39] with 2-methyl-2-butene catalyzed by Hoveyda-Grubbs second-generation catalyst (**39**) resulting in the desired product (1*S*,5*R*,7*S*,8*R*)-4-methoxy-8-methyl-8-(4-methyl-4-((triethylsilyl)oxy)pentyl)-5,7-bis(3-methylbut-2-en-1-yl) bicyclo[3.3.1]non-3-ene-2,9-dione (**40**; colorless oil). After silylation at the C-3 position, sequential bridgehead deprotonation-acylation [40] using LiTMP and *i*-PrC(O)CN gave ketone **41** with 44% yield over the two steps; one-pot desilylation and elimination of **41** was then accomplished through microwave irradiation with *p*-TsOH·H_2O resulting in the formation of 1-isobutyryl-4-methoxy-8-methyl-5,7-bis(3-methylbut-2-en-1-yl)-8-(4-methylpent-3-en-1-yl)bicyclo[3.3.1]non-3-ene-2,9-dione (**42**) in 65% yield. The final C-3 prenyl group was then installed utilizing a precedented sequence to afford hyperforin methyl ether **43**: (1) deprotonation of **42** with LDA, (2) transmetalation with 2-thienyl(cyano)copper lithium [Li(2-Th) CuCN], and (3) trapping with prenyl bromide. Finally, hyperforin (**1**) was obtained as colorless oil on demethylation of **43** by heating its DMSO solution with LiCl. The physical and spectral properties of this synthetic compound were found to be in excellent agreement with those for the natural product [1].

Ting and Maimone [41] reported a modified short route involving a 10-step total synthesis of (+)-hyperfolin starting from 2-methylcyclopent-2-en-1-one (**44**), based on diketene annulations reaction and oxidative ring expansion strategy (Scheme 3). Initially, a highly substituted cyclopentanone **47** was prepared as yellow oil from a copper-mediated conjugate addition of (3-methylbut-3-en-1-yl) magnesium bromide (**45**) with enone **44**, followed by lithium enolate generation and alkylation with homoprenyl iodide (**46**) after simple acid-catalyzed isomerization of the exocyclic olefin [42]. Cyclopentanone **47** was further alkylated with prenylbromide (**48**) in the presence of LDA affording (2*R*,3*S*)-2-methyl-3,5-bis(3-methylbut-2-en-1-yl)-2-(4-methylpent-3-en-1-yl)cyclopentanone (**49**). In the next step, the highly hindered lithium enolate of **49** (formed via LTMP-mediated deprotonation) engaged simple diketene in a formal *C*-acylation/ring annulation process leading to complex 5/6-fused bicycle **51** in 45% yield based on recovered starting material and as a single diastereomer. Diketone **51** then reacted smoothly with trimethylsilyldiazomethane (96%) to afford an easily separable 1:1 mixture of regioisomeric vinylogous esters **52** and **53**—regioisomer **52** was used for the next step, and **53** could be recycled to **51** via basic hydrolysis.

Vinylogous ester **52** on stirring with phenyliodine diacetate (PIDA) in basic methanol at room temperature effected a very clean rearrangement to bicyclo[3.3.1]nonane **55** with high yield, presumably via an intermediate similar to **54**. Based on a series of experimental findings, the investigators took the strategy first to chlorinate the C-3 position of **55** upon treating with TsCl/LTMP, thus blocking this site from competitive deprotonation as well as installing a handle for downstream functionalization. Chloride derivative **56** upon Shair's bridgehead functionalization [32] using LTMP as base and *i*-PrC(O)CN (**57**) as electrophile afforded **58** as yellow oil in 70% yield.

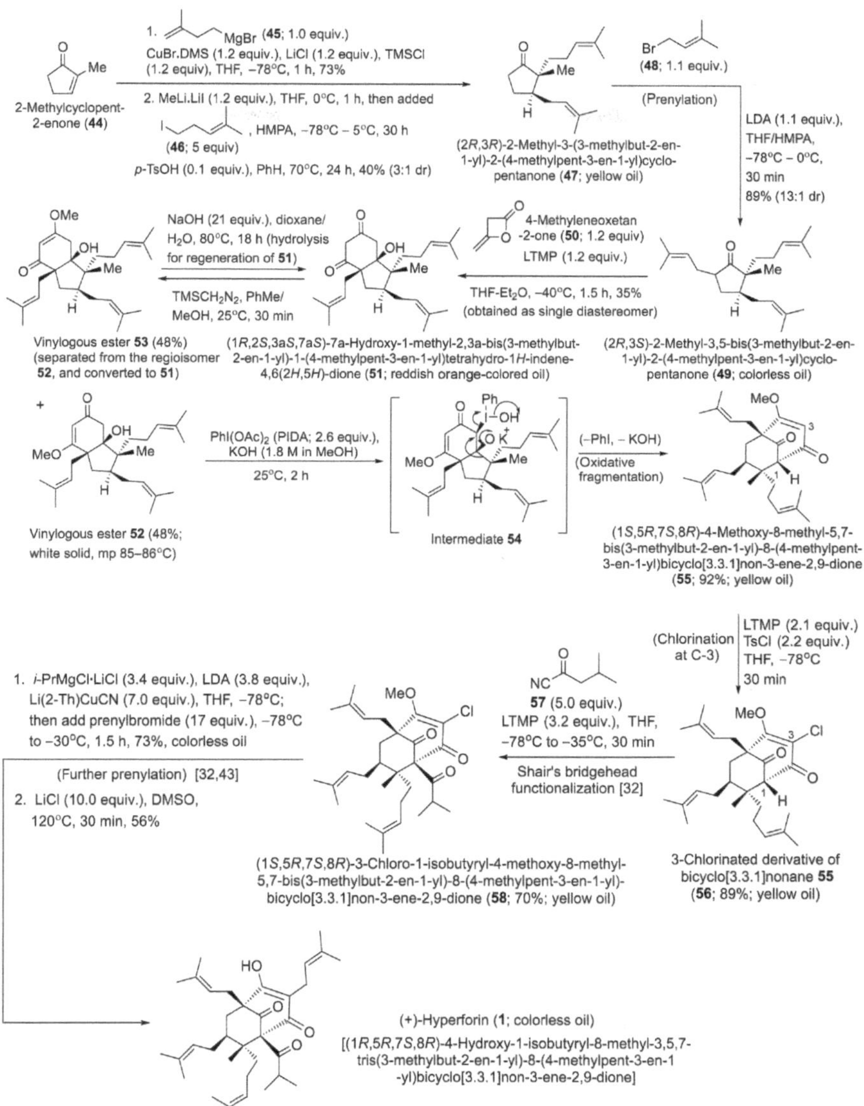

Scheme 3 Enantioselective total synthesis of (+)-hyperforin (**1**) [41].

The chloride derivative **58** thus obtained reacted with excess isopropylmagnesium chloride, and following transmetalation onto copper, the C-3 position could be readily functionalized with prenylbromide [32, 43]. On demethylation (LiCl/DMSO) of the resulting prenylated product, hyperforin (**1**) was obtained in the final step with good yield. The overall transformations are depicted in Scheme 3.

References

[1] A.I. Gurevich, V.N. Dobrynin, M.N. Kolosov, S.A. Popravko, I.D. Ryabova, B.K. Chernov, N.A. Derbentseva, B.E. Aizenman, A.D. Garagulya, Antibiotiki 16 (1971) 510.
[2] N.S. Bystrov, B.K. Chernov, V.N. Dobrynin, M.N. Kolosov, Tetrahedron Lett. 16 (1975) 2791.
[3] I. Brondz, T. Greibokk, P.A. Groth, A.J. Aasen, Tetrahedron Lett. 23 (1982) 1299.
[4] I. Brondz, T. Greibrokk, P. Groth, A.J. Aasen, Acta Chem. Scand. A 37 (1983) 263.
[5] J. Barnes, L.A. Anderson, J.D. Phillipson, J. Pharm. Pharmacol. 53 (2001) 583.
[6] L. Beerhues, Phytochemistry 67 (2006) 2201. M.A. Medina, B. Martínez-Poveda, M.I. Amores-Sánchez, A.R. Quesada, Life Sci. 79 (2006) 105; C. Quiney, C. Billard, C. Salanoubat, J.D. Fourneron, J.P. Kolb, Leukemia 20 (2006) 1519.
[7] S.S. Chatterjee, S.K. Bhattacharya, M. Wonnemann, A. Singer, W.E. Muller, Life Sci. 63 (1998) 499.
[8] W.E. Müller, A. Singer, M. Wonnemann, U. Hafner, M. Rolli, C. Schafer, Pharmacopsychiatry 1 (1998) 16.
[9] G. Laakmann, C. Schule, T. Baghai, M. Kieser, Pharmacopsychiatry 1 (1998) 54.
[10] S.T. Kaehler, C. Sinner, S.S. Chatterjee, A. Philippu, Neurosci. Lett. 262 (1999) 199.
[11] J.M. Greeson, B. Sanford, D.A. Monti, Psychopharmacology 153 (2001) 402.
[12] G. Di Carlo, F. Borrelli, E. Ernst, A.A. Izzo, Trends Pharmacol. Sci. 22 (2001) 292.
[13] W.E. Müller, Pharmacol. Res. 47 (2003) 101. K. Linde, Forsch. Komplementmed. 16 (2009) 146.
[14] K. Leuner, V. Kazanksi, M. Müller, K. Essin, B. Henke, M. Gollasch, C. Harteneck, W. E. Müller, FASEB J. 21 (2007) 4101.
[15] L.B. Moore, B. Goodwin, S.A. Jones, G.B. Wisely, C.J. Serabjit-Singh, T.M. Willson, J.L. Collins, S.A. Kliewer, Proc. Natl. Acad. Sci. U.S.A. 97 (2000) 7500. R.E. Watkins, J.M. Maglich, L.B. Moore, G.B. Wisely, S.M. Noble, P.D. Davis-Searles, M.H. Lambert, S.A. Kliewer, M.R. Redinbo, Biochemistry 42 (2003) 1430 (for crystal structure of human PXR in complex with hyperforin).
[16] C.M. Schempp, V. Kirkin, B. Simon-Haarhaus, A. Kersten, J. Kiss, C.C. Termeer, B. Gilb, T. Kaufmann, C. Borner, J.P. Sleeman, J.C. Simon, Oncogene 21 (2002) 1242.
[17] C.M. Schempp, K. Pelz, A. Wittmer, E. Schopf, J.C. Simon, Lancet 353 (1999) 2129.
[18] M. Uwamori, M. Kakada, Tetrahedron Lett. 54 (2013) 2022.
[19] E.C. Ashby, S.A. Noding, J. Org. Chem. 44 (1979) 4792.
[20] C.R. Graves, B.-S. Zeng, S.T. Nguyen, J. Am. Chem. Soc. 128 (2006) 12596.
[21] R.L. Danheiser, R.F. Miller, R.G. Brisbois, S.Z. Park, J. Org. Chem. 55 (1990) 1959.
[22] M. Uwamori, A. Satio, M. Nakada, J. Org. Chem. 77 (2012) 5098.
[23] D.L. Comins, A. Dehghani, Tetrahedron Lett. 33 (1992) 6299.
[24] H. Hagiwara, K. Hamano, M. Nozawa, T. Hoshi, T. Suzuki, F. Kido, J. Org. Chem. 70 (2005) 2250.
[25] R.H. Crabtree, P.C. Demou, D. Eden, J.M. Mihelcic, C.A. Parnell, J.M. Quirk, G.E. Morris, J. Am. Chem. Soc. 104 (1982) 6994. R.H. Crabtree, M.W. Davis, J. Org. Chem. 51 (1986) 2655.
[26] J.-Q. Yu, E.J. Corey, J. Am. Chem. Soc. 125 (2003) 3232.
[27] C. Tsukano, D.R. Siegel, S.J. Danishefsky, Angew. Chem. Int. Ed. 46 (2007) 8840.
[28] V. Rodeschini, N.S. Simpkins, C. Wilson, J. Org. Chem. 72 (2007) 4265.
[29] N.M. Ahmad, V. Rodeschini, N.S. Simpkins, S.E. Ward, A.J. Blake, J. Org. Chem. 72 (2007) 4803.
[30] A. Krasovskiy, R. Kopp, P. Knochel, Angew. Chem. Int. Ed. 45 (2006) 497.
[31] J.T. Njardarson, Tetrahedron 67 (2011) 7631. J.-A. Richard, R.H. Pouwer, D.Y.-K. Chen, Angew. Chem., Int. Ed. 51 (2012) 4536.

[32] B.A. Sparling, D.C. Moebius, M.D. Shair, J. Am. Chem. Soc. 125 (2013) 644.
[33] R.M. Hanson, K.B. Sharpless, J. Org. Chem. 51 (1986) 1922.
[34] R.C. Gash, F. MacCorquodale, J.C. Walton, Tetrahedron 45 (1989) 5531.
[35] E. Piers, J.R. Grierson, J. Org. Chem. 42 (1977) 3755.
[36] Y. Guidon, C. Yoakim, H.E. Morton, Tetrahedron Lett. 24 (1983) 2969. Y. Guidon, C. Yoakim, H.E. Morton, J. Org. Chem. 49 (1984) 3912.
[37] D.H.R. Barton, J.C. Jaszberenyi, Tetrahedron Lett. 30 (1989) 2619.
[38] G.E. Keck, J.H. Byers, J. Org. Chem. 50 (1985) 5442. G.E. Keck, E.J. Enholm, J.B. Yates, M.R. Wiley, Tetrahedron 41 (1985) 4079.
[39] S.B. Garber, J.S. Kingsbury, B.L. Gray, A.H. Hoveyda, J. Am. Chem. Soc. 122 (2000) 8168.
[40] N. Biber, K. Möws, B. Plietker, Nature Chem. 3 (2011) 938.
[41] C.P. Ting, T.J. Maimone, J. Am. Chem. Soc. 137 (2015) 10516.
[42] B.H. Lipshutz, C. Hackmann, J. Org. Chem. 59 (1994) 7437.
[43] G. Bellavance, L. Barriault, Angew. Chem. Int. Ed. 53 (2014) 6701.

CHAPTER THIRTY

Integric Acid

Abbreviations

Ac_2O acetic anhydride
CH_2Cl_2 dichloromethane
DIC diisopropyl carbodiimide
DMAP 4-dimethylaminopyridine
DMF *N*,*N*-dimethylformamide
DMPU 1,3-dimethyl-3,4,5,6-tetrahydro-2-pyrimidinone
Et_3N triethylamine
LDA lithium diisopropylamide
***m*-CPBA** *meta*-chloroperoxybenzoic acid
MS molecular sieves
RAMP (*R*)-1-amino-2-methoxymethylpyrrolidine
rt room temperature
TBAF tetrabutylammonium fluoride
TBSCl *tert*-butyldimethylsilyl chloride
THF tetrahydrofuran
TMSCl trimethylsilyl chloride

Systematic name: (1*S*,4*R*,7*R*,8a*R*)-4-(((*S*,*E*)-2,4-Dimethyloct-2-enoyl)oxy)-8a-methyl-6-oxo-7-(2-oxoacetyl)-1,2,3,4,6,7,8,8a-octahydronaphthalene-1-carboxylic acid

Compound class: Sesquiterpenoid (eremophilane type)

Structure:

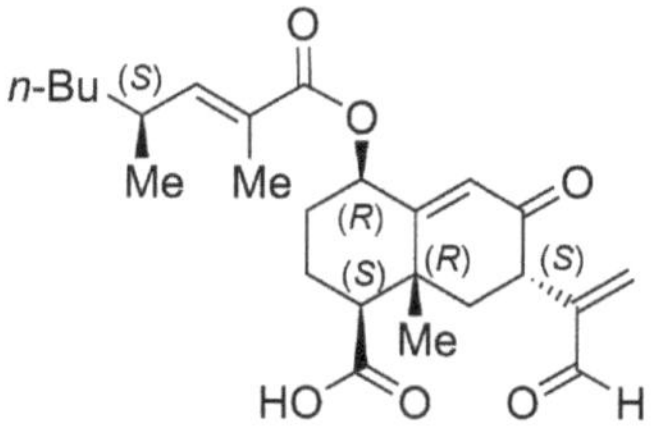

Integric acid (**1**)

Natural source: *Xylaria* sp. (MF6254) [1, 2]

Pharmaceutical potential: Anti-HIV [1–3]

Total Synthesis of Bioactive Natural Products
https://doi.org/10.1016/B978-0-08-102822-3.00030-4

Synthetic route:

The first total synthesis of integric acid (**1**) was achieved by Rutjes and his group [4]; their retrosynthetic strategy is depicted in Scheme 1.

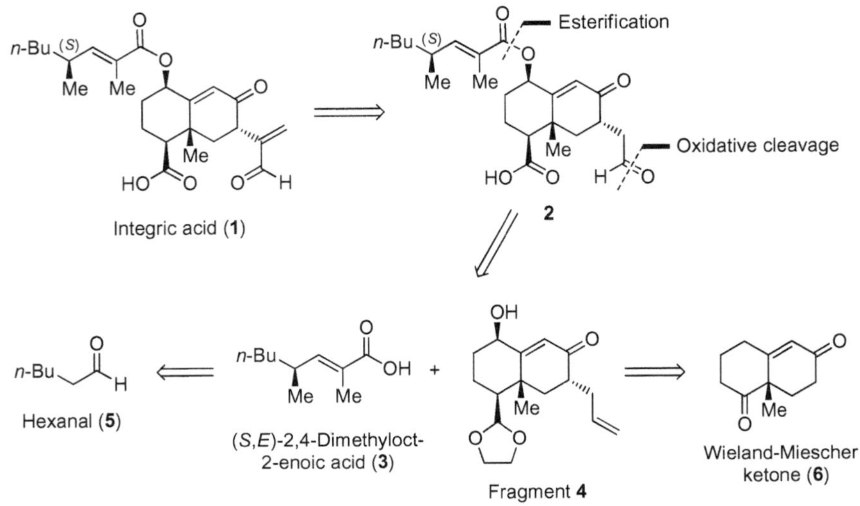

Scheme 1 Retrosynthetic approach to integric acid (**1**).

Preparation of (*S*,*E*)-2,4-dimethyloct-2-enoic acid (**3**) starting from hexanal (**5**) in five steps is described in Scheme 2. The RAMP-hydrazone derivative **7** was diastereoselectively alkylated with methyl iodide to give **8** as pale yellow oil. Compound **8** underwent oxidative cleavage with ozone [5] to produce an aldehyde, which was immediately converted to (*S*,*E*)-ethyl 2,4-dimethyloct-2-enoate (**9**; pale yellow oil; 60%; 98% *ee*) via Wittig reaction. Basic hydrolysis of this ester with methanolic NaOH ultimately afforded synthon **3** in 42% overall yield in five steps.

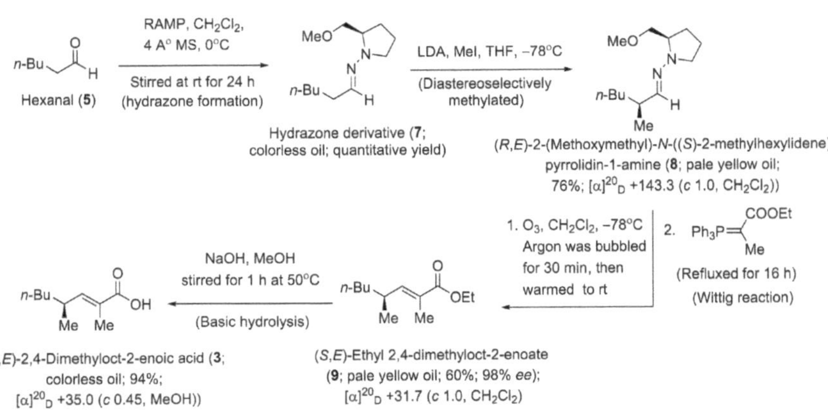

Scheme 2 Synthesis of (*S*,*E*)-2,4-dimethyloct-2-enoic acid (**3**).

Synthesis of fragment **4** from the Wieland-Liescher ketone **6** is shown in Scheme 3. First, ketone **6** was transformed into aldehyde **10** following a known process by Paquette et al. [6]. Deprotection of the dithioacetal with $Hg(ClO_4)_2$ and simultaneous dioxolane protection of the aldehyde in the presence of glycol gave (4a*R*,5*S*)-5-(1,3-dioxolan-2-yl)-4a-methyl-4,4a,5,6,7,8-hexahydronaphthalen-2(3*H*)-one (**11**) in a single step in 67% yield. Formation of dienyl acetate **12** was achieved via reaction of **11** with TMSCl and NaI in acetic anhydride and pyridine. Treatment of **12** with *m*-CPBA afforded the alcohols **4** and **4′** in 76% yield in a 4.4:1 ratio. Alcohol **4** was separated out and then protected as TBS ether in 96% yield.

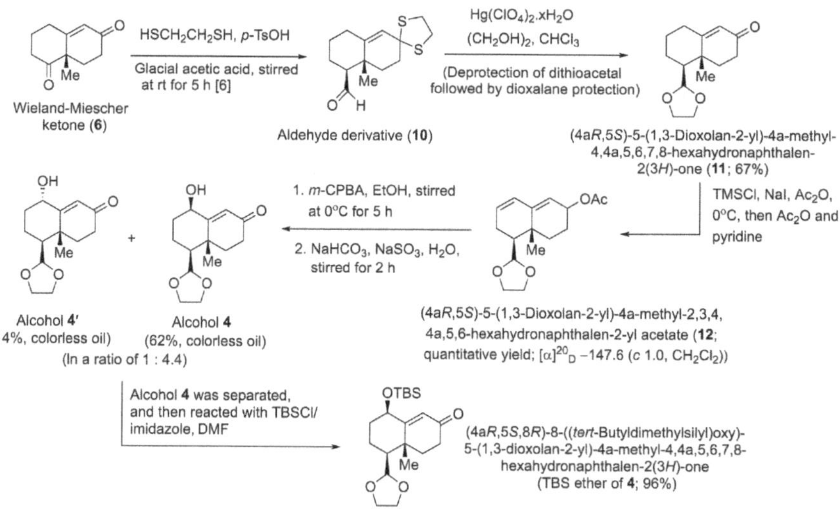

Scheme 3 Synthesis of TBS ether of alcohol **4**.

The next task was to assemble fragments **3** and **4** to arrive at the target molecule **1** (Scheme 4). The investigators then carried out alkylation of the enantiopure TBS ether of **4** with prenyl bromide in a diastereoselective manner to produce **13** in 83% yield as a colorless oil. After deprotection with TBAF, the resulting compound **14** was coupled with (*S*,*E*)-2,4-dimethyloct-2-enoic acid (**3**) in the presence of DIC and DMAP to furnish (*S*)-C4′ ester **15** (colorless oil; 88%), which eventually generated the carboxylic acid derivative **17** (colorless oil) on hydrolysis of the dioxolane followed by a Pinnick oxidation in a satisfactory overall yield. Ozonolysis of compound **17** in the presence of 4 equiv. of pyridine in a 1:1 mixture of methanol-dichloromethane, followed by reduction with Ph_3P, produced

(*S*)-C4′ aldehydic fragment **2** in 60% yield. In the final step, the unsaturated double bond was introduced by reaction with Eschenmoser's salt in the presence of triethylamine affording integric acid **1** as white powder as acceptable yield. The synthetic compound was found to have comparable physical and spectral properties with those reported for the natural molecule [1].

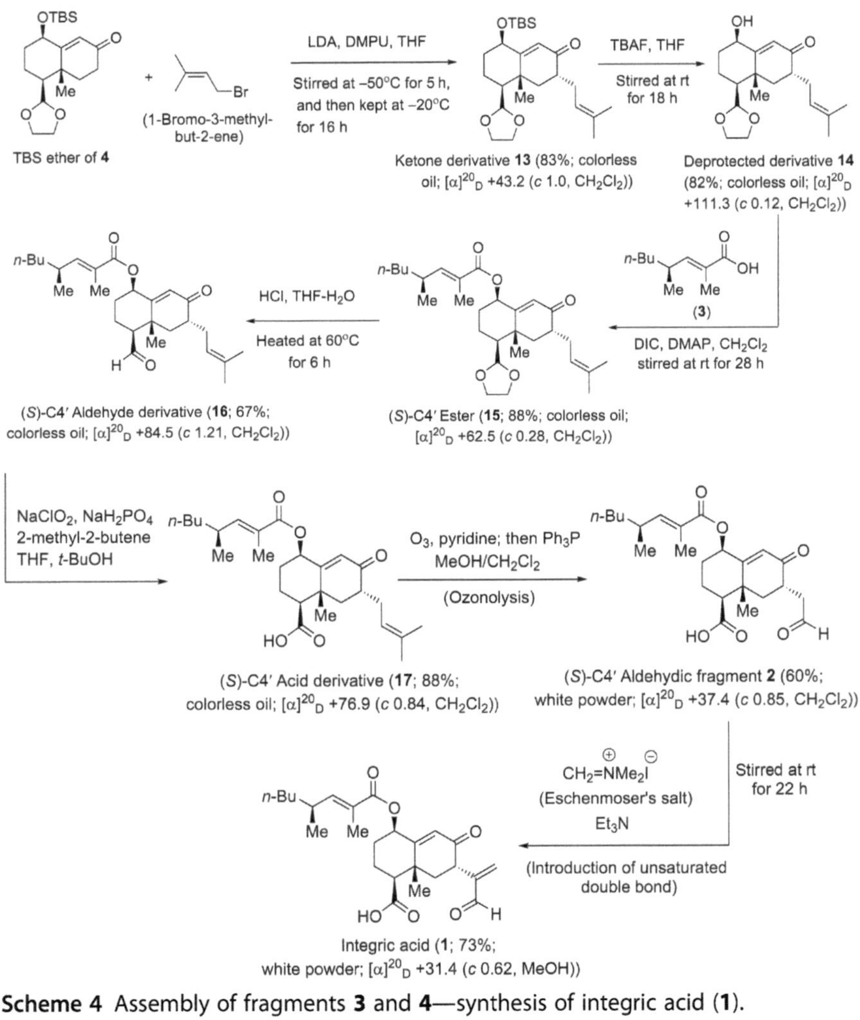

Scheme 4 Assembly of fragments **3** and **4**—synthesis of integric acid (**1**).

References

[1] S.B. Singh, D. Zink, J. Polishook, D. Valentino, A. Shafiee, K. Silverman, P. Felock, A. Teran, D. Vilella, D.J. Hazuda, R.B. Lingham, Tetrahedron Lett. 40 (1999) 8775.
[2] S.B. Singh, P. Felock, D.J. Hazuda, Bioorg. Med. Chem. Lett. 10 (2000) 235.
[3] D.J. Hazuda, C.U. Uncapher, P. Felock, J. Hastings, B. Pramanik, A. Wolfe, E. Bushman, C. Farnet, M. Goetz, M. Williams, K. Silverman, R. Lingham, S. Singh, Antiviral Chem. Chemother. 10 (1999) 63.
[4] D.C.J. Waalboer, H.A. van Kalkeren, M.C. Schaapman, F.L. van Delft, F.P.J.T. Rutjes, J. Org. Chem. 74 (2009) 8878.
[5] A. Job, C.F. Janeck, W. Bettray, R. Peters, D. Enders, Tetrahedron 58 (2002) 2253.
[6] L.A. Paquette, T.-Z. Wang, C.M.G. Philippo, S. Wang, J. Am. Chem. Soc. 116 (1994) 3367.

CHAPTER THIRTY ONE

Jineol

Abbreviations

$(CH_3)_3SiI$ trimethylsilyl iodide
$B(OMe)_3$ trimethyl borate
LAH lithium aluminum hydride
LTMP lithium tetramethylpiperidide
PDC pyridium dichromate
$POCl_3$ phosphorus oxychloride
Py pyridine
THF tetrahydrofuran

Systematic name: 3,8-Dihydroxyquinoline/quinoline-3,8-diol
Compound class: Quinoline alkaloid
Structure:

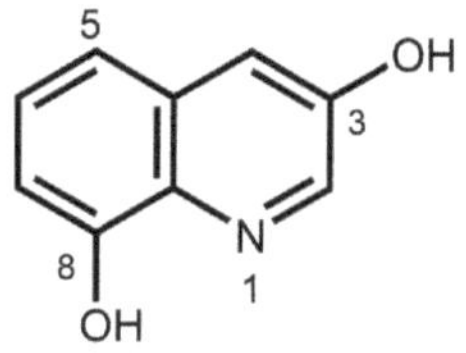

Jineol (**1**)

Natural source: *Scolopendra subspinipes mutilans* L. Koch (a centipede; family: Scolopendridae) [1]

Pharmaceutical potential: Cytotoxic (antitumor) [1, 2]; antibacterial and antifungal [2]

Synthetic routes:

In 1998, Goto and his group [3] first reported a convenient synthetic route for the cytotoxic alkaloid jineol (**1**) employing directed *ortho*-lithiation of a chloroquinoline derivative as the key step (Scheme 1). The starting material, 2-methoxycinnamanilide (**4**), was prepared by Schotten-Baumann reaction of commercially available 2-methoxyaniline (**2**) with cinnamoyl chloride (**3**); compound **4** on intramolecular Friedel-Crafts reaction in the presence of anhydrous aluminum chloride in chlorobenzene gave

https://doi.org/10.1016/B978-0-08-102822-3.00031-6

8-hydroxy-2(1*H*)-quinoline (**5**), which on selective *O*-methylation furnished the *O*-methylated derivative **6** in 95% yield. Compound **6** on treatment with phosphorus oxychloride ($POCl_3$) in the presence of a catalytic amount of pyridine in refluxing chlorobenzene afforded 2-chloro-8-methoxyquinoline (**7**) in quantitative yield. The electrophilic substitution reaction of compound **7** with trimethyl borate in the presence of lithium tetramethyl piperidide (LTMP) resulted in the formation of 2-chloro-8-methoxyquinolin-3-boronic acid (**8**; 82%), which was subsequently oxidized by 32% peracetic acid to give 2-chloro-3-hydroxy-8-methoxyquinoline (**9**) in 90% yield.

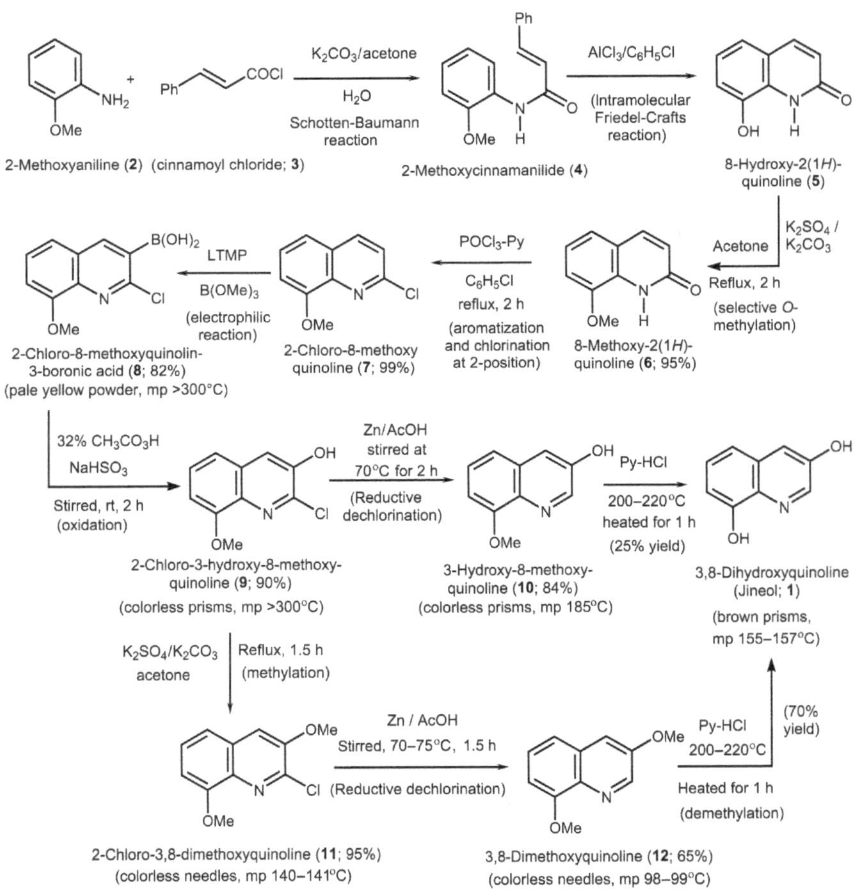

Scheme 1 Goto's total synthesis of jineol (**1**) [3].

Compound **9** was then subjected to reductive dechlorination by stirring with Zn-AcOH at 70°C for 2 h to obtain the 3-hydroxy-8-methoxyquinoline (**10**; 84%), which ultimately produced jineol (**1**; 3,8-dihydroxyquinoline) on demethylation by heating with pyridine hydrochloride at 200–220°C in 25% yield. In an alternative way, the investigators selectively methylated compound **9** to obtain 2-chloro-3,8-dimethoxyquinoline (**11**; 95%), which was then converted to the 3,8-dimethoxy derivative (**12**; 65%) via reductive dechlorination; compound **12**, in turn, afforded jineol (**1**; 70%) on its demethylation by heating with pyridine hydrochloride (Scheme 1) as brown prisms, mp 155–157°C [3].

Later on, another synthetic route (Scheme 2) for jineol (**1**) starting from 2-amino-3-methoxybenzoic acid (**13**) was developed by Moon and his group [2]; they also synthesized a number of jineol derivatives, and studied their antibacterial, antifungal, and anticancer activities. Compound **13** was first converted into 2-acetamido-3-methoxybenzyl acetate (**14**) in 71% yield on reduction with lithium aluminum hydride followed by acetylation with acetic anhydride-triethylamine. The investigators then carried out selective deacetylation of the acetoxyl moiety within **14** with K_2CO_3 in aqueous MeOH to obtain the benzyl alcohol derivative **15** (95%), which on subsequent oxidation with PDC in dichloromethane afforded the key intermediate 2-acetamido-3-methoxybenzaldehyde (**16**; 89%; unstable at room temperature). The aldehyde **16** underwent a modified Friedlander condensation [4] with benzyloxyacetaldehyde in the presence of sodium hydroxide in ethanol to furnish 3-benzyloxy-8-methoxyquinoline (**17**) in 81% yield. Hydrogenation of compound **17** with hydrogen in the presence of Pd/C or demethylation of the methoxy group at C-8 of compound **17** with trimethylsilyl iodide in dry methylene chloride afforded jineol 8-methyl ether (**10**) or jineol 3-benzyl ether (**18**), respectively. Both demethylation of ether **10** with trimethylsilyl iodide and debenzylation of ether **18** with hydrogen gas resulted in the production of jineol (**1**) in 30% and quantitative yields, respectively, as brown solid. Spectroscopic data of synthetic jineol were found to be closely comparable to those of the natural product [1].

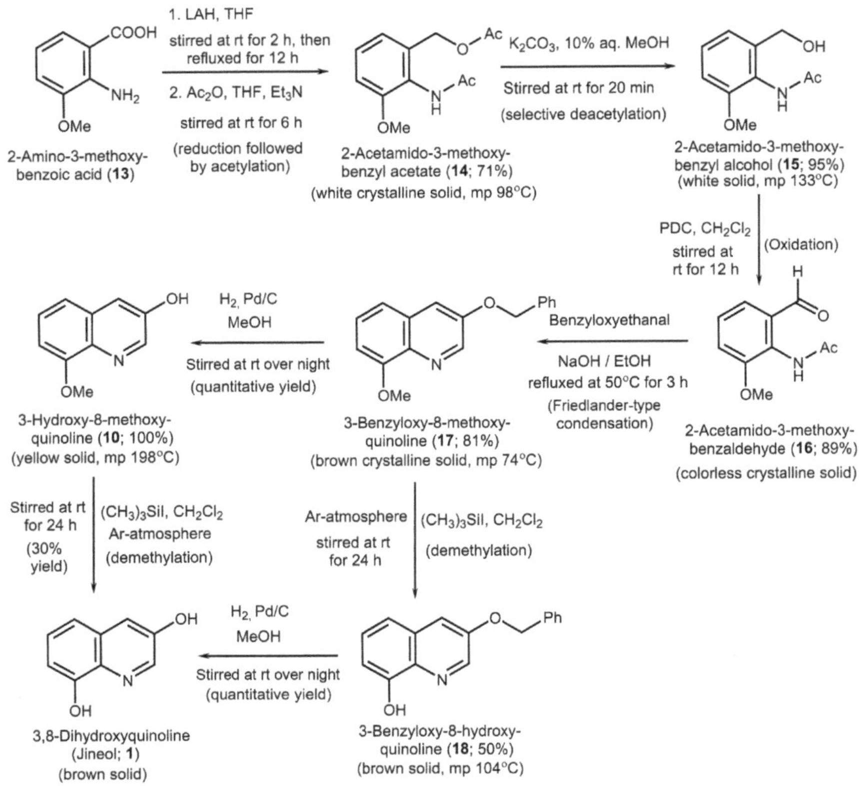

Scheme 2 Moon's total synthesis of jineol (**1**) [2].

References

[1] S.-S. Moon, N. Cho, J. Shin, Y. Seo, C.O. Lee, S.U. Choi, J. Nat. Prod. 59 (1996) 777.
[2] S.-C. Cho, M.Z. Sultan, S.-S. Moon, Bull. Kor. Chem. Soc. 29 (2008) 1587.
[3] Y. Tagawa, H. Yamashita, M. Nomura, Y. Goto, Heterocycles 48 (1998) 2379.
[4] D.L. Boger, J.H. Chen, J. Org. Chem. 60 (1995) 7369.

CHAPTER THIRTY TWO

Karalicin

Abbreviations

Ac acetyl
CH_2Cl_2 dichloromethane
DMAP 4-dimethylaminopyridine
DMF *N*,*N*-dimethylformamide
MeOH methanol
NMO *N*-methyl morpholine
PMB *p*-methoxybenzylidene
PTSA *para*-toluenesulfonic acid
TEA triethylamine
THF tetrahydrofuran
THP tetrahydro-2*H*-pyran-2-yl

Systematic name: 1,2,4-Trihydroxy-5-(4-methoxyphenyl)pentan-3-yl acetate

Compound class: Antibiotic

Structure:

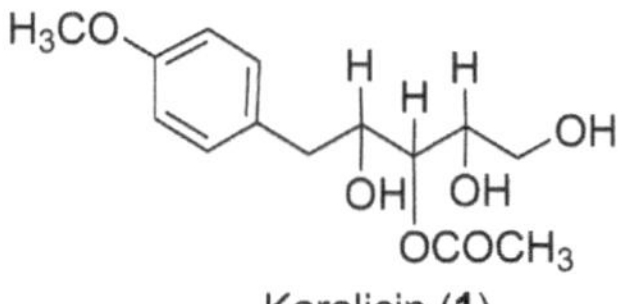

Karalicin (**1**)

Natural source: *Pseudomonas fluorescens/putida* strain SS-3 (CCM4430) (fermentation broth) [1, 2]

Pharmaceutical potential: Antiviral, cytotoxic, and antifungal [1, 2]

Synthetic route:

Rao [3] for the first time reported a stereo flexible total synthesis of the antiviral antibiotic karalicin **1** in six steps starting from *p*-methoxyphenylacetaldehyde (**2**). Aldehyde **2** was reacted with THP-protected propargyl alcohol **3** in the presence of *n*-BuLi/THF to give 1-(4-methoxyphenyl)-5-(tetrahydro-2*H*-pyran-2-yl)pent-3-yn-2-ol (**4**). This on stereospecific partial hydrogenation under Lindlar's conditions [4] afforded the olefin **5**. Dihydroxylation of **5** using OsO_4/NMO [5] gave the triol derivative 1-(4-methoxyphenyl)-5-((tetrahydro-2*H*-pyran-2-yl)oxy) pentane-2,3,4-triol (**6**). The 1,3-diol system within **6** was then protected [6] to generate **7**, the secondary hydroxyl group of which was derivatized as acetate with Ac_2O/TEA to afford 4-(4-methoxybenzyl)-2-(4-methoxyphenyl)-6-(((tetrahydro-2*H*-pyran-2-yl)oxy)methyl)-1,3-dioxan-5-yl acetate (**8**) [7]. In the final step, compound **8** underwent deprotection of THP and PMB (*p*-methoxybenzylidene) on prolonged stirring at ambient temperature in the presence of 1% HCl/MeOH furnishing karalicin **1**, which was then chromatographed over a silica gel column (hexane-ethyl acetate = 3:1) to give a pure sample of **1** (Scheme 1). Physical properties and spectral data of this synthetic compound were in close agreement with those reported for natural compound [1, 2].

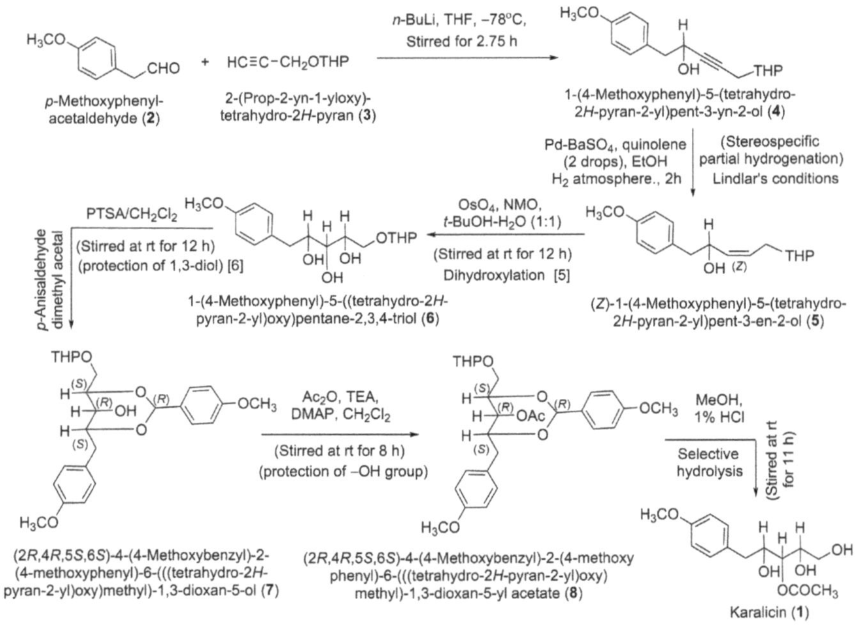

Scheme 1 Rao's total synthesis of karalicin (**1**) [3].

References

[1] G. Lampis, D. Deidda, C. Maullu, S. Petruzzelli, R. Pompei, F. Delle Monache, G. Satta, J. Antibiot. 49 (1996) 260.

[2] G. Lampis, D. Deidda, C. Maullu, S. Petruzzelli, R. Pompei, F. Delle Monache, G. Satta, J. Antibiot. 49 (1996) 263.

[3] T.P. Rao, Syn. Commun. 27 (1997) 3853.

[4] H. Lindlar, Helv. Chim. Acta 35 (1952) 446.
[5] V. Van Rheenen, R.C. Kelly, D.Y. Cha, Tetrahedron Lett. (1976) 1973.
[6] R. Johansson, B. Samuelsson, J. Chem. Soc., Perkin Trans I (1984) 2371.
[7] T. Katsuki, V.S. Martin, Org. React. (1996) 48.

CHAPTER THIRTY THREE

(−)-Kunstleramide

Abbreviations

DBU 1,8-diazabicyclo[5.4.0]undec-7-ene
DCM dichloromethane
DIBAL-H diisobutylaluminum hydride
DIPEA *N,N*-diisopropylethylamine (Hünig's base)
DIPEA *N,N*-diisopropylethylamine
HATU *O*-(7-azabenzotriazol-1-yl)-*N,N,N′,N′*-tetramethyluronium hexafluorophosphate
IBX 2-iodoxybenzoic acid
***p*-TSA** *p*-toluene sulfonic acid
TBS *tert*-butyldimethylsilyl
THF tetrahydrofuran

Systematic name: (6*R*,2*E*,4*E*)-7-(3,4-Dimethoxyphenyl)-*N*-ethyl-6-hydroxyhepta-2,4-dienamide

Compound class: Dienamide

Structure:

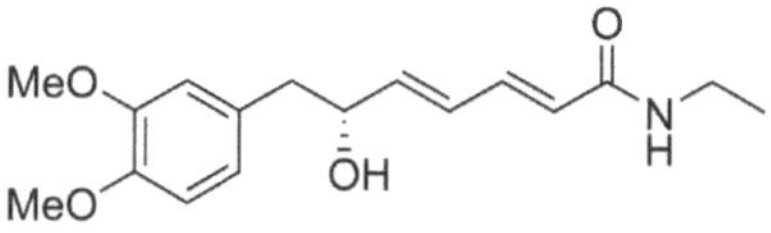

(−)-Kunstleramide (**1**)

Natural source: *Beilschmiedia kunstleri* gamble (barks; Lauraceae) [1]

Pharmaceutical potential: Antioxidant and cytotoxic [1]

Synthetic route:

Reddy and Venkataswarlu [2] reported the first stereoselective synthesis of the cytotoxic dienamide (−)-kunstleramide (1) in five steps in an overall yield of 31% from a commercially available starting material 3,4-dimethoxyphenylpropanol (2) using MacMillan α-hydroxylation [3], Horner-Wardsworth-Emmons (HWE) olefination, and amide-Wittig olefination [4] as key steps (Scheme 1).

Total Synthesis of Bioactive Natural Products
https://doi.org/10.1016/B978-0-08-102822-3.00033-X

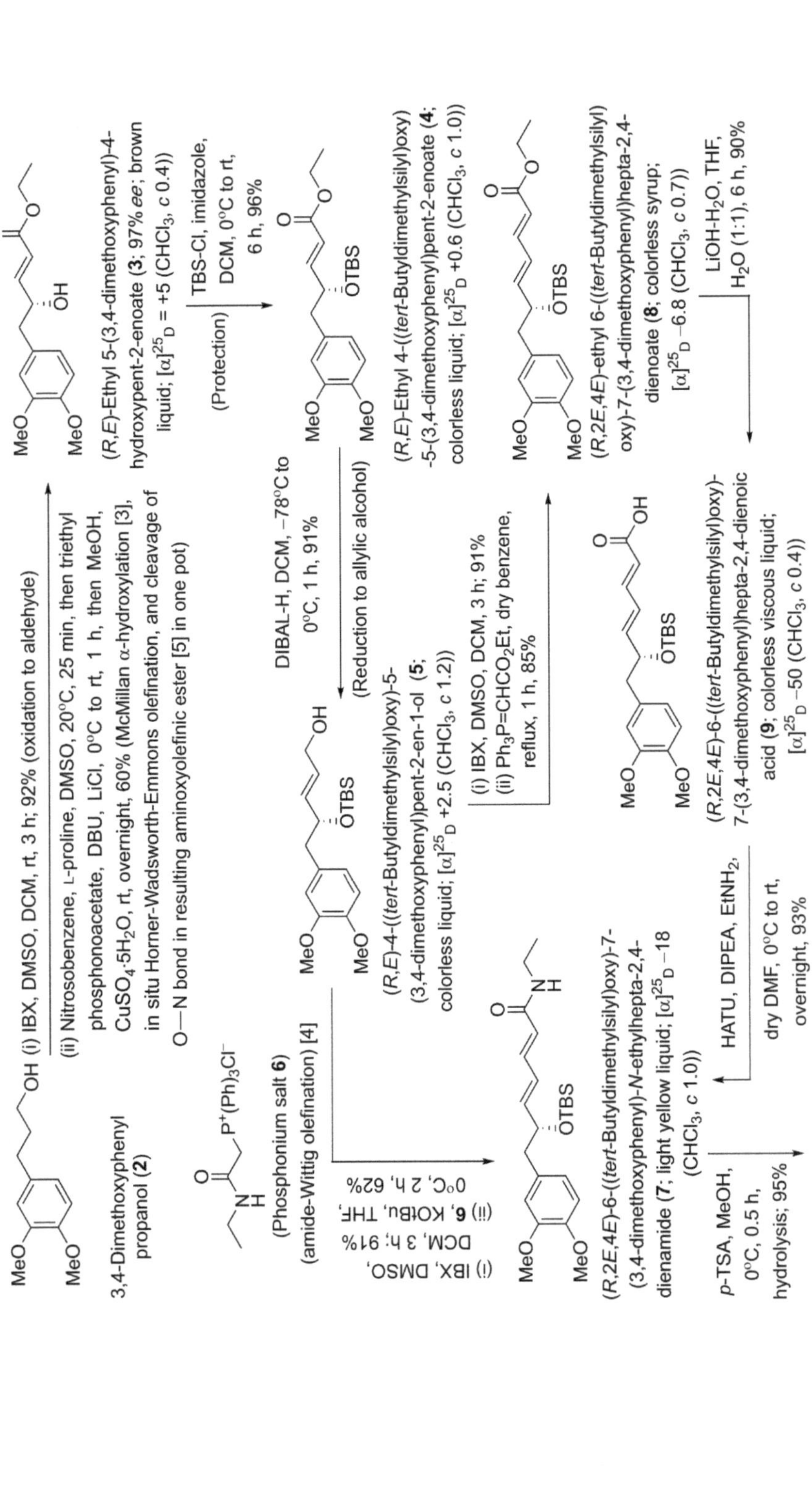

Scheme 1—Cont'd

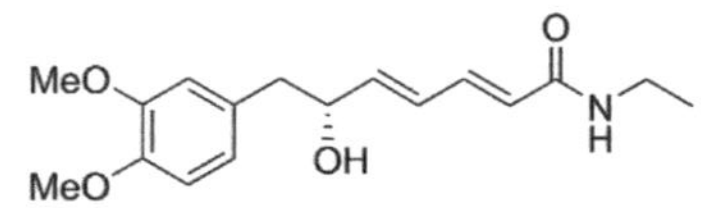

(−) Kunstleramide (**1**; white solid; $[\alpha]^{25}_{D}$ −50.8 (MeOH, *c* 0.2))

Scheme 1 Total synthesis of (−)-kunstleramide (**1**) [2].

3,4-Dimethoxyphenylpropanol (**2**) was first oxidized to the corresponding aldehyde using 2-iodoxybenzoic acid (IBX), and the crude aldehyde was subjected to MacMillan α-hydroxylation [4] using nitrosobenzene and 40 mol% of L-proline in DMSO, followed by rapid in situ HWE olefination with triethyl phosphonoacetate and DBU as base resulting in the formation of aminoxyolefinic ester. The cleavage of the O–N bond in aminoxyolefinic ester [5] using 20 mol% of copper sulfate in methanol at room temperature afforded (4*R*,2*E*)-ethyl 5-(3,4-dimethoxyphenyl)-4-hydroxypent-2-enoate (**3**; 97% *ee*) as brown liquid in 60% yield; the hydroxyl group of **3** was then protected as silyl ether **4**. The silyl ester was converted into allylic alcohol **5** using DIBAL-H in 91% yield. Compound **5** was oxidized with IBX to the corresponding aldehyde, which was subjected to amide-Wittig olefination [4] using phosphonium salt **6** and KO^tBu to afford (*R*,2*E*,4*E*)-6-((*tert*-butyldimethylsilyl)oxy)-7-(3,4-dimethoxyphenyl)-*N*-ethylhepta-2,4-dienamide (**7**) in 62% yield along with 6% of *Z*-isomer (separated through column chromatography). This amide derivative **7** was also prepared from compound **5** via another route as shown in Scheme 1. Finally, deprotection of TBS ether in compound **7** using *p*-TSA in MeOH furnished the target molecule (−)-kunstleramide (**1**) in 95% yield. The physical and spectral properties of the synthetic compound are in good agreement with the reported data of natural product [1].

References

[1] A. Mollataghi, A.H.A. Hadi, S.-C. Cheah, Molecules 17 (2012) 4197.

[2] P.S. Reddy, Y. Venkateswarlu, Tetrahedron Lett. 54 (2013) 4617.

[3] (a) S.P. Brown, M.P. Brochu, C.J. Sinz, D.W.C. MacMillan, J. Am. Chem. Soc. 125 (2003) 10808; (b) J.S. Yadav, U.V.S. Reddy, B.V.S. Reddy, Tetrahedron Lett. 50 (2009) 5984.

[4] N.J. Matovic, P.Y. Hayes, K. Penman, R.P. Lehmann, J.J. De voss, J. Org. Chem. 76 (2011) 4467.

[5] (a) G. Zhong, Y. Yu, Org. Lett. 6 (2004) 1637; (b) I.K. Mangion, D.W.C. MacMillan, J. Am. Chem. Soc. 127 (2005) 3967; (c) U. Ramulu, D. Ramesh, S. Rajaram, S.P. Reddy, K. Venkatesham, Y. Venkateswarlu, Tetrahedron Asymmetry 23 (2012) 117.

CHAPTER THIRTY FOUR

(±)-Limonin

Abbreviations

Ac_2O acetic anhydride
DBU 1.8-diazabicyclo[5.4.0]undec-7-ene
DIBAL diisobutylaluminum hydride
DMAP *N,N'*-dimethylaminopyridine
DME dimethylether
DMF *N,N*-dimethylformamide
DTBMP 2,6-di-*tert*-butyl-4-methylpyridine
LiHMDS lithium hexamethyldisilazide
***m*-CPBA** *meta*-chloroperbenzoic acid
NMO 4-methylmorpholine *N*-oxide
rt room temperature
TBAF tetrabutylammonium fluoride
TBHP *tert*-butyl hydroperoxide
TBSCl *tert*-butyldimethylsilyl chloride
THF tetrahydrofuran
TMSCl trimethylsilyl chloride
TPAP tetrapropylammonium perruthenate

Systematic name: (4a*S*,6a*R*,8a*R*,8b*R*,9a*S*,12*S*,12a*S*,14a*R*,14b*R*)-12-(Furan-3-yl)-6,6,8a,12a-tetramethyldecahydrooxireno[2,3-*d*]pyrano[4',3':3,3a]isobenzofuro[5,4-*f*]isochromene-3,8,10(1*H*,6*H*,8a*H*)-trione

Compound class: Limonoid (tetranortriterpenoid)

Structure:

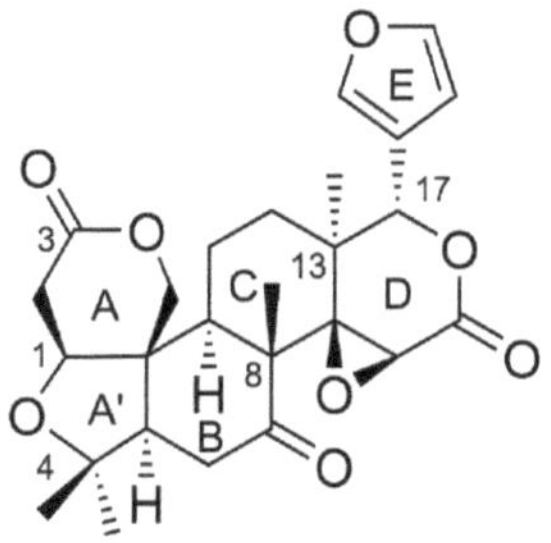

(±)-Limonin (**1**)

Total Synthesis of Bioactive Natural Products
https://doi.org/10.1016/B978-0-08-102822-3.00034-1

Natural sources: Citrus fruits (higher concentrations in seeds, for example orange and lemon seeds) [1, 2]. *Dictamnus dasycarpus* (family: Rutaceae) [3].

Pharmaceutical potential: Antioxidant, anticancer, cholesterol lowering, antiviral, neuroprotective [3–8]

Synthetic route:

Yamashita and coworkers [9] reported the first total synthesis of (±)-limonin (**1**); the investigators accomplished the total synthesis of this limonoid molecule in 35 steps from geraniol (**4**). They envisioned a retrosynthetic pathway for the construction of the limonoid skeleton as depicted in Scheme 1.

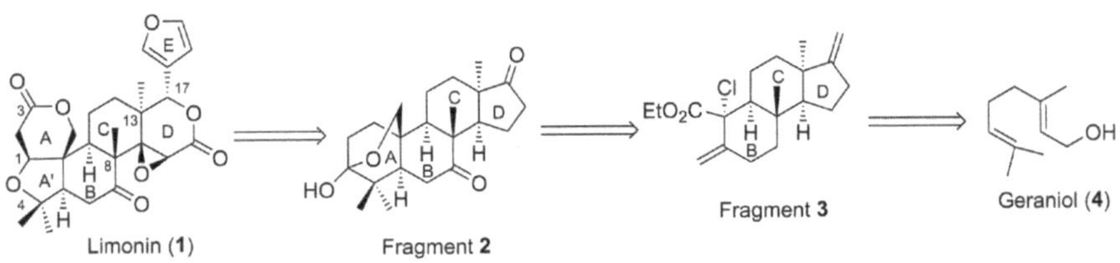

Scheme 1 Retrosynthetic analysis for (±)-limonin (**1**) [9].

The 35-step synthesis of the target compound **1** is shown in Scheme 2. First, the investigators synthesized the tricyclic fragment compound **3**. Chlorination of **4** followed by regioselective epoxidation producing an epoxygeranyl chloride, which was successively treated with lithiated 1-(trimethylsilyl) propyne [10, 11] and aluminum isopropoxide [12] to obtain allylic alcohol **6** in 75% overall yield. Compound **6** was then subsequently treated with thionyl chloride and the dianion of ethyl 2-chloroacetoacetate [13, 14], and then with TBAF for desilylation of **7** to generate the requisite alkynyl β-ketoester **8** as a colorless oil. Compound **8** was subjected to a manganese-mediated tandem radical cyclization [15–18] leading to the synthesis of the tricyclic BCD ring system **3** with C-13α configuration as a major product. On treatment with zinc dust in acetic acid at room temperature, the tricyclic ketoester **9** was obtained on reductive removal of the chloride, which underwent Robinson annulation with methyl vinyl ketone in the presence of a catalytic amount of potassium *tert*-butoxide to furnish the tetracyclic enone **10** (colorless crystals with 62% yields over two steps) with excellent diastereoselectivity at C-10. Lastly, installation of a dimethyl group at the C-4 position of **10** was achieved by treating the compound with iodomethane and *tert*-BuOK to give **11** in 80% yield [19].

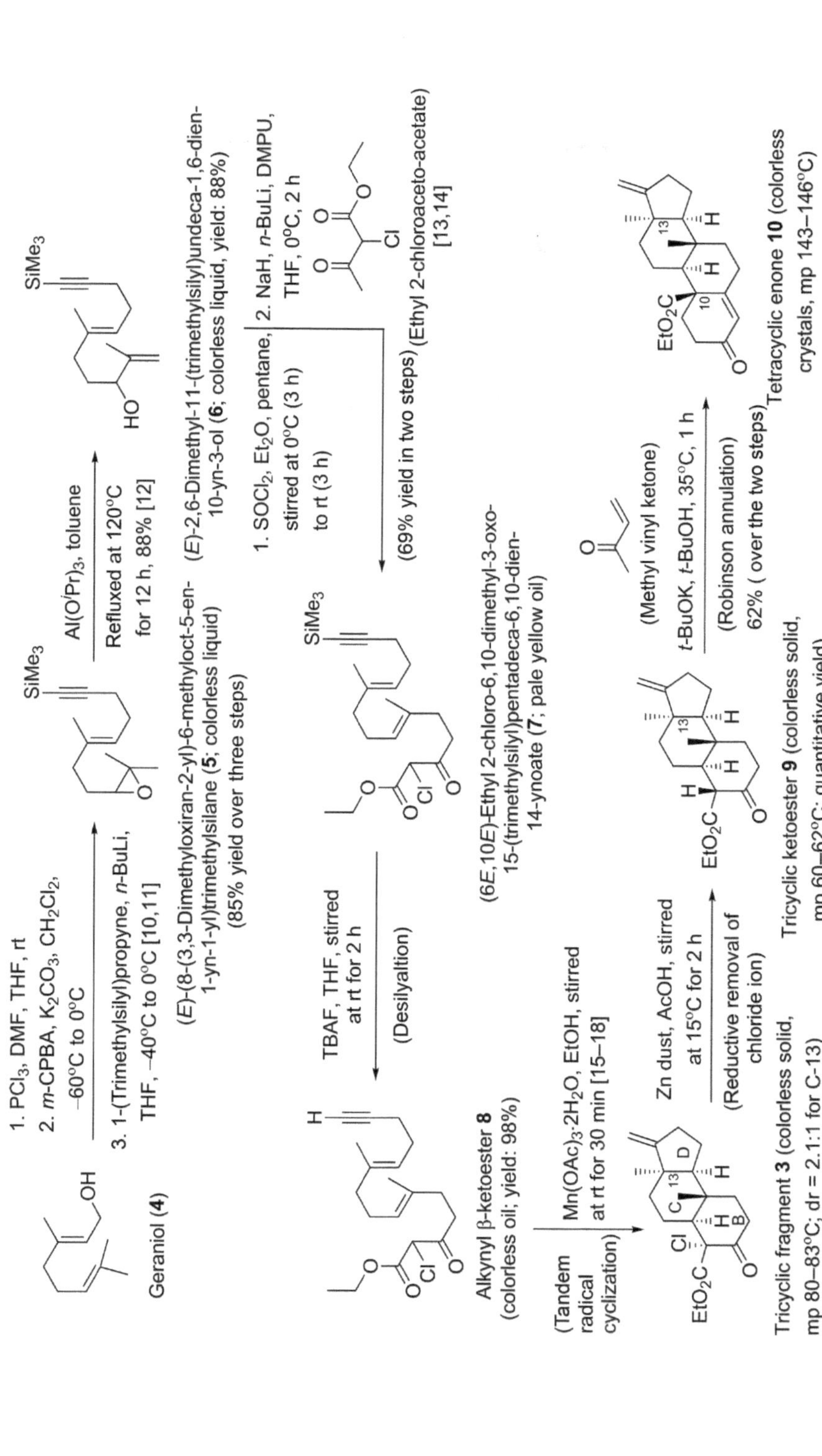

Scheme 2—Cont'd

(Continued)

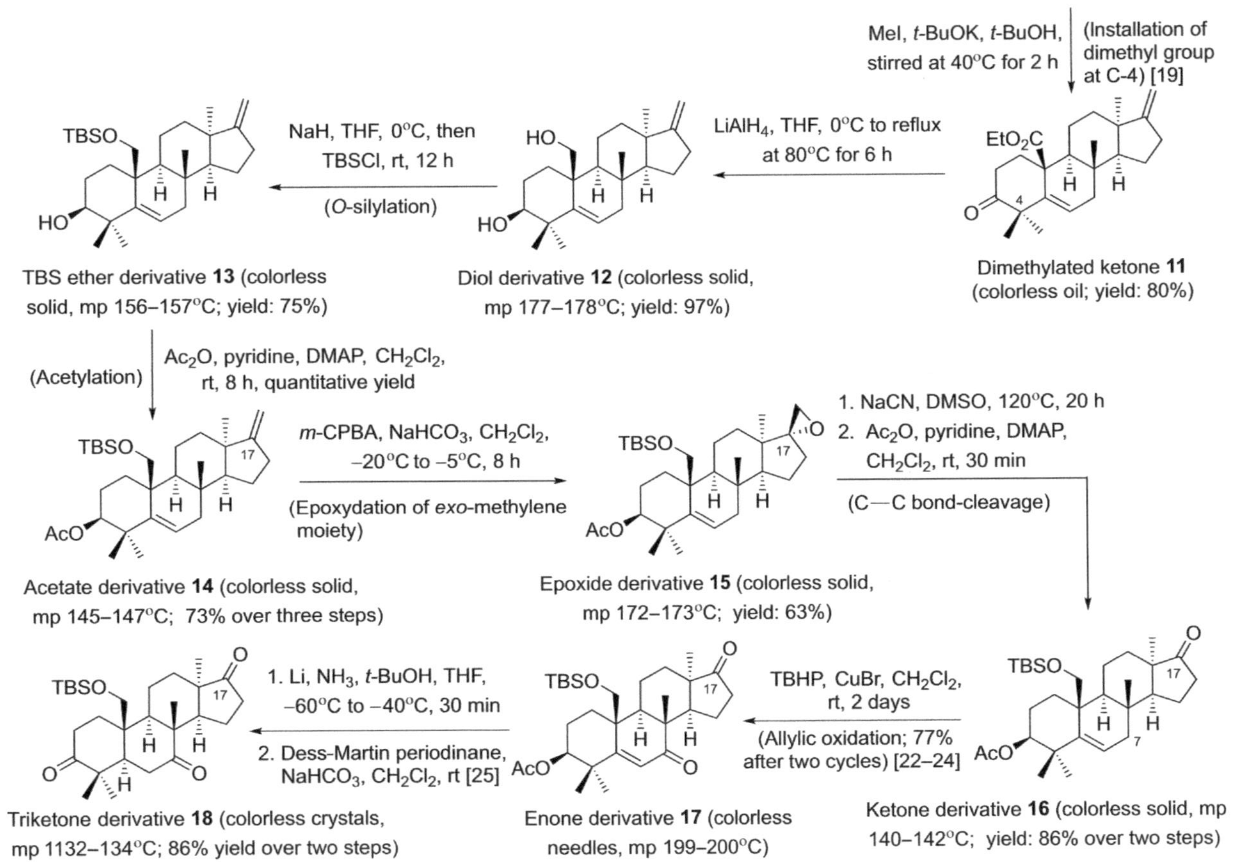

Scheme 2—Cont'd

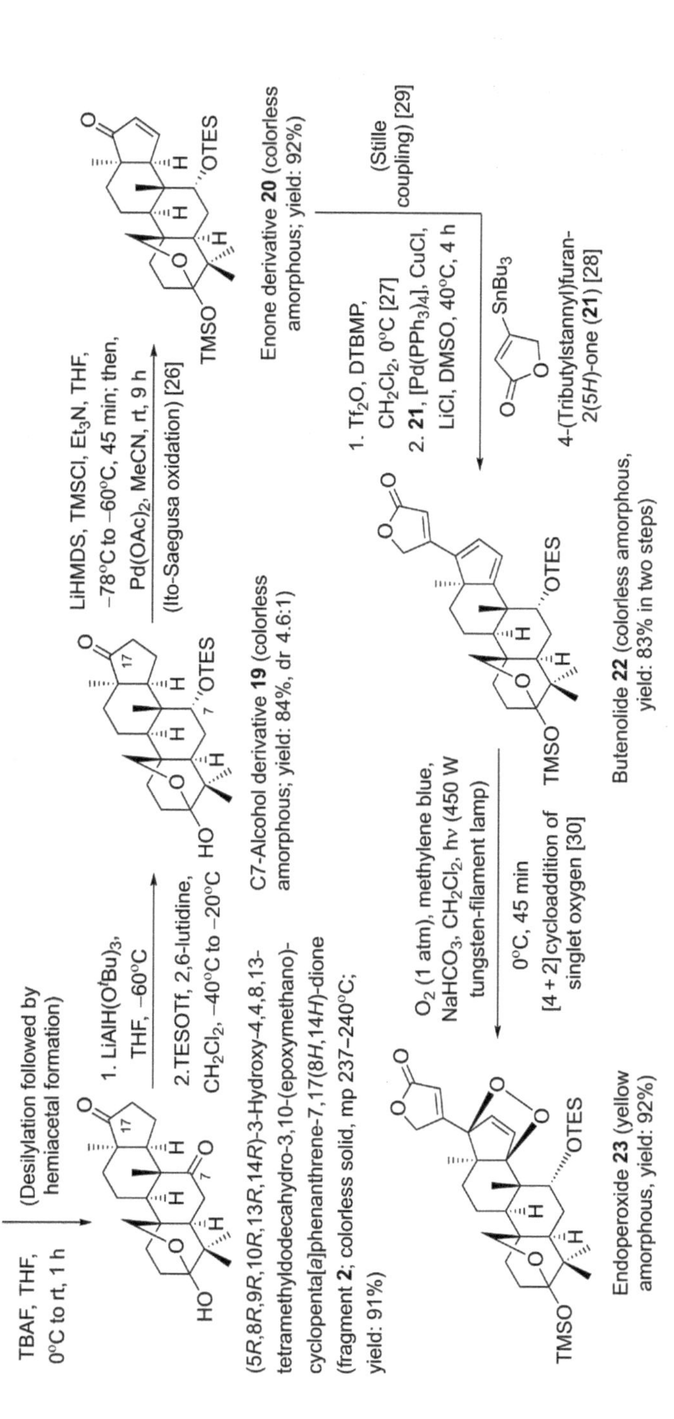

Scheme 2—Cont'd

(Continued)

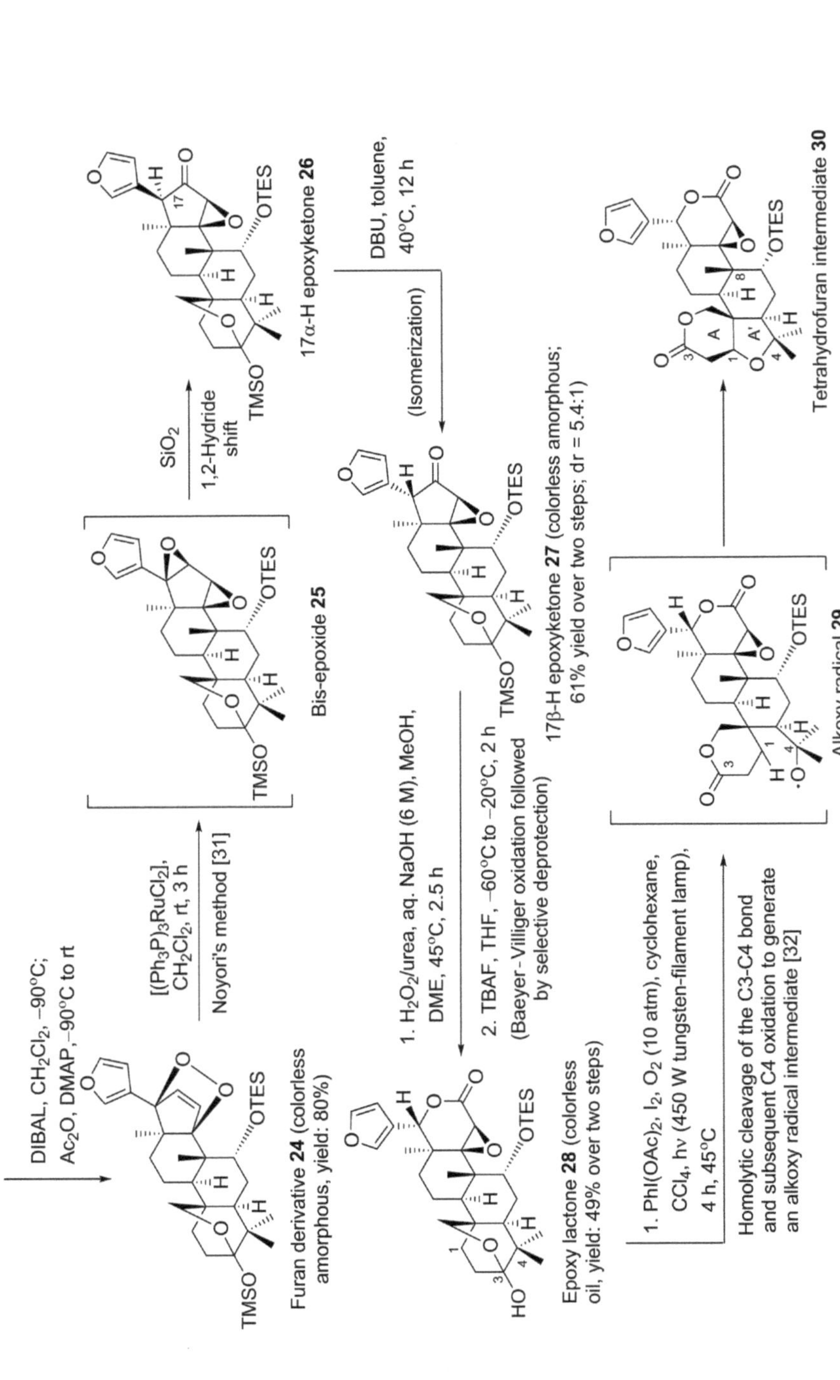

Scheme 2—Cont'd

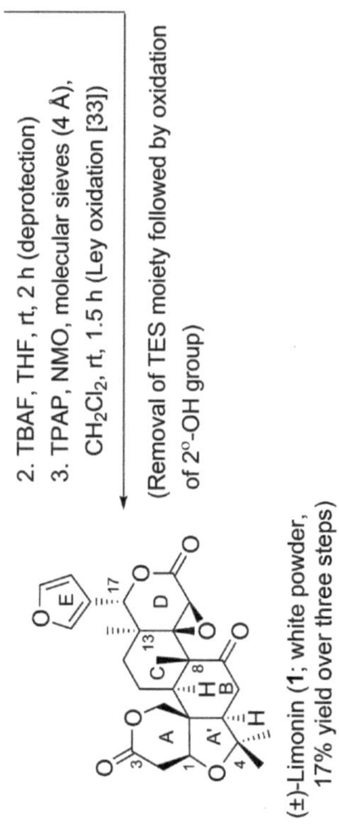

Scheme 2 Total synthesis of (±)-limonin (**1**) [9].

Reduction of **11** with lithium aluminum hydride afforded the corresponding diol **12**, and the two hydroxy groups were protected as a TBS ether and an acetate, respectively, to give compound **14** with 73% yield over three steps. Chemo- and stereoselective epoxidation of the *exo*-methylene group of **14** was performed with *m*-CPBA to generate epoxide **15**. Epoxide **15** on heating at 120°C in DMSO in the presence of NaCN underwent an interesting C—C bond cleavage, possibly due to nitrile addition to the epoxide followed by elimination of acetonitrile [20, 21], thereby leading to the formation of the desired ketone **16** along with the corresponding deacetylated compound, which was reacetylated to give **16**. Allylic oxidation of **16** by means of *tert*-butyl hydroperoxide (TBHP) in the presence of a catalytic amount of CuBr [16] furnished enone **17** in 77% yield after two cycles. Enone **17** yielded hemiacetal fragment **2** on successive reactions such as Birch reduction, Dess-Martin oxidation, and then desilylation with TBAF in 78% overall yield. $LiAlH(OtBu)_3$ reduced the C-7 ketone of **2** chemoselectively, and TES protection of the hydroxy groups then afforded compound **19** in 69% yield (along with the C7 epimer in 15% yield). Ito-Saegusa oxidation [26] of ketone **19** proceeded smoothly to give enone **20**.

The investigators then designed the installation of the furan moiety and the construction of the epoxylactone D ring in the next part. Treatment of **20** with Tf_2O in the presence of 2,6-di-*tert*-butyl-4-methylpyridine (DTBMP) [27] afforded the corresponding vinylogous enol triflate; butenolide **21** [28] was then attached to this intermediate in a Stille coupling [29] to give butenolide **22** in 83% overall yield. The [4+2] cycloaddition of singlet oxygen to diene **22** smoothly provided the endoperoxide **23** in a stereoselective manner [30]. Compound **23** on sequential treatment with diisobutylaluminum hydride (DIBAL) and acetic anhydride/DMAP led to the formation of furan **24** as colorless amorphous solid in 80% yield. Ruthenium-catalyzed isomerization of the endoperoxide in **24** via Noyori's method [31] resulted in the intermediate formation of a bis-epoxide **25**, which smoothly underwent a 1,2-hydride shift on silica gel to produce 17α-H epoxyketone **26**. This compound was isomerized by heating at 40°C in the presence of 1,8-diazabicyclo[5.4.0]undec-7-ene (DBU) to provide 17β-H epoxyketone **27** (colorless amorphous; 61% yield over two steps; dr = 5.4:1). Baeyer-Villiger oxidation of **27** under basic conditions followed by selective deprotection of the TES acetal afforded epoxylactone **28**.

In the final stage, the investigators installed a AA′ ring system within the intermediate **30**, which was accomplished by a Suárez reaction in a single operation [32], possibly through the homolytic cleavage of the C3–C4 bond and subsequent C-4 oxidation generating an alkoxy radical intermediate **29**—this radical eventually underwent C-1 hydrogen abstraction and iodination followed by formation of a tetrahydrofuran ring to give the compound **30**. Removal of the TES group within the compound **30** on treating with TBAF/THF followed by a Ley oxidation [33] afforded the target compound (±)-limonin (**1**) (white powder, 17% over three steps), the spectral properties of which were found to be identical to those for the natural compound [9]. The overall synthesis is summarized in Scheme 2.

References

[1] Bernays, Justus Liebigs Ann. Chem. 40 (1841) 317.
[2] C. Liu, J. Liu, Y. Rong, N. Liang, L. Rong, Czech J. Food Sci. 30 (2012) 364.
[3] J.S. Yoon, H. Yang, S.H. Kim, S.H. Sung, Y.C. Kim, J. Mol. Neurosci. 42 (2010) 9.
[4] E.M. Kurowska, N.M. Borradaile, J.D. Spence, K.K. Carroll, Nutrition Res. 20 (2000) 121.
[5] E.M. Kurowska, J.D. Spence, J. Jordan, S. Wetmore, D.J. Freeman, L.A. Piche, P. Serratore, Am. J. Clin. Nutr. 72 (2000) 1095.
[6] L. Battinelli, F. Mengoni, M. Lichtner, G. Mazzanti, A. Saija, C.M. Mastroianni, V. Vullo, Planta Med. 69 (2003) 910.
[7] E.G. Miller, J.L. Porter, W.H. Binnie, I.Y. Guo, S. Hasegawa, J. Agric. Food Chem. 52 (2004) 4908.
[8] G.D. Manners, J. Agric. Food Chem. 55 (2007) 8285.
[9] S. Yamashita, A. Naruko, Y. Nakazawa, L. Zhao, Y. Hayashi, M. Hirama, Angew. Chem. Int. Ed. 54 (2015) 8538.
[10] B.H. Lipshutz, G. Bulow, F. Fernandez-Lazaro, S.-K. Kim, R. Lowe, P. Mollard, K.L. Stevens, J. Am. Chem. Soc. 121 (1999) 11664.
[11] E.J. Corey, C. Rucker, Tetrahedron Lett. 23 (1982) 719.
[12] E. Doron, K. Ehud, J. Am. Chem. Soc. 110 (1988) 4356.
[13] S.N. Huckin, L. Weiler, J. Am. Chem. Soc. 96 (1974) 1082.
[14] R.C.D. Brown, C.J. Bataille, R.M. Hughes, A. Kenney, T.J. Luker, J. Org. Chem. 67 (2002) 8079.
[15] B.B. Snider, J.J. Patricia, S.A. Kates, J. Org. Chem. 53 (1988) 2137.
[16] B.B. Snider, R.M. Mohan, S.A. Kates, J. Org. Chem. 50 (1985) 3659.
[17] B.B. Snider, Chem. Rev. 96 (1996) 339.
[18] S. Yamashita, A. Naruko, T. Yamada, Y. Hayashi, M. Hirama, Chem. Lett. 42 (2013) 220.
[19] M. Kato, M. Matsumura, K. Heima, N. Fukamiya, C. Kabuto, A. Yoshikoshi, Tetrahedron 43 (1987) 711.
[20] P.A. Wade, J.F. Bereznak, J. Org. Chem. 52 (1987) 2973.
[21] S. Scholz, H. Hofmeister, G. Neef, E. Ottow, C. Scheidges, R. Wiechert, Liebigs Ann. Chem. (1989) 151.
[22] M.S. Kharasch, G. Sosnovsky, J. Am. Chem. Soc. 80 (1958) 756.
[23] J.A.R. Salvador, M.L.S. Melo, A.S.C. Neves, Tetrahedron Lett. 38 (1997) 119.
[24] D. Li, T.A. Spencer, Steroids 65 (2000) 529.
[25] D.B. Dess, J.C. Martin, J. Org. Chem. 48 (1983) 4155.
[26] Y. Ito, T. Hirao, T. Saegusa, J. Org. Chem. 43 (1978) 1011.
[27] P.J. Stang, T.E. Fisk, Synthesis (1979) 438.

[28] G.J. Hollingworth, G. Perkins, J. Sweeney, J. Chem. Soc. Perkin Trans. 1 (1996) 1913.
[29] K. Mukai, D. Urabe, S. Kasuya, N. Aoki, M. Inoue, Angew. Chem. Int. Ed. 52 (2013) 5300.
[30] W. Li, P.L. Fuchs, Org. Lett. 5 (2003) 2849.
[31] M. Suzuki, H. Ohtake, Y. Kameya, N. Hamanaka, R. Noyori, J. Org. Chem. 54 (1989) 5292.
[32] A. Boto, R. Freire, R. Hernández, E. Suárez, J. Org. Chem. 62 (1997) 2975.
[33] S.V. Ley, J. Norman, W.P. Griffith, S.P. Marsden, Synthesis (1994) 639.

CHAPTER THIRTY FIVE

Linderaspirone A and Bi-linderone

Abbreviations

LiHMDS lithium hexamethyldislazide
rt room temperature
THF tetrahydrofuran

Compound class: Spirocyclopentenedione derivatives
Structures:

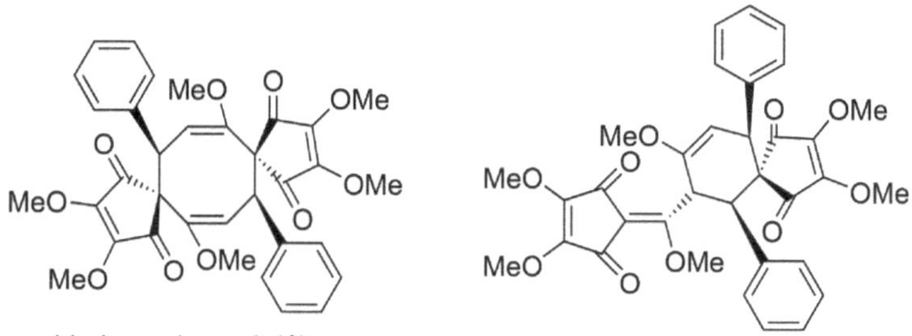

Linderaspirone A (**1**)

Bi-linderone (**2**)

Natural source: *Lindera aggregata* (Sims) kosterm (Lauraceae; roots) [1, 2]
Pharmaceutical potential: Antidiabetic (showed activity against glucosamine-induced insulin resistance in bioactivity tests with HepG2 cells) [1, 2]
Synthetic route:
Both the spirocyclopentenediones **1** and **2** bear highly congested eight- or six-membered ring skeletons, respectively. Significant bioactivity coupled with such unique architecture and very limited natural availability (15 mg **1** and 4 mg **2** from 800 g of air-dried powdered roots of the plant) provoked organic chemists to regard them as attractive synthetic targets. In 2011, Liu and coworkers [3] first developed a one-step biomimetic approach to **1** and **2**, and at almost the same time Wang and coworkers [4] also reported biomimetic total syntheses of **1** and **2** through

Total Synthesis of Bioactive Natural Products
https://doi.org/10.1016/B978-0-08-102822-3.00035-3

photochemical [2 + 2] cycloaddition/Cope rearrangement and photochemical [2 + 2] cycloaddition/radical rearrangement, respectively. Hu and his group [5] developed a concise and efficient total synthesis of linderaspirone A (**1**) through a Darzens cyclopentenedione synthesis and dioxygen-assisted photochemical dimerization; they also established the thermal isomerization of **1** as an efficient and biomimetic route for the synthesis of **2** [5] (Scheme 1).

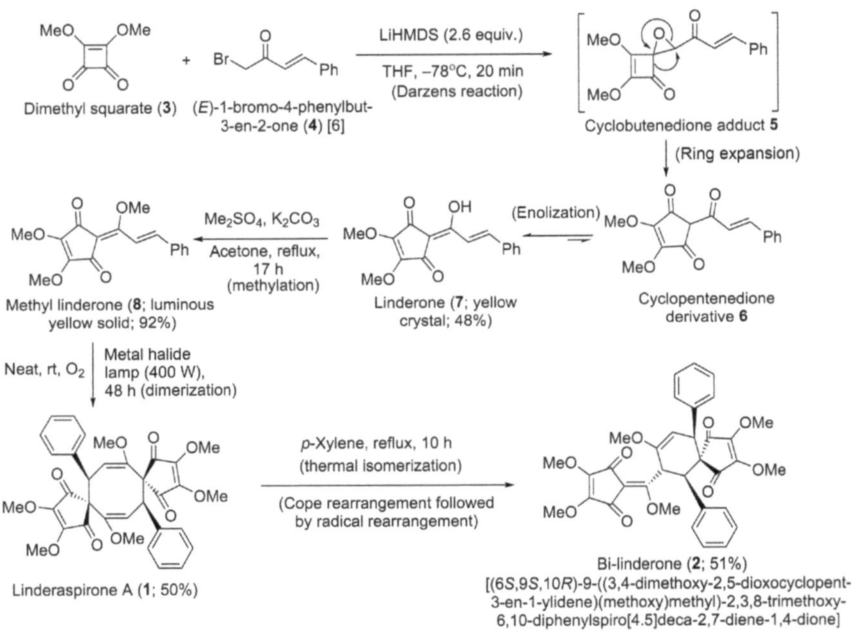

Scheme 1 Total synthesis of linderaspirone A (**1**) and bi-linderone (**2**) [5].

The investigators first synthesized methyl linderone (**8**), the precursor for the target molecule **1**, starting from dimethyl squarate (**3**) in only two steps with 57% overall yield. α-Bromoketone (**4**) took part in Darzens reaction with dimethyl squarate (**3**) in the presence of the strong base LiHMDS/THF at −78°C to form a cyclobutenedione adduct **5** that underwent rapid in situ ring expansion followed by enolization resulting in the formation of linderone (**7**) as yellow crystals in 48% yield. Subsequent methylation with dimethyl sulfate afforded methyl linderone (**8**) in good yield. Methyl linderone was then dissolved in dichloromethane in a silica tubular reactor and the solvent was blow dried with dioxygen. The reaction was then irradiated for 48 h at room temperature by using a 400 W metal halide lamp when photo-dimerization took place to give a crude product purified by chromatography on a silica gel column (acetone/petroleum ether = 1:6) to furnish pure linderaspirone A (**1**) in 50% yield; the spectroscopic data of the synthetic compound were found to be completely in agreement with those of the natural product [2]. Thermal isomerization of compound **1** on refluxing in *p*-xylene afforded bi-linderone (**2**; 51%), possibly with the involvement of an interesting thermal Cope/radical rearrangement cascade of linderaspirone A.

References

[1] F. Wang, Y. Gao, L. Zhang, J.K. Liu, Org. Lett. 12 (2010) 2354.
[2] F. Wang, Y. Gao, L. Zhang, B. Bai, Y.N. Hu, Z.J. Dong, Q.W. Zhai, H.J. Zhu, J.K. Liu, Org. Lett. 12 (2010) 3196.
[3] G.Q. Wang, K. Wei, L. Zhang, T. Feng, F. Wang, Q.A. Wang, J.K. Liu, Tetrahedron Lett. 52 (2011) 2719.
[4] H. Tan, C. Zheng, Z. Liu, D.Z. Wang, Org. Lett. 13 (2011) 2192.
[5] F. Xiao, W. Liu, Y. Wang, Q. Zhang, X. Li, X. Hu, Asian J. Org. Chem. 2 (2013) 216.
[6] Y. Onizawa, H. Kusama, N. Iwasawa, J. Am. Chem. Soc. 130 (2008) 802.

CHAPTER THIRTY SIX

Lucidone

Abbreviations

CH_2Cl_2 dichloromethane
DMSO dimethylsulfoxide
Et_2O diethyl ether
LiHMDS lithium bis(trimethylsilyl)amide
Me_2SO_4 dimethylsulfide
Me_3SiOTf trimethylsilyltriflate
rt room temperature
THF tetrahydrofuran

Systematic name: (*Z*)-2-((*E*)-1-Hydroxy-3-phenylallylidene)-4-methoxycyclopent-4-ene-1,3-dione

Compound class: Cyclopentenedione

Structure:

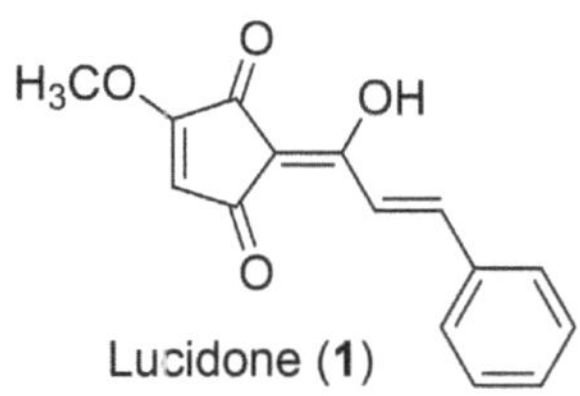

Lucidone (**1**)

Natural sources: *Lindera lucida* (dried fruits; family: Lauraceae) [1] and *L. erythrocarpa* (dried fruits) [2, 3]

Pharmaceutical potentials: Antioxidant, antiinflammatory, antitumor, hypolipidemic, hepatoprotective, melanin inhibitory, antihepatitis, and antidengue [4–9]

Synthetic routes:

The first total synthesis of lucidone (**1**) was accomplished by Lee and Que. in 1985 from (*E*)-3-(2-hydroxy-3,4,6-trimethoxyphenyl)-1-phenylprop-2-en-1-one (**2**) (Scheme 1) [10]. The chalcone derivative **2** on treatment with a mixture of glacial acetic acid and concentrated nitric acid was oxidized to form 2-cinnamoyl-3-hydroxy-5-methoxy-1,4-benzoquinone (**3**); the crude

https://doi.org/10.1016/B978-0-08-102822-3.00036-5

product was washed thoroughly with water and dried in vacuo over phosphorus pentaoxide, and the dried mass was recrystallized from benzene to afford the pure product of quinochalcone **3** as dark orange plates with just 39% yield. In the next step, the quinochalcone derivative **3** was stirred with acetic anhydride in dimethylsulfoxide under nitrogen atmosphere to form (*E*)-3-methoxy-5-((*E*)-2-oxo-4-phenylbut-3-en-1-ylidene)furan-2(5*H*)-one (**4**) as yellow rods. The butenolide derivative **4** was then refluxed with a solution of sodium methoxide either in dry benzene or dry methanol for 1–1.5 h under nitrogen atmosphere, followed by work-up with addition of 10% aqueous hydrochloric acid to furnish bright yellow needles of lucidone with respective yields of 46% and 67%. The physical and spectral properties of the synthetic compound were in full agreement with those of natural product [10].

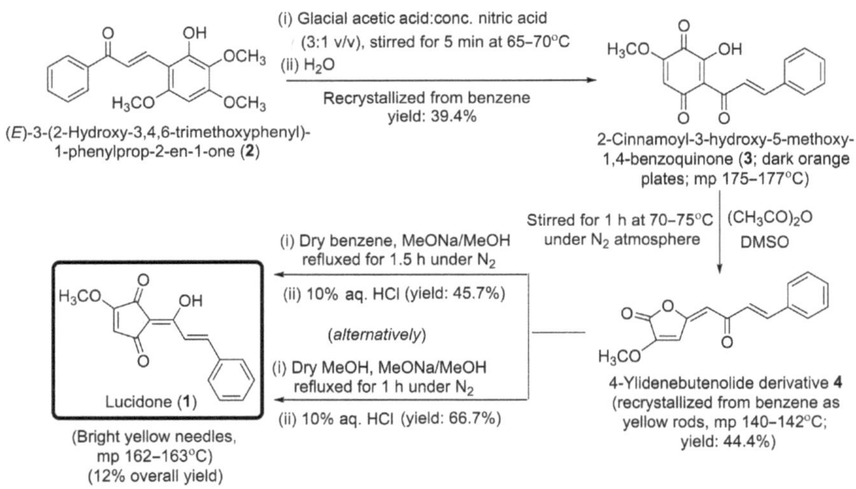

Scheme 1 Total synthesis of lucidone (**1**) [10].

Later on, the group of Prof. Langer [11] reported a formal total synthesis of lucidone (**1**) (Scheme 2). The key starting compound, 6-phenyl-2,4-bis (trimethysilyloxy)hexa-1,3,5-triene (**8**), was prepared from the condensation reaction of the 1,3-bis-silyl enol ether **5** with benzaldehyde (**6**), followed by silylation of the resulting 1,3-diketone **7** with trimethylsilyl triflate in triethylamine/diethyl ether medium. The Me_3SiOTf-catalyzed cyclization of **8** with oxalyl chloride (**9**) afforded (*E*)-3-hydroxy-5-((*E*)-2-oxo-4-phenylbut-3-en-1-ylidene)furan-2(5*H*)-one (**10**) as yellow solid in 64% yield. Compound **10** was then methylated with dimethyl sulfate to the methyl derivative **11**, which was finally transformed into the natural acylcyclopentenedione lucidone (**1**) as per the procedure adopted earlier by Clemo et al. [12], with an overall yield of 17%.

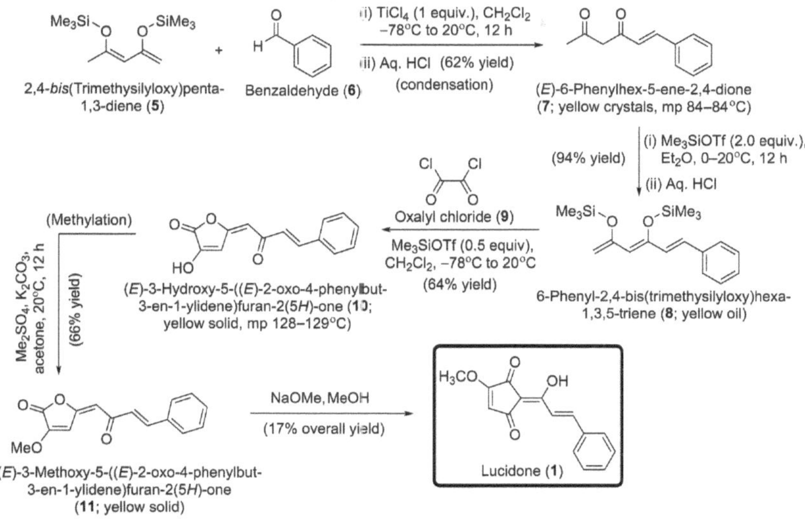

Scheme 2 Formal total synthesis of lucidone (**1**) [11].

In 2013, a concise synthesis of lucidone was achieved by Hu and his group [13] via a one-pot two-step reaction starting from commercially available dimethyl squarate (**12**) with somewhat high overall yield of 46% (Scheme 3). The transformation involves the one-pot reduction/rearrangement of dimethyl squarate and the Darzens/ring expansion of the monomethoxyl cyclobutenedione (**13**) upon reaction with (*E*)-1-bromo-4-phenylbut-3-en-2-one (**14**) in the presence of lithium *bis*(trimethylsilyl)amide (LiHMDS) as a basic reagent.

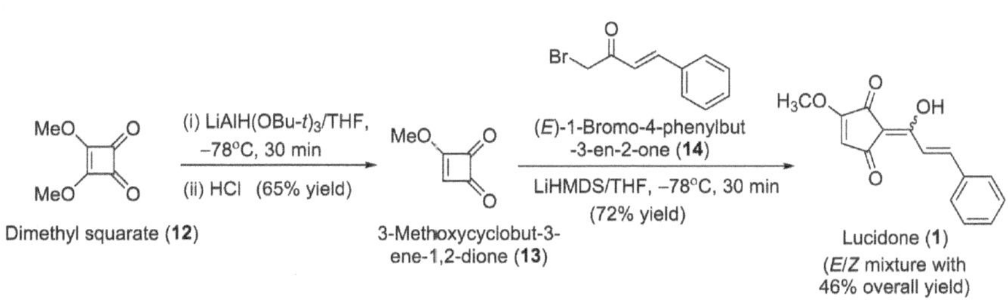

Scheme 3 Concise total synthesis of lucidone (**1**) [13].

References

[1] H.H. Lee, Tetrahedron Lett. 9 (1968) 4243.
[2] S.Y. Liu, S. Hisada, I. Inagaki, Phytochemistry 12 (1973) 472.
[3] S.-Y. Wang, X.-Y. Lan, J.-H. Xiao, J.-C. Yang, Y.-T. Kao, S.-T. Chang, Phytother. Res. 22 (2008) 213.
[4] H.M. Oh, S.K. Choi, S.K. Lee, H.Y. Kim, D.C. Han, H.M. Kim, K.H. Son, B.M. Kvon, Bioorg. Med. Chem. 13 (2005) 6182.
[5] K.J.S. Kumar, J.C. Yang, F.H. Chu, S.T. Chang, S.Y. Wang, Phytother. Res. 24 (2010) 1158.
[6] K.J.S. Kumar, H.W. Hsieh, S.-Y. Wang, Int. Immunopharmacol. 10 (2010) 385.
[7] K.J.S. Kumar, H.L. Yang, Y.C. Tsai, P.C. Hung, S.H. Chang, H.W. Lo, P.C. Shen, S.C. Chen, H.M. Wang, S.Y. Wang, C.W. Chou, Y.C. Hseu, Food Chem. Toxicol. 59 (2013) 55.
[8] W.-C. Chen, S.-Y. Wang, C.-C. Chiu, C.-K. Tseng, C.-K. Lin, H.-C. Wang, J.-C. Lee, Antimicrob. Agents Chemother. 57 (2013) 1180.
[9] W.-C. Chen, C.-K. Tseng, C.-K. Lin, S.-N. Wang, W.-H. Wang, S.-H. Hsu, Y.-H. Wu, L.-C. Hung, Y.-H. Chen, J.-C. Lee, Virulence 9 (2018) 588.
[10] H.-H. Lee, Y.-T. Que, J. Chem. Soc. Perkin Trans. 1 (1985) 453–454.
[11] I. Freifeld, G. Bose, T. Eckardt, P. Langer, Eur. J. Org. Chem. (2007) 351–355.
[12] N.G. Clemo, D.R. Gedge, G. Pattenden, J. Chem. Soc., Perkin Trans. 1 (1981) 1448.
[13] W. Liu, F. Xiao, X. Hu, Chin. J. Org. Chem. 33 (2013) 1587.

CHAPTER THIRTY SEVEN

(±)-Marinopyrrole A

Abbreviations

DBU 1,8-diazabicyclo[5.4.0]undec-7-ene
DIBAL-H diisobutylaluminum hydride
DMAP 4-*N*,*N*-dimethylaminopyridine
DMF *N*,*N*-dimethylformamide
DMSO dimethylsulfoxide
IBX 2-iodoxybenzoic acid
LDA lithium diisoproylamide
MeCN acetonitrile
NCS *N*-chlorosuccinimide
PPTS pyridinium *p*-toluenesulfonate
***p*-TsOH** *para*-toluenesulfonic acid
THF tetrahydrofuran

Systematic name: (4,4′,5,5′-Tetrachloro-1′*H*-1,3′-bipyrrole-2,2′-diyl)bis(2-hydroxyphenyl)methanone; alternatively, (2-hydroxyphenyl)-[2,3,4′,5′-tetrachloro-5-(2-hydroxybenzoyl)-1′*H*-[1,3′]bipyrrolyl-2′-yl]-methanone

Compound class: Alkaloid (a halogenated bipyrrole antibiotic)

Structure:

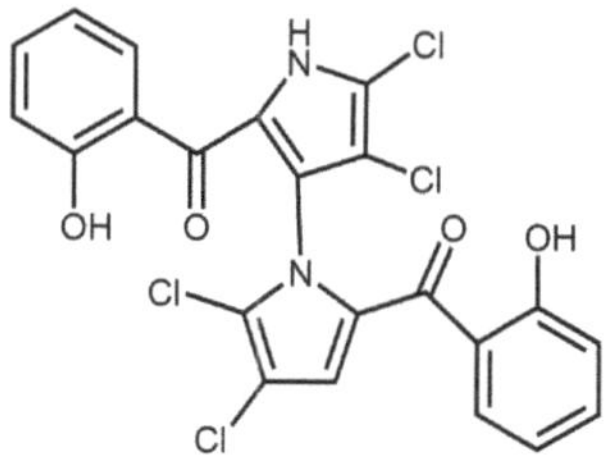

(±)-Marinopyrrole A (**1**)
(−)-Marinopyrrole A (**1a**; natural)
(+)-Marinopyrrole A (**1b**; unnatural)

Total Synthesis of Bioactive Natural Products
https://doi.org/10.1016/B978-0-08-102822-3.00037-7

Natural source: *Streptomyces* sp. strain CNQ-418 (an obligate marine culture broth) [1]

Pharmaceutical potentials: Antibacterial activity against methicillin-resistant *Staphylococcus aureus* (MRSA) [1–4]; anticancer [1, 2, 5, 6]

Synthetic routes:

The first total synthesis of the marine natural product (±)-marinopyrrole A (**1**), bearing an unprecedented bispyrrole moiety, was carried out by Li and his group via a nine-step synthesis in an overall yield of 30%. 2-Ethoxycarbonyl-3-aminopyrrole (**2**) [8, 9] underwent a one-pot condensation followed by cyclization reaction with ethyl 4-(1,3-dioxan-2-yl)-2-oxobutanoate (**3**) [10] in the presence of *p*-TsOH catalyst under reflux conditions in toluene to afford the bispyrrole derivative **4** in 82% yield. After tosylation to protect the nitrogen on the pyrrole ring in **4**, the two ester groups were reduced with DIBAL-H at room temperature in CH_2Cl_2 to furnish the diol **6** in 87% yield over these two steps. The hydroxyl groups in **6** were then oxidized with IBX in DMSO to form the dialdehyde **7** (90% yield). The crude diol intermediate **8** formed on addition with 2-methoxyphenylmagnesium bromide in THF was directly subjected to oxidation by CrO_3 in anhydrous pyridine at room temperature to furnish diketone **9** in 69% yield in two consecutive steps (to avoid the spontaneous conversion of **9** to oxazepine **10** in the presence of weak acid; although oxazepine **10** was isolated in 10% yield as by-product under these conditions, and separated out by column chromatography).

Compound **11** obtained on deprotection of the nitrogen on the pyrrole ring in **10** with KOH in MeOH/THF (1:1) was then treated with *N*-chlorosuccinimide (4.4 equiv) in anhydrous acetonitrile to give tetrachloro-substituted derivative **12** as a light yellow solid in 75% yield. Demethylation of phenolic methyl ethers in **12** with BBr_3 in CH_2Cl_2 at −78°C eventually resulted in the production of (±)-marinopyrrole A (**1**; yellow solid, mp 205–207°C) in 95% yield (Scheme 1); the overall yield of **1** in this nine-step synthesis was, however, found to be 30%. The spectroscopic properties of the synthetic sample of (±)-**1** were identical with those of the natural product (−)-**1** [1]. At the same time, the investigators also reported the synthesis of a good number of marinopyrrole derivatives as a focused library of potential antibiotic and anticancer agents in the same report [7].

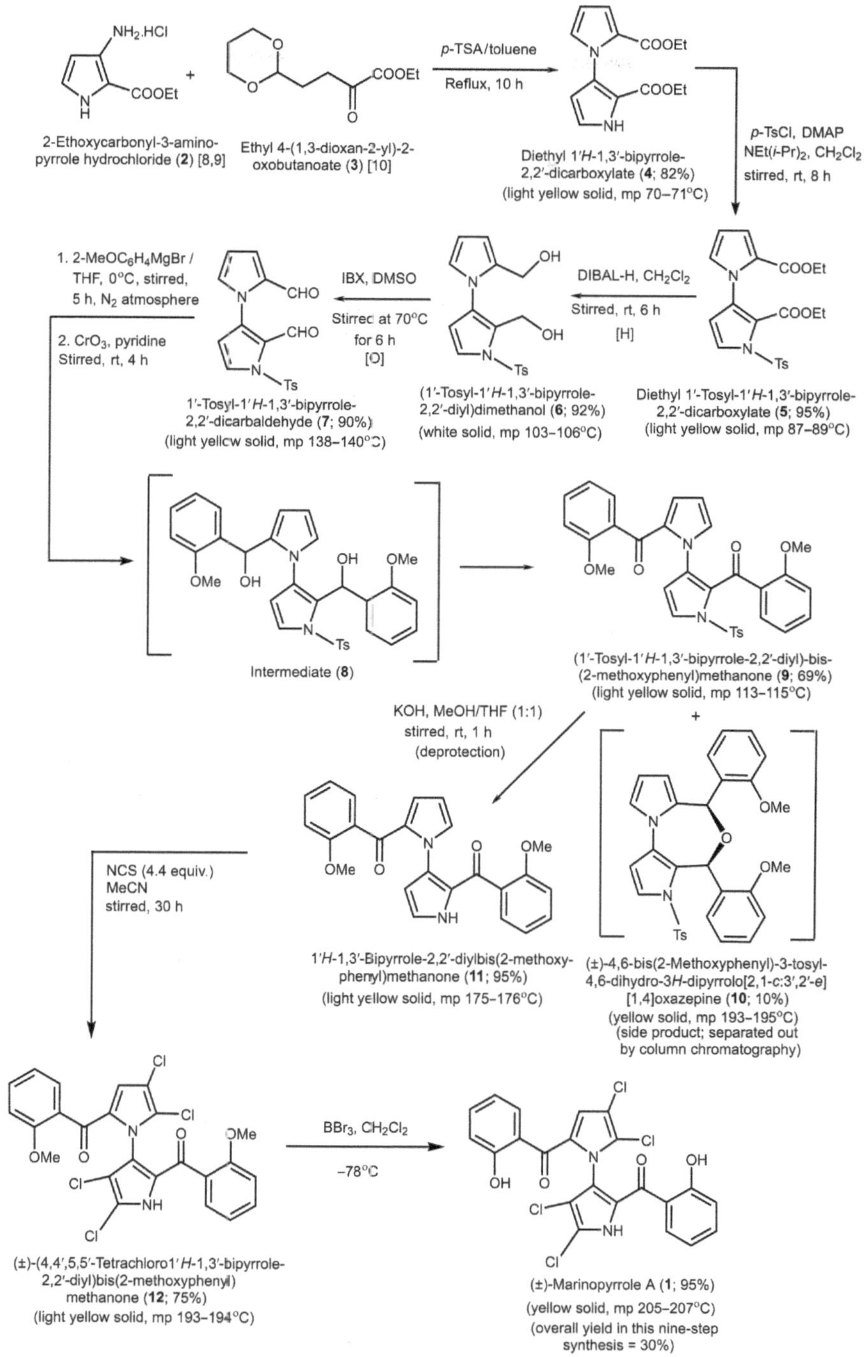

Scheme 1 Li's total synthesis of (±)-marinopyrrole A (**1**) [7].

Later on, Kanakis and Sarli [11] reported a six-step synthesis of racemic marinopyrrole A (**1**) in an overall yield of 22% (Scheme 2). Nicolaou and his group [4] also accomplished another synthetic route (Scheme 3) for this marine natural product with an overall yield of 16% in six steps. In addition, they synthesized a library of marinopyrrole A analogs and compared their antibacterial efficacies against the methicillin-resistant *Staphylococcus aureus*

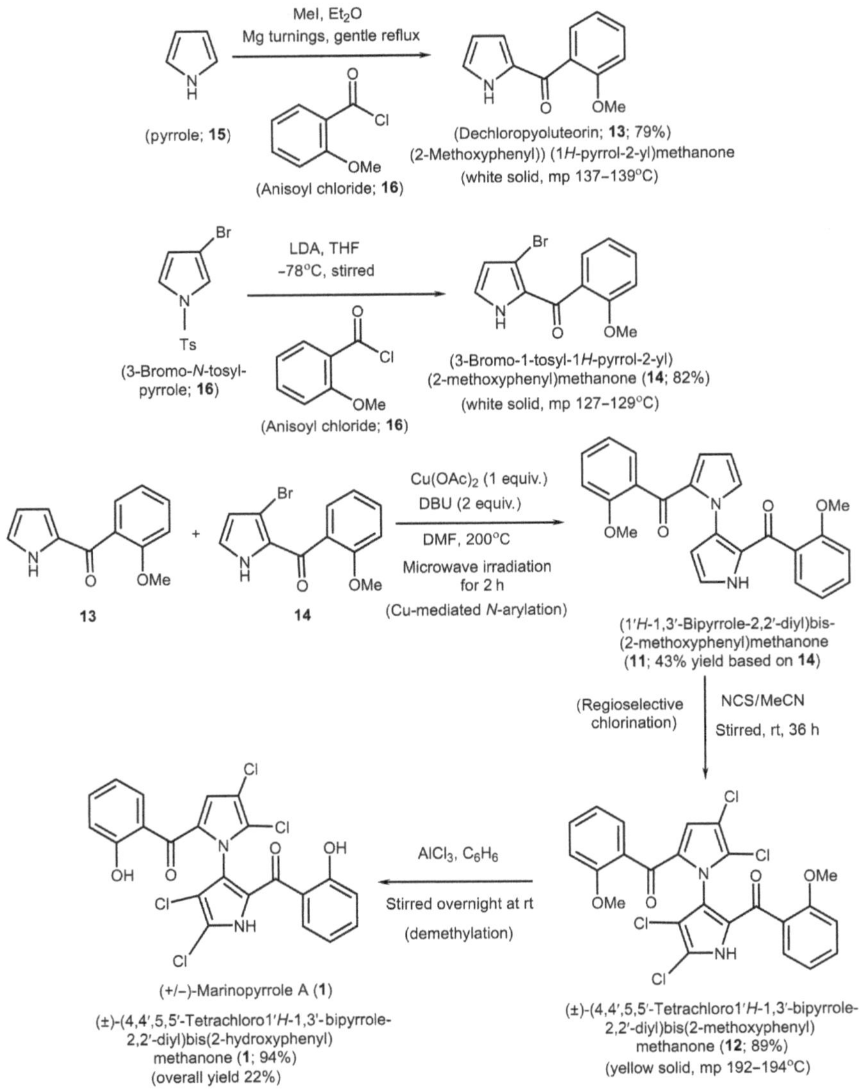

Scheme 2 Kanakis and Sarli's total synthesis of (±)-marinopyrrole A (**1**) [11].

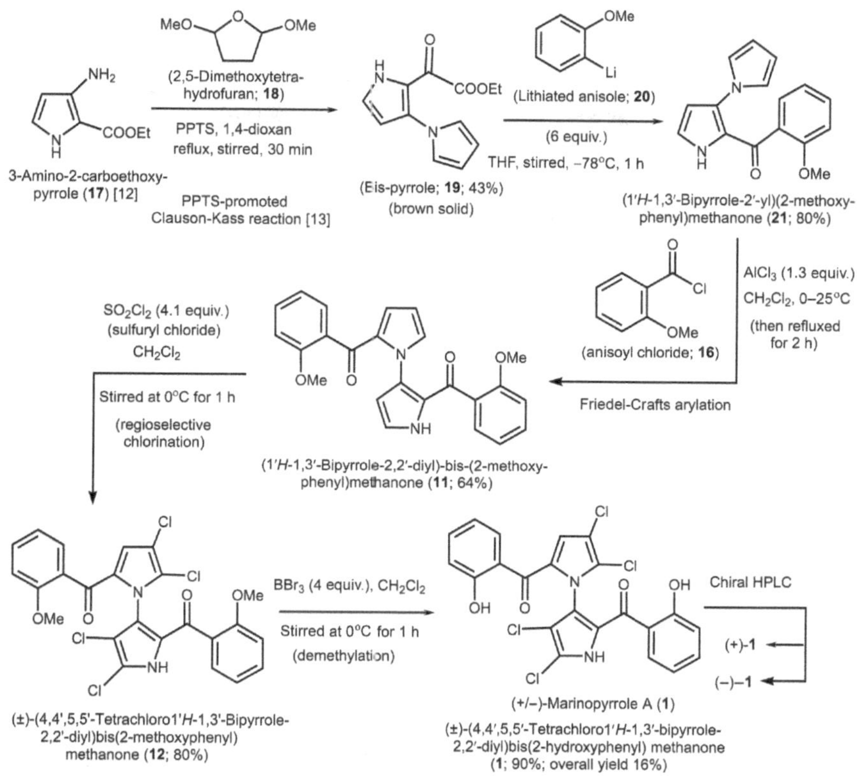

Scheme 3 Nicolaou's total synthesis of (±)-marinopyrrole A (**1**) [4].

(MRSA) TCH1516 strain [3]. The schematic representations for the synthetic routes of (±)-marinopyrrole A (**1**) developed by Kanakis and Sarli, and Nicolaou and his group, are described herein, respectively, in Schemes 2 and 3.

References

[1] C.C. Hughes, A. Prieto-Davo, P.R. Jensen, W. Fenical, Org. Lett. 10 (2008) 629.
[2] C.C. Hughes, W. Fenical, J. Org. Chem. 75 (2010) 3240.
[3] N.M. Haste, C.C. Hughes, D.N. Tran, W. Fenical, P.R. Jensen, V. Nizet, M.E. Hensler, Antimicrob. Agents Chemother. 55 (2011) 3305.
[4] K.C. Nicolaou, N.L. Simmons, J.C. Chen, N.M. Haste, V. Nizet, Tetrahedron Lett. 52 (2011) 2041.
[5] C.C. Hughes, Y.-L. yang, W.-T. Liu, P.C. Dorrestein, J.J. La Clair, W. Fenical, J. Am. Chem. Soc. 131 (2009) 12094.
[6] K. Doi, R. Li, S.-S. Sung, H. Wu, Y. Liu, W. Manieri, G. Krishnegowda, A. Awwad, A. Dewey, X. Liu, S. Amin, C. Cheng, Y. Qin, E. Schonbrunn, G. Daughdrill, T.P. Loughran Jr., S. Sebti, H.-G. Wang, J. Biol. Chem. 287 (2012) 10224.
[7] C. Cheng, L. Pan, Y. Chen, H. Song, Y. Qin, R. Li, J. Comb. Chem. 12 (2010) 541.

[8] J.C. Stowell, D.R. Keith, B.T. King, Org. Syn. 7 (1990) 59.
[9] U. Schmidt, B. Riedl, Synthesis 8 (1993) 809.
[10] J.A. Ragan, B.P. Jones, M.J. Castaldi, P.D. Hill, T.W. Makowski, Org. Synth. 10 (2004) 418.
[11] A.A. Kanakis, V. Sarli, Org. Lett. 12 (2010) 4872.
[12] R.H. Furneaux, P.C. Tyler, J. Org. Chem. 64 (1999) 8411.
[13] C. Rochais, V. Lisowski, P. Dallemagne, S. Rault, Bioorg. Med. Chem. 14 (2006) 8162.

CHAPTER THIRTY EIGHT

Martefragin A

Abbreviations

Bn benzyl
Boc_2O di-*tert*-butyl dicarbonate
DDQ 2,3-dichloro-5,6-dicyanobenzoquinone
DMF *N,N*-dimethylformamide
Et_3N triethylamine
$KMnO_4$ potassium permanganate
$KN(TMS)_2$ lithium bis(trimethylsilyl)amide
LAH lithium aluminum hydride
MsCl methanesulfonyl chloride
$NaIO_4$ sodium periodate
rt room temperature
$SOCl_2$ thionyl chloride

Systematic name: 2-((1*S*,3*S*)-1-(Dimethylammonio)-3-methylpentyl)-5-(1*H*-indol-3-yl)oxazole-4-carboxylate

Compound class: 3-Oxazolylindole alkaloid

Structure:

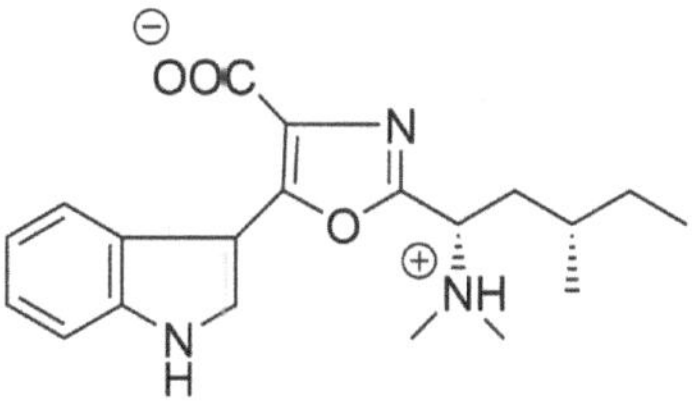

Martefragin A (**1**)

Natural source: *Martensia fragilis* Harvey (sea red alga; family: Pholadidae) [1]

Pharmaceutical potential: Inhibitor of lipid peroxidation (antioxidant) [1–3]

Synthetic route:

The first total synthesis of martefragin A (**1**) was accomplished by Nakagawa and coworkers staring from (*R*)-citronellol (**2**) (Scheme 1) [3].

Total Synthesis of Bioactive Natural Products
https://doi.org/10.1016/B978-0-08-102822-3.00038-9

Compound **2** was converted to (*S*)-4-methylhexanoic acid (**3**) via three-step reaction (methylsulfonylation, reduction with lithium aluminum hydride, and oxidative cleavage of the double bond by the successive addition of sodium periodate and potassium permanganate) in good yield. Acid chloride of **3** was then coupled with lithium salt of (*S*)-4-benzyl-2-oxo-1,3-oxazolidine (**4**) to generate (*S*)-4-benzyl-3-((*S*)-4-methylhexanoyl) oxazolidin-2-one (**5**; 86%), which on asymmetric azidation afforded (*S*)-3-((2*S*,4*S*)-2-azido-4-methylhexanoyl)-4-benzyloxazolidin-2-one (**6**) in a diastereomerically pure form. The azide **6** was converted to the protected (2*S*,4*S*)-homoisoleucine **7** in three steps, and the product **7** was condensed with *O*-benzyl-L-tryptophan hydrochloride to obtain the dipeptide (*S*)-benzyl 2-((2*S*,4*S*)-2-((*tert*-butoxycarbonyl)amino)-4-methylhexanamido)-3-(1*H*-indol-3-yl)propanoate (**8**; 97%). On oxidative cyclization in the presence of DDQ in THF at reflux temperature, the dipeptide **8** gave the oxazole derivative **9** (63%). *N*,*N*-Dimethylation followed by debenzylation of the oxazole derivative **9** ultimately furnished (1″*S*, 3″*S*)-martefragin A (**1**) in 47% yield.

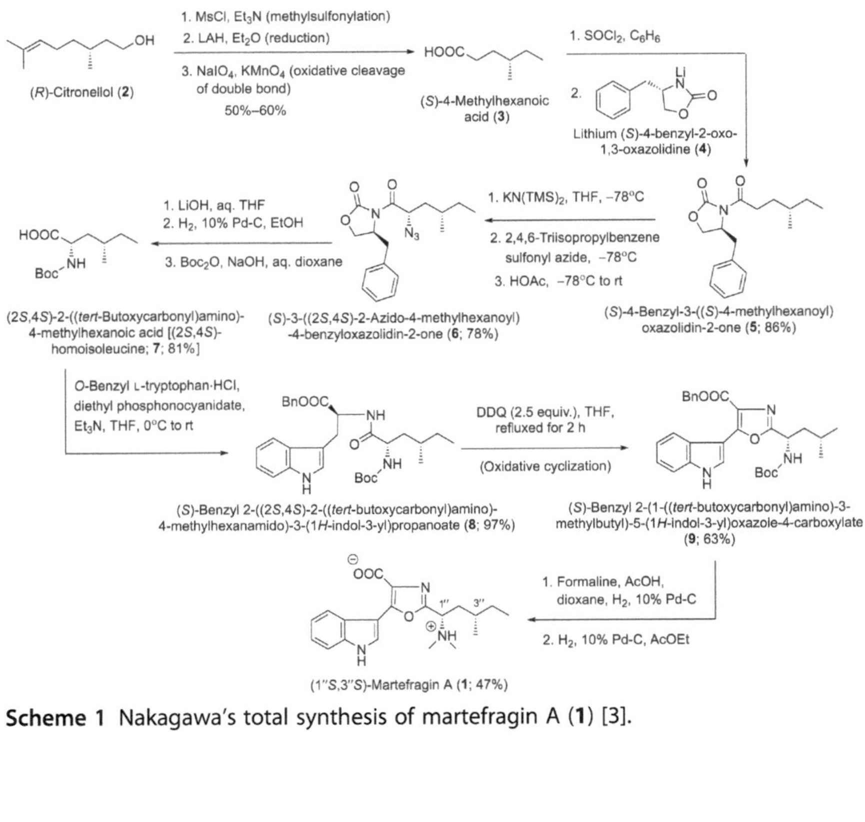

Scheme 1 Nakagawa's total synthesis of martefragin A (**1**) [3].

References

[1] S. Takahashi, T. Matsunaga, C. Hasegawa, H. Saito, D. Fujita, F. Kiuchi, Y. Tsuda, Toyama-ken Yakuji Kenkyusho Nenpo 24 (1997) 53.
[2] S. Takahashi, T. Matsunaga, C. Hasegawa, H. Saito, D. Fujita, F. Kiuchi, Y. Tsuda, Chem. Pharm. Bull. 46 (1998) 1527.
[3] A. Nishida, M. Fuwa, Y. Fujikawa, E. Nakahata, A. Furuno, M. Nakagawa, Tetrahedron Lett. 39 (1998) 5983.

CHAPTER THIRTY NINE

(−)-Melotenine A

Abbreviations

Boc *tert*-butoxycarbonyl
DBU 1,8-diazabicyclo[5.4.0]undec-7-ene
DDQ 2,3-dichloro-5,6-dicyano-1,4-benzoquinone
DEAD diethylazodicarboxylate
DMP Dess-Martin periodinane
Et_3N triethylamine
HG-II second-generation Hoveyda-Grubbs catalyst
HMPA hexamethylphosphoramide
LDA lithium diisopropylamide
$LiAlH_4$ lithium aluminum hydride
MS molecule sieves
rt room temperature
Tf trifluoromethanesulfonyl
THF tetrahydrofuran
TMS trimethylsilyl
Ts 4-toluenesulfonyl

Systematic name: ($4a^1S$,11b*R*)-Methyl 4-methyl-1,$4a^1$,5,7,12,13-hexahydroazepino[3′,2′,1′:7,1]indolo[4,3a-*b*]indole-6-carboxylate

Compound class: Monoterpenoid indole alkaloid

Structure:

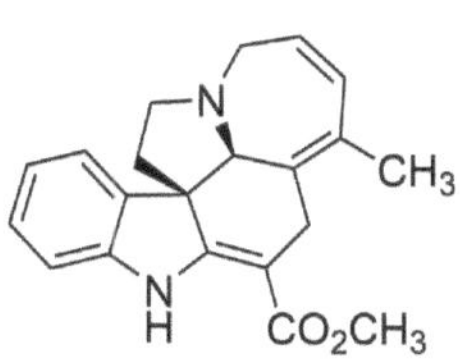

(–)-Meletonine A (**1**)

Natural source: *Melodinus tenuicaudatus* (Apocynaceae; whole plants) [1]

Pharmaceutical potential: Anticancer [1]

https://doi.org/10.1016/B978-0-08-102822-3.00039-0

Synthetic route:

Andrade and his group [2] accomplished the first asymmetric total synthesis of the potent cytotoxic alkaloid (−)-melotenine A (**1**) in 14 steps with an overall yield of 1% from commercially available starting materials (Scheme 1). Condensation of *N*-tosyl indole-3-carboxaldehyde (**2**) with (*R*)-*N*-*tert*-butanesulfinamide (**3**) in the presence of $Ti(OEt)_4$ and In(0) gave *N*-sulfinylimine intermediate that subsequently underwent stereoselective Barbier coupling reaction with allyl bromide to afford homoallylic sulfinamide **4** in 87% yield (dr = 10:1) [3]. It was then transformed into

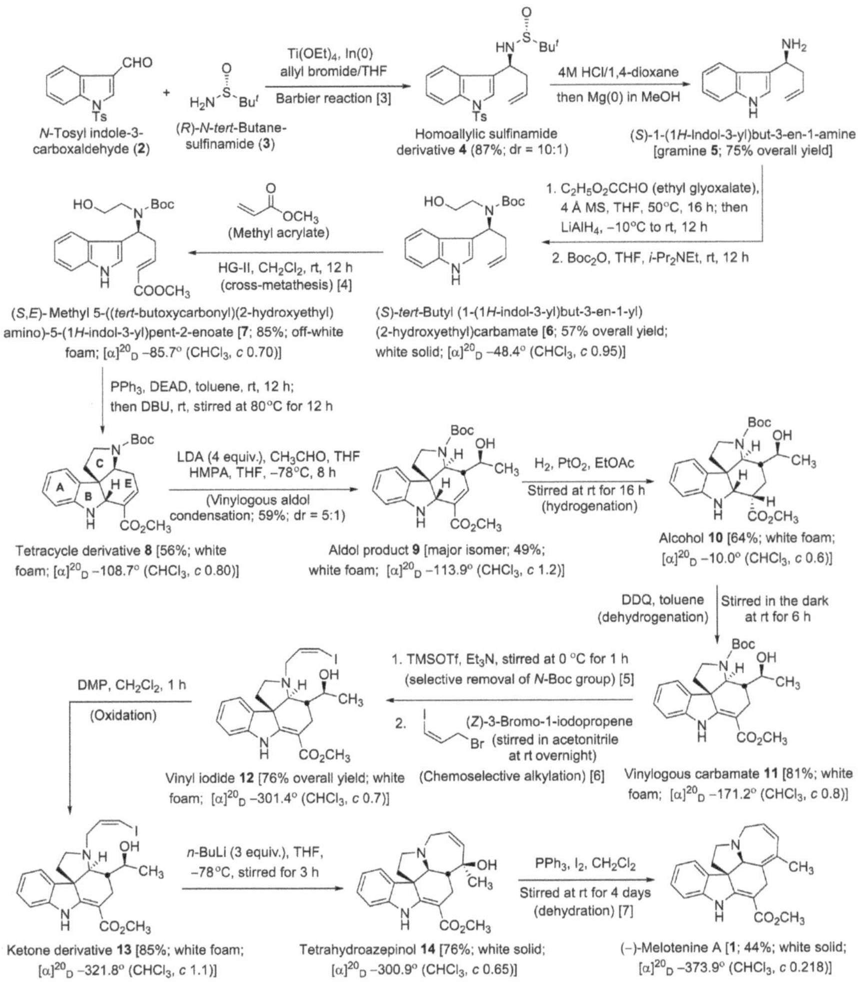

Scheme 1 Total synthesis of (−)-melotenine A (**1**) [2].

the unsubstituted amine (*S*)-1-(1*H*-indol-3-yl)but-3-en-1-amine (gramine **5**) in good overall yield. Hydroxyethylation of gramine **5** was done by stepwise condensation with ethyl glyoxaldehyde and reduction with $LiAlH_4$, followed by protection of the secondary amine with a *tert*-butoxycarbonyl (Boc) to form derivative **6** in 57% overall yield. Cross-metathesis of compound **6** with methyl acrylate in the presence of the second-generation Hoveyda-Grubbs catalyst (HG-II) [4] afforded the enoate **7** in 85% yield.

The investigators then treated the enolate **7** with DEAD and PPh_3 in toluene for a smooth spirocyclization to constitute the pyrrolidine C ring, followed by heating the reaction mixture in the presence of DBU at 80° C for 12 h when the tetracyclic derivative **8** was obtained in 56% yield. It is worth mentioning that this method was found to yield a single stereoisomer and was easily scaled to 10 g. At this stage, a vinylogous aldol reaction was envisioned for installation of the D-ring, and for this purpose the tetracyclic derivative **8** was treated with four equivalents of LDA in THF and HMPA followed by acetaldehyde to obtain the aldol adduct **9** as the major diastereomer in 59% yield (dr = 5:1). Isomerization of the olefinic double bond in **9** to the requisite β-anilinoacrylate position was accomplished by sequential platinum-catalyzed hydrogenation and DDQ-mediated oxidation resulting in the formation of vinylogous carbamate **11**. The Piers annulation substrate **13** was prepared following three steps. First, treatment of **11** with TMSOTf and Et_3N selectively removed the *N*-Boc group [5]. Second, chemoselective alkylation of the C-ring nitrogen atom with (*Z*)-3-bromo-1-iodopropene [6] in the presence of the C-19 alcohol installed the remaining three carbon atoms. Third, the C-19 alcohol **12** was oxidized with the Dess-Martin periodinane (DMP) to obtain **13**. After then the D-ring was introduced smoothly within tetrahydroazepinol **14** from the reaction of **13** with three equivalents of *n*-BuLi at −78°C for 3 h in 76% yield. Regioselective dehydration of the tertiary, allylic C-19 carbinol **14** using Appel protocol [7] ultimately furnished (−)-melotenine A (**1**) in 44% yield; spectroscopic data of the synthetic (−)-melotenine A were found to be completely in agreement with those reported for the natural compound [1].

References

[1] T. Feng, Y. Li, Y.-P. Liu, X.-H. Cai, Y.-Y. Wang, X.-D. Luo, Org. Lett. 12 (2010) 968.
[2] S. Zhao, G. Sirasani, S. Vaddypally, M.J. Zdilla, R.B. Andrade, Angew. Chem. Int. Ed. 52 (2013) 8309.
[3] J.C. Gonzalez-Gomez, M. Medjahdi, F. Foubelo, M. Yus, J. Org. Chem. 75 (2010) 6308.

[4] S.B. Garber, J.S. Kingsbury, B.L. Gray, A.H. Hoveyda, J. Am. Chem. Soc. 122 (2000) 8168.
[5] M.E. Kuehne, F. Xu, J. Org. Chem. 63 (1998) 9434.
[6] E. Piers, J. Renaud, S.J. Rettig, Synthesis (1998) 590.
[7] E.J. Alvarez-Manzaneda, R. Chahboun, E.C. Torres, E. Alvarez, R. Alvarez-Manzaneda, A. Haidour, J. Ramos, Tetrahedron Lett. 45 (2004) 4453.

CHAPTER FORTY

Microcin SF608

Abbreviations

Bn benzyl
Boc *tert*-butoxycarbonyl
CH_2Cl_2 dichloromethane
Et_3N triethylamine
LS-Selectride lithium trisiamylborohydride solution
NMM *N*-methylmorpholine
TFA trifluoroacetic acid
THF tetrahydrofuran

Systematic name: (2*S*,3a*S*,6*R*,7a*S*)-*N*-(4-Guanidinobutyl)-6-hydroxy-1-((*S*)-2-((*S*)-2-hydroxy-3-(4-hydroxyphenyl)propanamido)-3-phenylpropanoyl)octahydro-1*H*-indole-2-carboxamide

Compound class: Peptide

Structure:

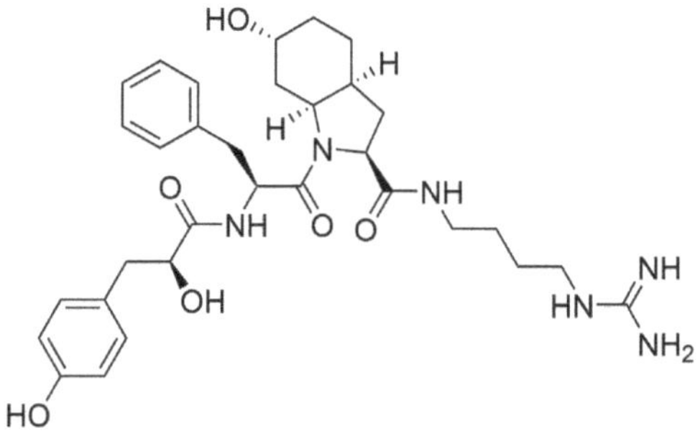

Microcin SF608 (**1**)

Natural source: *Microcystis aeruginosa* (cyanobacterium) [1]

Pharmaceutical potential: Serine protease trypsin inhibitor [1]

Synthetic route:

The first total synthesis of the aquatic peptide microcin SF608 (**1**) was achieved by Bonjoch and coworkers (Scheme 2) [2]. The peptide molecule is composed of four units such as L-phenylalanine (L-Phe; amino acid),

Total Synthesis of Bioactive Natural Products
https://doi.org/10.1016/B978-0-08-102822-3.00040-7

2-carboxy-6-hydroxyoctahydroindole (L-Choi; an unusual bicyclic α-amino acid core), α-hydroxy-*p*-hydroxyphenyllactic acid (L-Hpla), and agmatine (Agma; the decarboxylated arginine). The investigators framed the total synthesis by coupling of L-Hpla with the dipeptide L-Phe-L-Choi followed by coupling with agmatine and a deprotection step. For this purpose, they first synthesized (2*S*,3a*S*,6*R*,7a*S*)-1-*tert*-butyl 2-methyl 6-hydroxyoctahydro-1*H*-indole-1,2-dicarboxylate [*N*-Boc-L-Choi(OMe) (**8**)] starting from *N*-acetyl-L-tyrosine according to their earlier reported protocol (Scheme 1) [3].

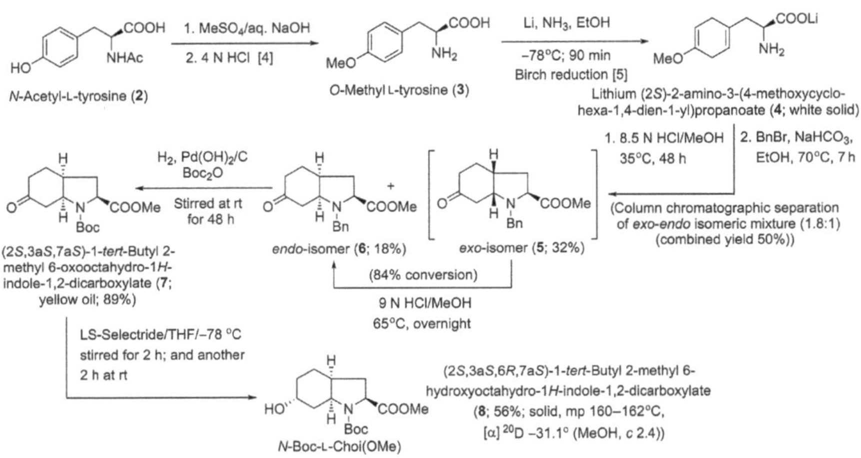

Scheme 1 Synthesis of *N*-Boc-L-Choi(OMe) (**8**) [3].

Then the investigators proceeded to assemble the structure of microcin SF608 (**1**) with these four starting units, namely, L-Hpla, L-Phe, L-Choi, and Agma, as their appropriate protected forms. The total synthesis is outlined in Scheme 2. Removal of the Boc group from L-Choi **8** with TFA followed by coupling with (*S*)-2-amino-3-phenylpropanoic acid (Boc-Phe-OH), using PyBOP in the presence of NMM, furnished the dipeptide Boc-L-Phe-L-Choi-OMe (**9**) **8** as a separable mixture of alcohol **9a** and its trifluoroacetate **9b** in 90% combined yield. Boc group removal from **9** followed by coupling with (*S*)-2-acetoxy-3-(4-(benzyloxy)phenyl)propanoic acid [AcO-L-Hpla (Bn)-OH] (**11**) under the same reaction conditions gave tripeptide AcO-L-Hpla(Bn)-L-Leu-L-Choi-OMe (**12**). Treatment with LiOH effected smooth saponification of the methyl ester in **12** along with concomitant deprotection of the secondary alcohol hydroxyl groups yielding carboxylic acid in 86% yield, which in turn underwent coupling reaction with di-Boc-agmatine (**13**) [6] giving rise to tetrapeptide derivative HO-L-Hpla (Bn)-L-Phe-L-Choi-diBoc-Agma (**14**) in good yield (77%). Cleavage of the orthogonal protecting groups in **14** in two high-yielding steps afforded microcin SF608 (**1**) that showed identical physical and spectral properties to those of the natural product.

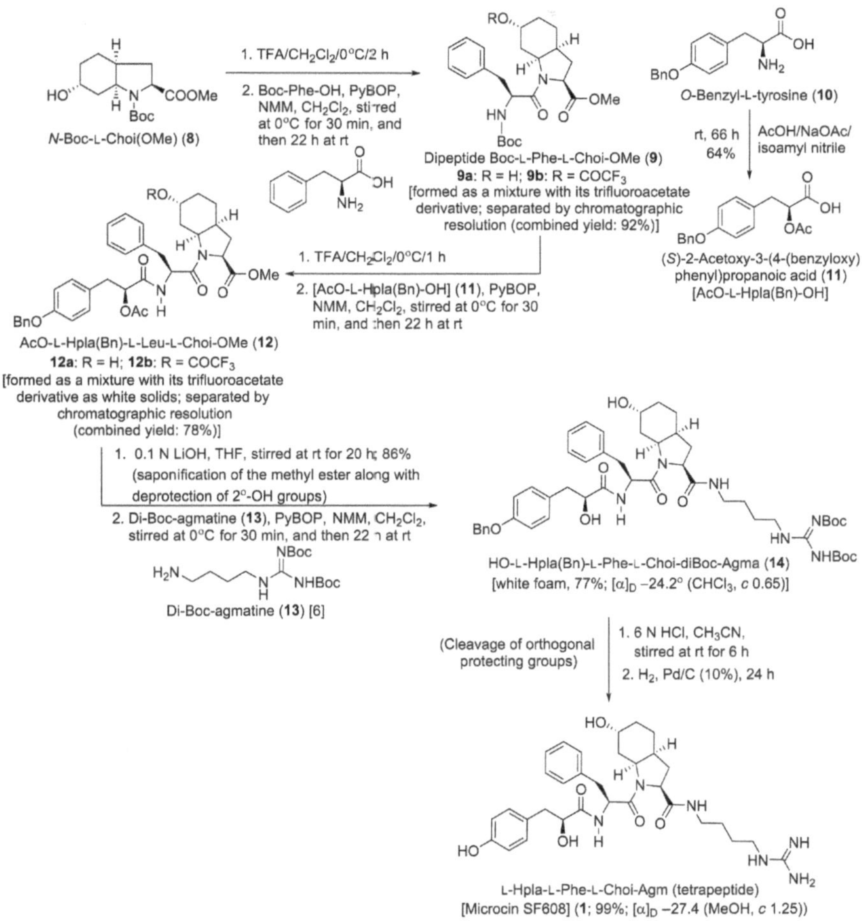

Scheme 2 Total synthesis of microcin SF608 (**1**) [2].

Later on, Carreira and his group [7] also reported another route for the total synthesis of this natural peptide **1** utilizing TMSOTf-mediated nucleophilic opening of an oxabicyclo[2.2.1]heptane building block to rapidly assemble the Choi core common to the aeruginosin members of this family of serine protease inhibitors.

References

[1] (a) R. Banker, S. Carmeli, Tetrahedron 55 (1999) 10835; (b) V. Reshef, S. Carmeli, Tetrahedron 57 (2001) 2885.
[2] N. Valls, M. Vallribera, M. López-Canet, J. Bonjoch, J. Org. Chem. 67 (2002) 4945.
[3] N. Valls, M. López-Canet, M. Vallribera, J. Bonjoch, Chem. Eur. J. 7 (2001) 3446.
[4] (a) B. Wünsch, M. Zott, Liebigs Ann. Chem. (1992) 39; (b) W. Siedel, K. Sturm, R. Geiger, Chem. Ber. 96 (1963) 1636.
[5] B.C. Laguzza, B. Ganem, Tetrahedron Lett. 22 (1981) 1483.
[6] (a) A. Expósito, M. Fernández-Suárez, T. Iglesias, L. Muñoz, R. Riguera, J. Org. Chem. 66 (2001) 4206; (b) E.J. Iwanovicz, M.A. Poss, J. Lin, Synth. Commun. 23 (1993) 1443.
[7] S. Diethelm, C.S. Schindler, E.M. Carreira, Org. Lett. 12 (2010) 3950.

CHAPTER FORTY ONE

Nemorosone

Abbreviations

Ac_2O acetic anhydride
AIBN azobisisobutyronitrile
CH_2Cl_2 dichloromethane
$(DHQD)_2PHAL$ hydroquinidine 1,4-phthalazinediyl diether
DIBAL-H diisobutylaluminum hydride
DMAP 4-methylaminopyridine
DMF *N*,*N*-dimethylformamide
DMP 2,2-dimethoxypropane
DMSO dimethylsulfoxide
Et_3N triethylamine
HMPA hexamethylphosphoramide
KHMDS potassium hexamethyldisilazide
LDA lithium diisopropylamide
LTMP/LiTMP lithium 2,2,6,6-tetramethylpiperidide
MS molecular sieves
rt room temperature
TBABr tetrabutylammonium bromide
TBAF tetrabutylammonium fluoride
TBHP *tert*-butyl hydroperoxide
TBS *tert*-butyldimethylsilyl
TESCl triethylsilyl chloride
THF tetrahydrofuran
TIPSOTf triisopropylsilyl trifluoromethanesulfonate
TMS trimethylsilyl

Systematic name: (1*R*,5*S*,6*R*,7*S*)-5-Benzoyl-4-hydroxy-6-methyl-1,3,7-tris(3-methylbut-2-en-1-yl)-6-(4-methylpent-3-en-1-yl)bicyclo [3.3.1]non-3-ene-2,9-dione

Compound class: Polycyclic polyprenylated acylphloroglucinol

Total Synthesis of Bioactive Natural Products
https://doi.org/10.1016/B978-0-08-102822-3.C0041-9

Structure:

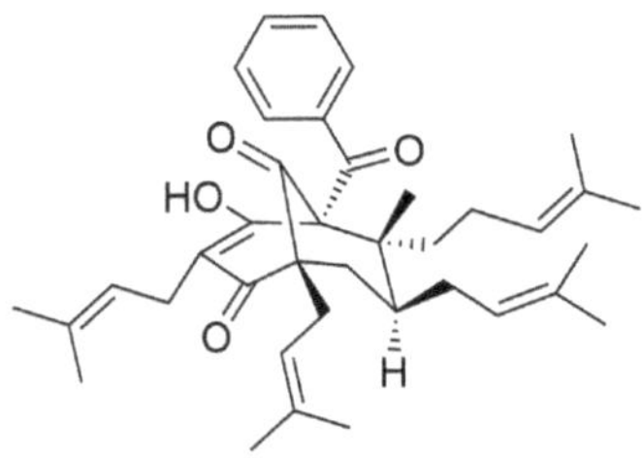

Nemorosone (**1**)

Natural source: *Clusia rosea* (flowers; family: Clusiaceae) [1–3]

Pharmaceutical potentials: Anti-HIV and antitumor activity, possibly associated with telomerase inhibition as well as inhibition of ERK-1/2 [1–5]

Synthetic routes:

Danishefsky and his group [6] accomplished the first total synthesis of nemorosone (**1**) starting from commercially available 3,5-dimethoxyphenol (**2**) by employing allylative de-aromatization and iodinative cyclization as the key skeleton-building steps (Scheme 1). 4,4-Diallyl-3,5-dimethoxy-cyclohexa-2,5-dienone (**3**) obtained from twofold allylation of **2** [7] subsequently underwent twofold cross-metathesis with 2-methyl-2-butene in the presence of Grubbs second generation catalyst [8] to afford 3,5-dimethoxy-4,4-bis(3-methylbut-2-en-1-yl)cyclohexa-2,5-dienone (**4**) as white solid, cleavage of a single methoxy function of which yielded 5-methoxy-6,6-bis(3-methylbut-2-en-1-yl)cyclohex-4-ene-1,3-dione (**5**) [9]. Iodine-induced carbocyclization of **5** gave only 32% of **6** along with an almost 1:1 mixture of two undesired products, both of which could be converted into **5** on treatment with Zn/THF/H_2O. Compound **6** was converted into the novel bridgehead-fused cyclopropane **7** by reductive elimination; the cyclopropane derivative **7** then underwent nucleophilic ring opening through the agency of TMSI to afford (1*R*,5*S*,7*R*)-7-iodo-4-methoxy-8,8-dimethyl-5-(3-methylbut-2-en-1-yl)bicyclo[3.3.1]non-3-ene-2,9-dione (**8**) as colorless oil. The key intermediate **9** was prepared from compound **8** following the highly elegant allylation protocol of Keck and Yates [10].

In the next step, compound **9** was treated with excess LDA in the presence of excess TMSCl followed by oxidative quenching with iodine to

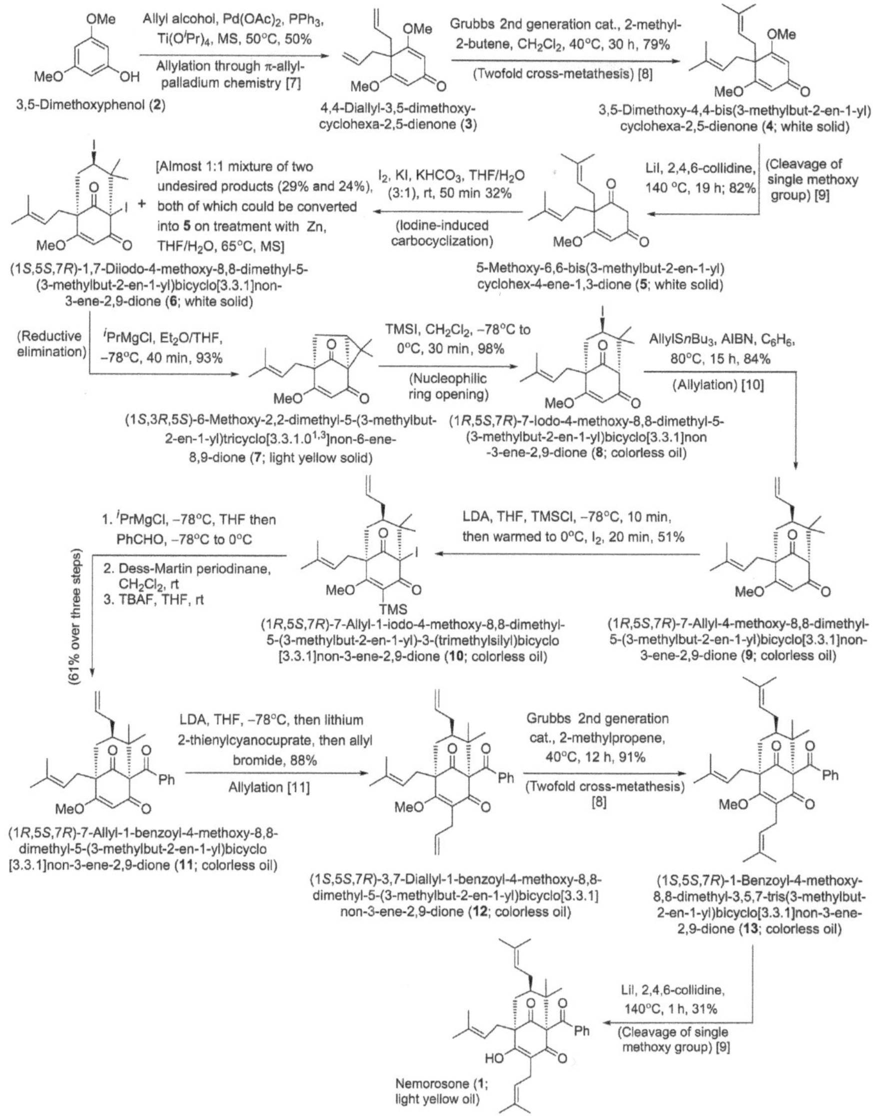

Scheme 1 Total synthesis of nemorosone (**1**) [6].

obtain (1*R*,5*S*,7*R*)-7-allyl-1-iodo-4-methoxy-8,8-dimethyl-5-(3-methylbut-2-en-1-yl)-3-(trimethylsilyl)bicyclo[3.3.1]non-3-ene-2,9-dione (**10**) directly from **9** with 51% yield. The investigators hereafter introduced the benzoyl group at C-1 in three successive steps. For this purpose they carried out reductive deiodination of **10** in the presence of isopropylmagnesium chloride followed by quenching of the presumed metalated intermediate with benzaldehyde; work-up afforded a benzaldehyde adduct, which upon Dess-Martin oxidation and subsequent cleavage of the C_3-vinylsilane linkage eventually furnished (1*R*,5*S*,7*R*)-7-allyl-1-benzoyl-4-methoxy-8,8-dimethyl-5-(3-methylbut-2-en-1-yl)bicyclo[3.3.1]non-3-ene-2,9-dione (**11**) with 61% yield over the three steps. The C_3-allyl derivative **12** then underwent twofold cross-metathesis [8] in the presence of 2-methylpropene under Grubbs second generation catalysis to generate the prenylated derivative (1*S*,5*S*,7*R*)-1-benzoyl-4-methoxy-8,8-dimethyl-3,5,7-tris(3-methylbut-2-en-1-yl)bicyclo[3.3.1]non-3-ene-2,9-dione (**13**). Demethylation of 4-methoxy function of **13** on treatment with LiI/2,4,6-collidine [9] ultimately afforded the target compound nemorosone (**1**) with 31% yield (Scheme 1).

Later on, Simpkins et al. [12] performed the synthesis of nemorosone (**1**) via a bridgehead substitution process, involving initial iodination and subsequent lithium–iodine exchange followed by acylation. Nakada et al. [13] also reported the highly stereoselective total synthesis of nemorosone (**1**) starting from commercially available methyl 2,6-dimethoxybenzoate; the total synthetic route as adopted by this group is illustrated in Scheme 2.

Methyl 2,6-dimethoxybenzoate (**14**) was subjected to Birch reduction, followed by a one-pot reaction with allylbromide, reduction of the methyl ester with lithium aluminum hydride, and protection of the primary hydroxyl group as a TIPS ether to afford ((1-allyl-2,6-dimethoxycyclohexa-2,5-dien-1-yl)methoxy)triisopropylsilane (**15**) as an oil; selective dihydroxylation of **15** followed by 1,2-diol cleavage afforded aldehyde **16**, which was then converted to 1-[2,6-dimethoxy-1-((triisopropylsilyloxy)methyl) cyclohexa-2,5-dienyl]propan-2-one (**18**; oil) using one-pot methylation—Oppenauer oxidation protocol in good yield [14, 15]. In the following steps, 1-diazo-3-(2,6-dimethoxy-1-(((triisopropylsilyl)oxy)methyl)cyclohexa-2,5-dien-1-yl)propan-2-one (**20**, yellow powder) was successfully prepared from the methyl ketone **18** using Danheiser's protocol [16]. Intramolecular cyclopropanation (IMCP) of **18** was carried out in the presence of a chiral ligand to obtain bicyclic ketone **21**, which in turn was converted into (1*S*,5*R*)-6-methoxy-4,4-dimethyl-5-(((triisopropylsilyl)oxy)methyl)bicyclo[3.3.1]non-6-ene-3,9-dione (**22**). Conversion of **22** to the corresponding enol triflate and subsequent palladium-mediated carbonylation afforded **23** in good yield [17], which then converted smoothly at room temperature into (1*S*,3*S*,5*R*)-methyl 6-methoxy-4,4-dimethyl-9-oxo-5-(((triisopropylsilyl)oxy)methyl)bicyclo

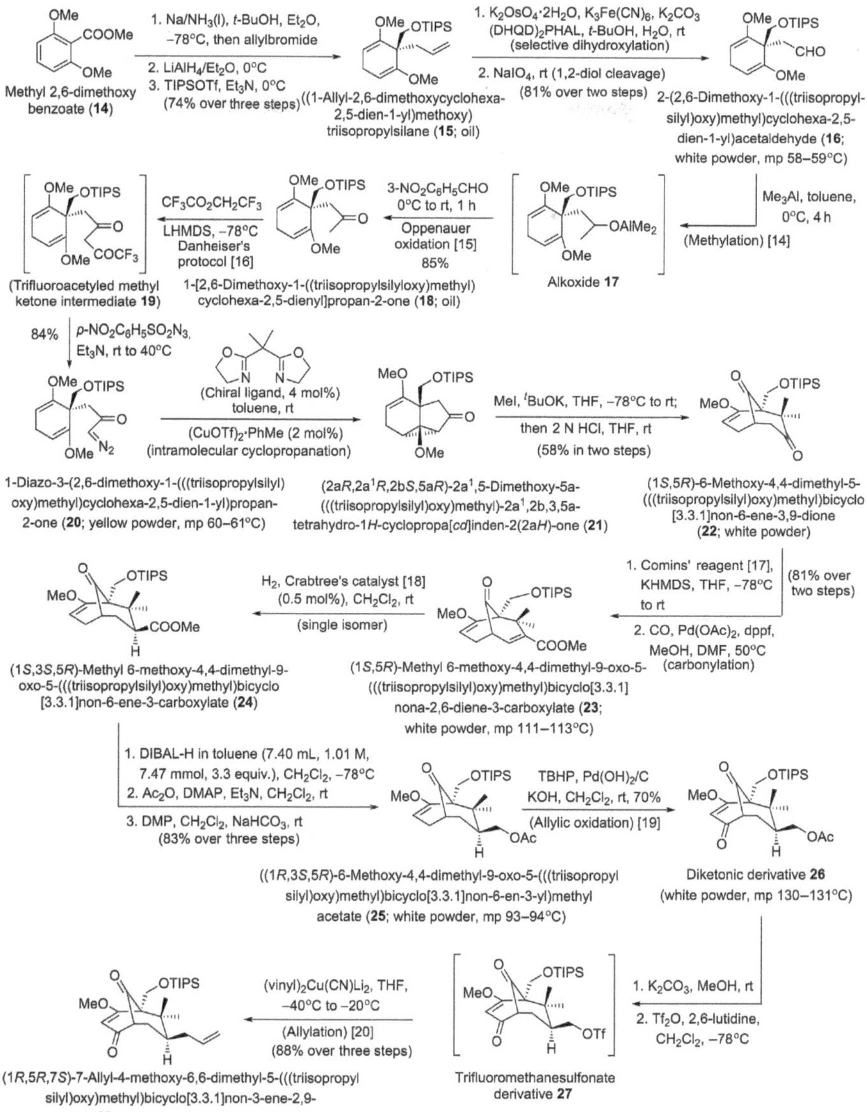

Scheme 2—Cont'd

(Continued)

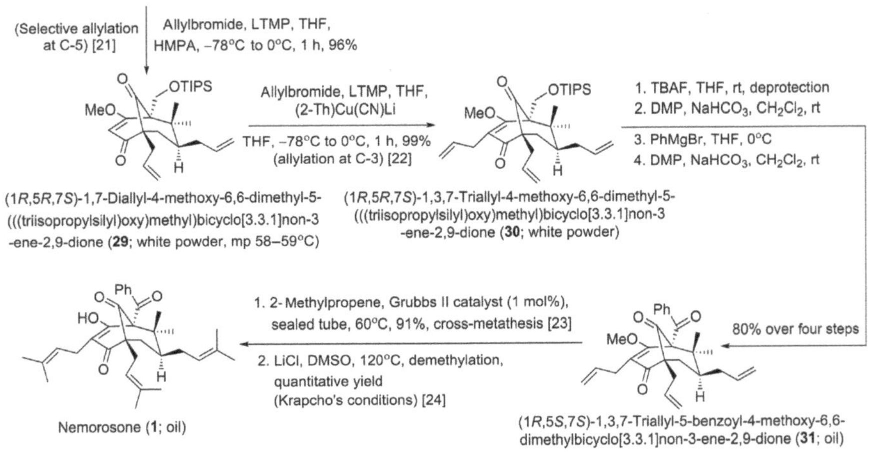

Scheme 2 Total synthesis of nemorosone (**1**) [13].

[3.3.1]non-6-ene-3-carboxylate (**24**) as the single isomer under Crabtree's catalysis [18].

Reduction of **24** with DIBAL-H, subsequent selective acetylation of the primary hydroxyl group, and oxidation of the secondary hydroxyl group afforded ((1*R*,3*S*,5*R*)-6-methoxy-4,4-dimethyl-9-oxo-5-(((triisopropylsilyl) oxy)methyl)bicyclo[3.3.1]non-6-en-3-yl)methylacetate (**25**) with 83% yield over the three steps. On allylic oxidation at the C-4 position, compound **25** successfully produced diketonic derivative **26** [19], the acetate moiety of which was then removed, and the resultant alcohol was converted to triflate **27** so as to introduce an allyl group at the C-7 position via a coupling reaction with divinyl cuprate having **28** smoothly without involving 1,4-addition in the enone system. At this stage, the investigators introduced two more allyl groups at the C-5 and C-3 positions to obtain compound **30** [21, 22]. Thereafter, removal of the TIPS group of 22, Dess-Martin oxidation, reaction with phenylmagnesium bromide, and further Dess-Martin oxidation afforded (1*R*,5*S*,7*S*)-1,3, 7-triallyl-5-benzoyl-4-methoxy-6,6-dimethylbicyclo[3.3.1]non-3-ene-2,9-dione (**31**; oil) with 80% yield over the four steps. Lastly, it was planned to convert all three allyl groups into prenyls by cross-metathesis; the task was performed cleanly with 2-methylpropene under Grubbs II catalysis [23] and the methyl ether was cleaved under Krapcho's conditions [24] to furnish nemorosone (**1**). All the relevant spectroscopic properties of the synthetic compound were found to be matched in all respects with those reported for naturally occurring nemorosone [1–5].

References

[1] O. Cuesta-Rubio, H. Velez-Castro, B.A. Frontana-Uribe, J. Cardenas, Phytochemistry 57 (2001) 279.
[2] O. Cuesta-Rubio, B.A. Frontana-Uribe, T. Ramirez-Apan, J.Z. Cardenas, Z. Naturforsch. C 57 (2002) 372.
[3] S. Seeber, R.A. Hilger, D. Diaz-Carballo, PCT Int. Appl. (2003) WO-2003043622 A1.
[4] C.M.A. de Oliveira, A.L.M. Porto, V. Bittrichc, A.J. Marsaioli, Phytochemistry 50 (1999) 1073.
[5] J. Lokvama, J.F. Braddocka, P.B. Reichardtb, T.P. Clausenc, Phytochemistry 55 (2000) 29.
[6] C. Tsukano, D.R. Siegel, S.J. Danishefsky, Angew. Chem. Int. Ed. 46 (2007) 8840.
[7] T. Satoh, M. Ikeda, M. Miura, M. Nomura, J. Org. Chem. 62 (1997) 4877.
[8] Chatterjee, A. K., Sanders, D. P., and Grubbs R. H. (2002), Org. Lett., 4, 1939; S.J. Spessard, B.M. Stoltz, Org. Lett. 4 (2002) 1943.
[9] A.P. Krapcho, J.F. Weimaster, J.M. Eldridge, E.G.E. Jahngen Jr., A.J. Lovey, W.P. Stephens, J. Org. Chem. 43 (1978) 138 A.P. Krapcho, Synthesis (1982) 805; A.P. Krapcho, Synthesis (1982) 893.
[10] G.E. Keck, J.B. Yates, J. Am. Chem. Soc. 104 (1982) 5829.
[11] B.H. Lipshutz, M. Koerner, D.A. Parker, Tetrahedron Lett. 28 (1987) 945.
[12] N.S. Simpkins, J.D. Taylor, M.D. Weller, C.J. Hayes, Synlett (2010) 639.
[13] M. Uwamori, A. Satio, M. Nakada, J. Org. Chem. 77 (2012) 5098.
[14] E.C. Ashby, S.A. Noding, J. Org. Chem. 44 (1979) 4792.
[15] C.R. Graves, B.-S. Zeng, S.T. Nguyen, J. Am. Chem. Soc. 128 (2006) 12596.
[16] R.L. Danheiser, R.F. Miller, R.G. Brisbois, S.Z. Park, J. Org. Chem. 55 (1990) 1959.
[17] D.L. Comins, A. Dehghani, Tetrahedron Lett. 33 (1992) 6299.
[18] R.H. Crabtree, P.C. Demou, D. Eden, J.M. Mihelcic, C.A. Parnell, J.M. Quirk, G.E. Morris, J. Am. Chem. Soc. 104 (1982) 6994 R.H. Crabtree, M.W. Davis, J. Org. Chem. 51 (1986) 2655.
[19] R. Tagaki, Y. Inoue, K. Ohkata, J. Org. Chem. 73 (2008) 9320 J.-Q. Yu, E.J. Corey, J. Am. Chem. Soc. 125 (2003) 3232.
[20] B.H. Lipshutz, R.S. Wilhelm, J. Kozlowski, Tetrahedron Lett. 23 (1982) 3755.
[21] V. Rodeschini, N.S. Simpkins, C. Wilson, J. Org. Chem. 72 (2007) 4265 J.C. Hayes, N.S. Simpkins, D.T. Kirk, L. Mitchell, J. Baudoux, A.J. Blake, C. Wilson, J. Am. Chem. Soc. 131 (2009) 8196.
[22] B.H. Lipshutz, M. Koener, D.A. Parker, Tetrahedron Lett. 28 (1987) 945 C. Tsukano, D.R. Siegel, S.J. Danishefsky, Angew. Chem., Int. Ed. 46 (2007) 8840; N.M. Ahmad, V. Rodeschini, N.S. Simpkins, S.E. Ward, A.J. Blake, J. Org. Chem. 72 (2007) 4803.
[23] A. Krasovskiy, R. Kopp, P. Knochel, Angew. Chem. Int. Ed. 45 (2006) 497 J. Qi, J.A. Porco Jr., J. Am. Chem. Soc. 129 (2007) 12682; N.S. Simpkins, J.D. Taylor, M.D. Weller, C.J. Hayes, Synlett 4 (2010) 639.
[24] J.T. Njardarson, Tetrahedron 67 (2011) 7631 J.-A. Richard, R.H. Pouwer, D.Y.-K. Chen, Angew. Chem. Int. Ed. 51 (2012) 4536.

CHAPTER FORTY TWO

Nicotlactone A

Abbreviations

BAIB bis(acetoxy)iodobenzene
CH_2Cl_2 dichloromethane
$(COCl)_2$ oxalyl chloride
DDQ 2,3-dichloro-5,6-dicyano-1,4-benzoquinone
DIAL-H diisobutylaluminum hydride
(+)-DIPT D-(+)-diisopropyl tartarate
2,2-DMP 2,2-dimethoxypropane
DMSO dimethylsulfoxide
Et_3N triethylamine
PMB *p*-methoxybenzyl
PPTS pyridinium *p*-toluenesulfonate
PTSA *p*-toluenesulfonic acid
rt room temperature
TBHP *tert*-butylhydroperoxide
TEMPO tetramethylpiperdinyloxy free radical
THF tetrahydrofuran

Systematic name: (3*R*,4*R*,5*S*)-5-(Benzo[*d*][1,3]dioxol-5-yl)-3-hydroxy-3,4-dimethyldihydrofuran-2(3*H*)-one

Compound class: Lignan derivative

Structure:

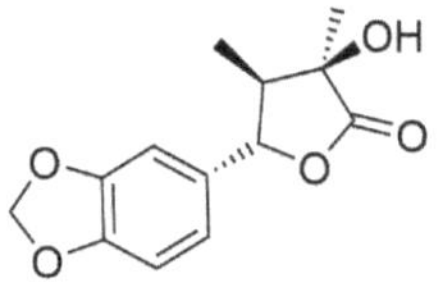

Nicotlactone A (**1**)

Natural source: *Nicotiana tabacum* (leaves; family: Solanaceae) [1]

Pharmaceutical potential: Anti-TMV; anti-HIV [1]

Synthetic route:

Radha Krishna and his coworkers reported the first stereoselective total synthesis of nicotlactone A (**1**) via acid-catalyzed acetonide deprotection

Total Synthesis of Bioactive Natural Products
https://doi.org/10.1016/B978-0-08-102822-3.00042-0

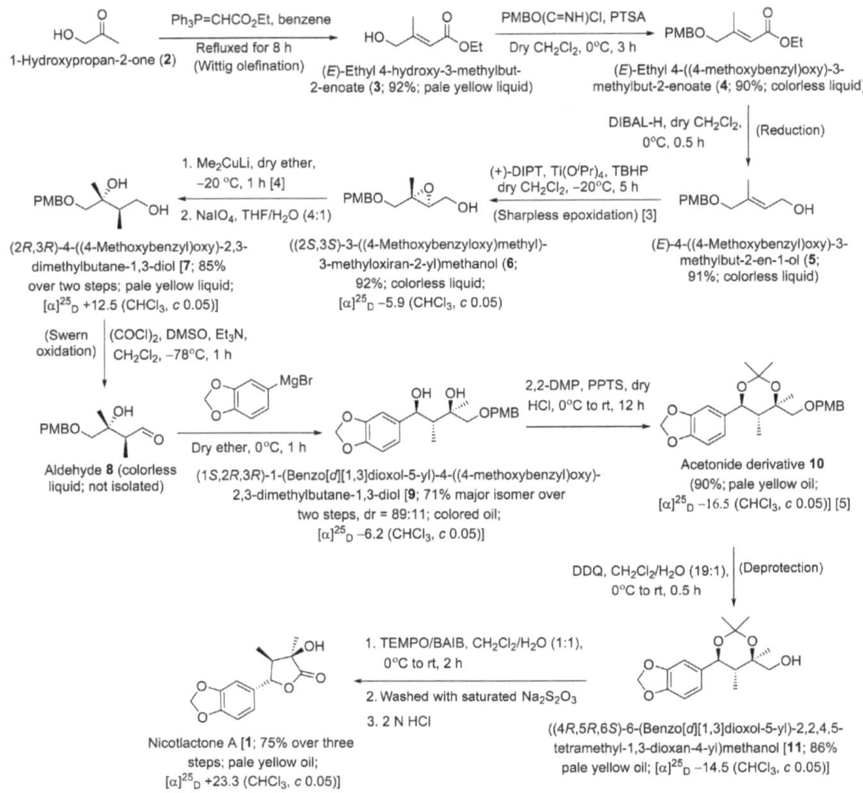

Scheme 1 Total synthesis of nicotlactone A (**1**) [2].

followed by intramolecular lactonization in one pot as the key step with 24% overall yield (Scheme 1) [2]. The investigators initially prepared the key precursor (2*R*,3*R*)-4-((4-methoxybenzyl)oxy)-2,3-dimethylbutane-1,3-diol (**7**) starting from the inexpensive commercially available starting material 1-hydroxypropan-2-one (**2**). Wittig olefination of **2** with (ethoxycarbonylmethylene)triphenyl phosphorane under reflux conditions in benzene gave the desired hydroxy ester **3** in 92% yield as an exclusive *E*-isomer. The alcohol functionality in **3** was then protected as its PMB ether (**4**). Next, the α,β-conjugated ester was reduced with DIBAL-H to the corresponding allylic alcohol **5** as colorless liquid (91%). Subsequently, allylic alcohol **5** was converted into chiral epoxy alcohol **6** under Sharpless conditions [3]. On regioselective ring-opening reaction with Gilman's reagent (Me_2CuLi/ether) [4], the epoxy alcohol **6** eventually furnished the desired precursor (2*R*,3*R*)-4-((4-methoxybenzyl)oxy)-2,3-dimethylbutane-1,3-diol (**7**) in 85% yield as the major isomer.

In the next step, the primary alcohol functionality of precursor **7** was oxidized under Swern oxidation conditions to furnish the corresponding aldehyde **8**, which was immediately subjected to Grignard reaction with 4-bromo-1,2-(methylenedioxy)benzene in THF to afford compound **9** in 71% yield as a major isomer (dr = 89:11). To a stirred solution of diol **9** in dry CH_2Cl_2 at 0°C, PPTS (cat.) was added and the solution stirred for 20 min, and then 2,2-DMP was added and the solution stirred for an additional 12 h. The reaction mixture was quenched with Et_3N and concentrated in vaccuo. The crude residue was purified by column chromatography (silica gel, 60–120 mesh, EtOAc:*n*-hexane = 4:96) to afford acetonide derivative **10** (90%) as a pale yellow oil [5]. Then the PMB group of compound **10** was deprotected under standard DDQ oxidation conditions giving rise to alcohol **11** (86%). Finally, the primary alcohol was oxidized to its carboxylic acid under TEMPO/BAIB conditions; washing the crude product with saturated $Na_2S_2O_3$ followed by work-up with 2 N HCl afforded the target compound nicotlactone A (**1**; 75%) in one pot via sequential reactions such as acetonide deprotection followed by the intramolecular lactonization reactions. The data of the synthetic sample matched the reported values of the natural product [1].

References

[1] X. Gao, X. Li, X. Yang, H. Mu, Y. Chen, G. Yang, G. Hu, Heterocycles 85 (2012) 147.
[2] P. Radha Krishna, S. Prabhakar, C. Sravanthi, Tetrahedron Lett. 54 (2013) 669.
[3] K.B. Sharpless, H.C. Behrens, T. Katsuki, M.W.A. Lee, S.V. Martin, M. Takatani, M.S. Viti, J.F. Walker, S.S. Woodard, Pure Appl. Chem. 55 (1983) 589.
[4] K. Komatsu, K. Tanino, M. Miyashita, Angew. Chem. 116 (2004) 4441.
[5] S.D. Rychnovsky, D.J. Skalitzky, *Tetrahedron Lett.* 31 (1990) 945 S.D. Rychnovsky, G. Yang, J. Org. Chem. 58 (1993) 3511.

CHAPTER FORTY THREE

(−)-6-*O*-Desmethylantofine

Abbreviations

Ac_2O acetic anhydride
CF_3COOH trifluoroacetic acid
$(COCl)_2$ oxalyl chloride
ee enantiomeric excess
Et_3SiH triethylsilane
$LiAlH_4$ lithium aluminum hydride
$NaBH_4$ sodium borohydride
PBr_3 phosphorus tribromide
rt room temperature
THF tetrahydrofuran

Systematic name: (*R*)-2,3-Dimethoxy-9,11,12,13,13a,14-hexahydrodibenzo[*f,h*]pyrrolo[1,2-*b*]isoquinolin-6-ol

Compound class: Alkaloid (phenanthroindolizidine alkaloid)

Structure:

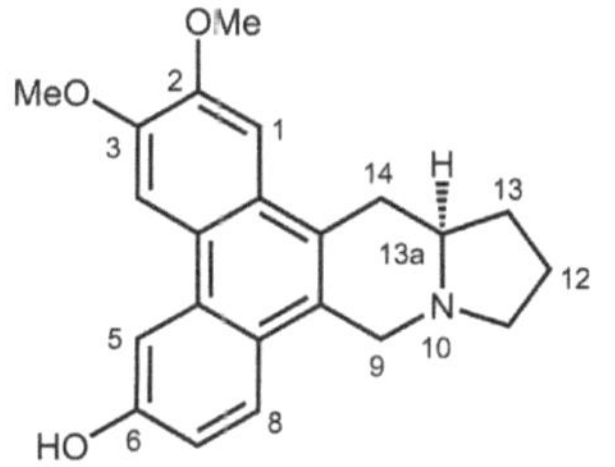

(−)-(*R*)-6-*O*-Desmethylantofine (**1**)

Natural sources: *Cynanchum vincetoxicum* (L.) Pers. (*syn. Vincetoxicum officinale* Moench, *V. hirundinaria* Medic.; leaves; family: Asclepiadaceae) [1–5]; *C. hancockianum* (Maxim.) Al. Iljinski [6]; *C. komarovii* [7]

Pharmaceutical potential: Cytotoxic (anticancer) [5, 8–10]; anti-tobacco mosaic virus [7]

Synthetic route:

Natural abundance of potentially bioactive (−)-6-*O*-desmethylantofine (**1**) is very low (~0.003%) as reported so far [7], and an efficient approach to

https://doi.org/10.1016/B978-0-08-102822-3.00043-2

its synthesis is demanded. Only a few syntheses of racemic (±)-6-*O*-desmethylantofine have been reported [11,12]; however, Wang and coworkers [13] offered the first total synthetic route for (−)-6-*O*-desmethylantofine (**1**) (and also for its unnatural enantiomer) in moderate overall yield of 9.68% (*ee* 90%) (Scheme 1).

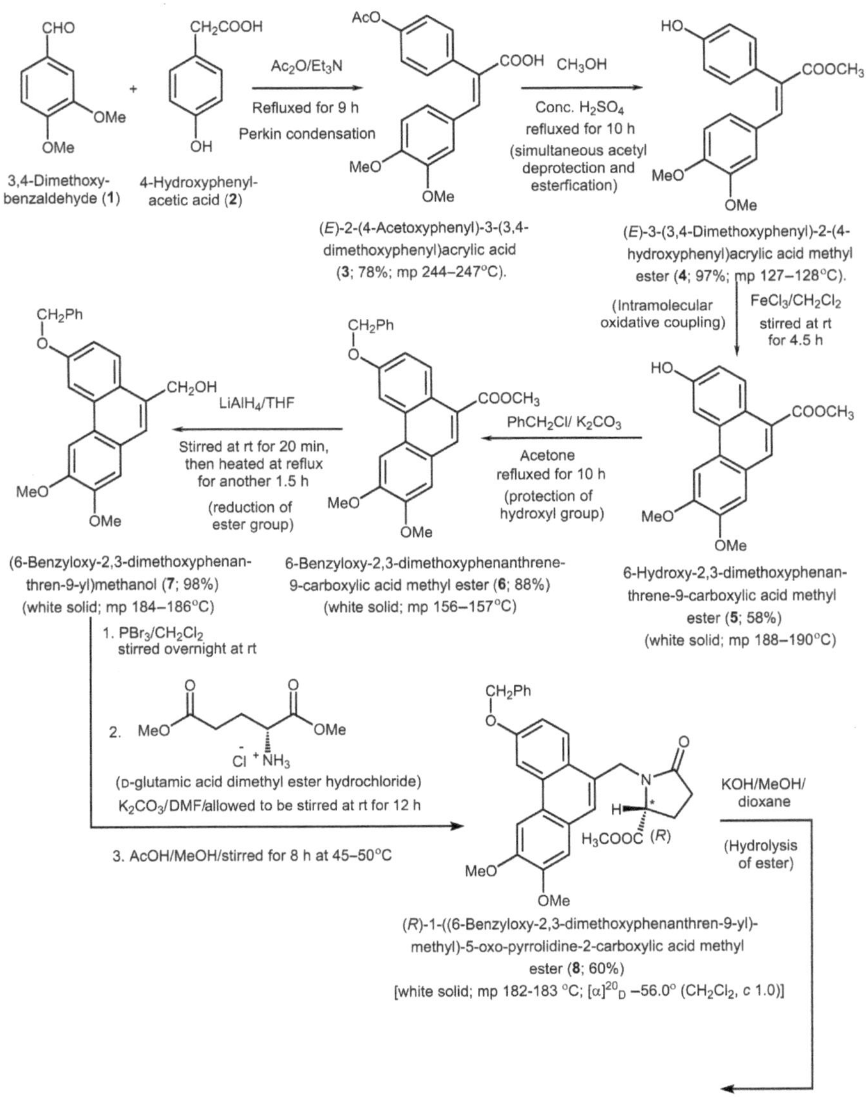

Scheme 1—Cont'd

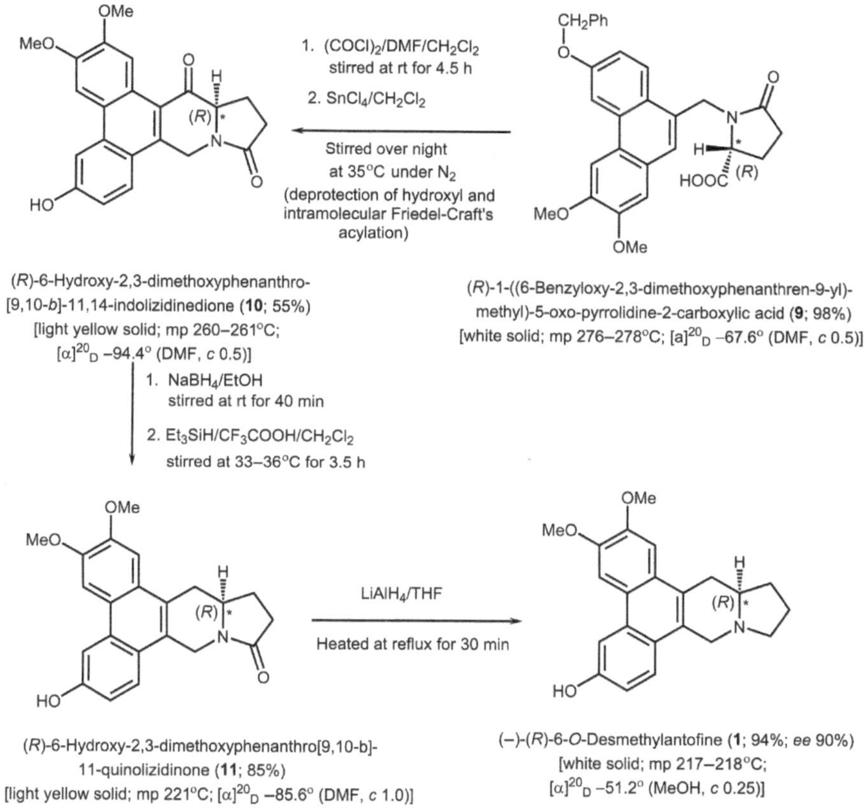

Scheme 1 Wang's total synthesis of (−)-6-*O*-desmethylantofine (**1**) [13].

The investigators started their synthetic process with Perkin condensation of commercially available 3,4-dimethoxybenzaldehyde (**1**) and 4-hydroxyphenylacetic acid (**2**) to prepare (*E*)-2-(4-acetoxyphenyl)-3-(3,4-dimethoxyphenyl)acrylic acid (**3**), which underwent both acetyl deprotection and esterification reaction in one pot on reaction with methanol in the presence of concentrated sulfuric acid to afford the methyl ester **4** in 97% yield. To construct the phenanthrene ring system, the investigators carried out intramolecular oxidative coupling of compound **4** using iron(III) chloride ($FeCl_3$) as an effective reagent to yield 6-hydroxy-2, 3-dimethoxyphenanthrene-9-carboxylic acid methyl ester (**5**; 58%). Ester **6**, obtained on protection of the hydroxyl group in ester **5** with benzyl bromide, was then converted to its corresponding alcohol **7** using lithium aluminum hydride in THF. The alcohol **7** was treated with phosphorus tribromide, and then reacted with D-glutamic acid dimethyl ester hydrochloride followed by cyclization in warm methanol and acetic acid to afford the lactam ester **8** successfully in three steps with 60% yield.

Ester **8** underwent hydrolysis to the corresponding carboxylic acid **9** (98% yield) with potassium hydroxide. It was noted interestingly that use of $SnCl_4$ as Lewis acid effected both deprotection of benzyl group and intramolecular Friedel-Crafts acylation in one pot smoothly to afford indolizidine derivative **10** from compound **9**. The decarbonylation within indolizidine derivative **10** was carried out on treating the compound with sodium borohydride followed by immediate conversion of the resulting alcohol (reduced product) using triethylsilane and trifluoroacetic acid into compound **11**. Finally, **11** was further reduced with lithium aluminum hydride to afford (−)-6-*O*-desmethylantofine (**1**) with an overall yield of 9.68% (*ee* 90%), white solid, mp 217–218°C (lit. 226–228°C) [4]; $[\alpha]_D^{20}$ −51.2° (MeOH, *c* 0.25) (lit. $[\alpha]_D^{20}$ −54.5° (MeOH, *c* 0.88) [9,10].

By the same procedure, the investigators synthesized (+)-(*S*)-6-*O*-desmethylantofine, the unnatural enanatiomer of **1**, using L-glutamic acid dimethyl ester hydrochloride during the reaction with compound **7** instead of D-glutamic acid dimethyl ester hydrochloride in an overall yield of 9.32% (*ee* 91%), white solid, mp 219–220°C, $[\alpha]_D^{20}$ +66.4° (MeOH, *c* 0.25) [13].

References

[1] A. Haznagy, K. Szendrei, L. Toth, Pharmazie 20 (1965) 541.
[2] A. Haznagy, L. Toth, K. Szendrei, Pharmazie 20 (1965) 649.
[3] L. Ferenczy, J. Zsolt, A. Haznagy, L. Toth, A. Szendrei, Acta Microbiol. Acad. Sci. Hung. 66 (1965) 337.
[4] W. Wiegrebe, L. Faber, H. Brockman, K.U. Bidzikiewicz, Justus Liebigs Ann. Chem. 721 (1969) 154.
[5] D. Stærk, A.K. Lykkeberg, J. Christensen, B.A. Budnik, F. Abe, J.W. Jaroszewski, J. Nat. Prod. 65 (2002) 1299.
[6] X. Li, J. Peng, M. Onda, Y. Konda, M. Iguchi, Y. Harigaya, Heterocycles 29 (1989) 1797.
[7] Wang, Q.M., Wang, K.L., Huang, Z.Q., Liu, Y.X., Li, H., Hu, T.S., Jin, Z., Fan, Z.J., and Huang, R.Q. (2008) *CN* 101189968, 2008 [*Chem. Abstr.* 149, 97630]; Z.Q. Huang, Y.X. Liu, Z.J. Fan, Q.M. Wang, G.R. Li, Y.C. Yao, X.S. Yu, R.Q. Huang, Fine Chemical Intermediates (Ch.) 37 (2007) 20.
[8] S.K. Lee, K.-A. Nam, Y.-H. Heo, Planta Med. 69 (2003) 21.
[9] A. Toribio, A. Bonfils, E. Delannay, E. Prost, D. Harakat, E. Henon, B. Richard, M. Litaudon, J.-M. Nuzillard, J.-H. Renault, Org. Lett. 8 (2006) 3825.
[10] Y. Fu, S.K. Lee, H.Y. Min, T. Lee, J.W. Lee, M.S. Cheng, S.H. Kim, Bioorg. Med. Chem. Lett. 17 (2007) 97.
[11] W. Wiegrebe, L. Faber, H. Budzikiewicz, Justus Liebigs Ann. Chem. 733 (1970) 125.
[12] D.S. Bhakuni, P.K. Gupta, Indian J. Chem. 21B (1982) 393.
[13] M. Wu, L. Li, B. Su, Z. Liu, Q. Wang, Org. Biomol. Chem. 9 (2011) 141.

CHAPTER FORTY FOUR

Oleocanthal

Abbreviations

DBU 1,8-diazabicyclo[5.4.0]undec-7-ene
DEAD diethylazodicarboxylate
DMF *N*,*N*-dimethylformamide
HMPA hexamethylphosphoramide
KHMDS potassium bis(trimethylsilyl)amide
LDA lithium diisopropylamide
LHMDS lithium bis(trimethylsilyl)amide
rt room temperature
TBAF tetra-*n*-butylammonium fluoride
TBAI tetra-*n*-butylammonium iodide
TBS tetrabutylsilyl
THF tetrahydrofuran
TMEDA tetramethylethylenediamine

Systematic name: (3*S*,4*E*)-4-Hydroxyphenethyl 4-formyl-3-(2-oxoethyl)hex-4-enoate

Compound class: Phenolic compound (a phenolic moiety is linked by an ester bond to a monoterpenoid secoiridoid unit)

Structure:

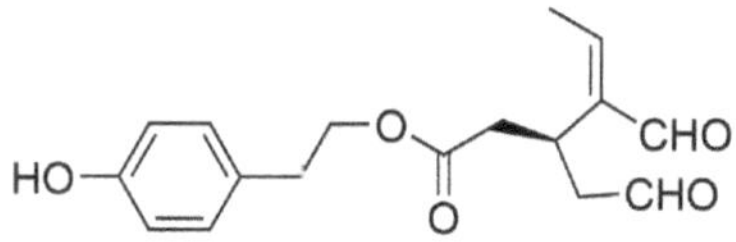

(−)-Oleocanthal (**1**)

Natural source: Extra virgin olive oil [1, 2]

Pharmaceutical potential: Nonsteroidal antiinflammatory drug similar to ibuprofen [3, 4]; antioxidant [4]; anti-Alzheimer's [5–8]; anticancer and antiangiogenic [9]

Synthetic routes:

The first total synthesis of the naturally occurring levorotatory (−)-oleocanthal (**1**) was accomplished by Smith III and collaborators starting

https://doi.org/10.1016/B978-0-08-102822-3.00044-4

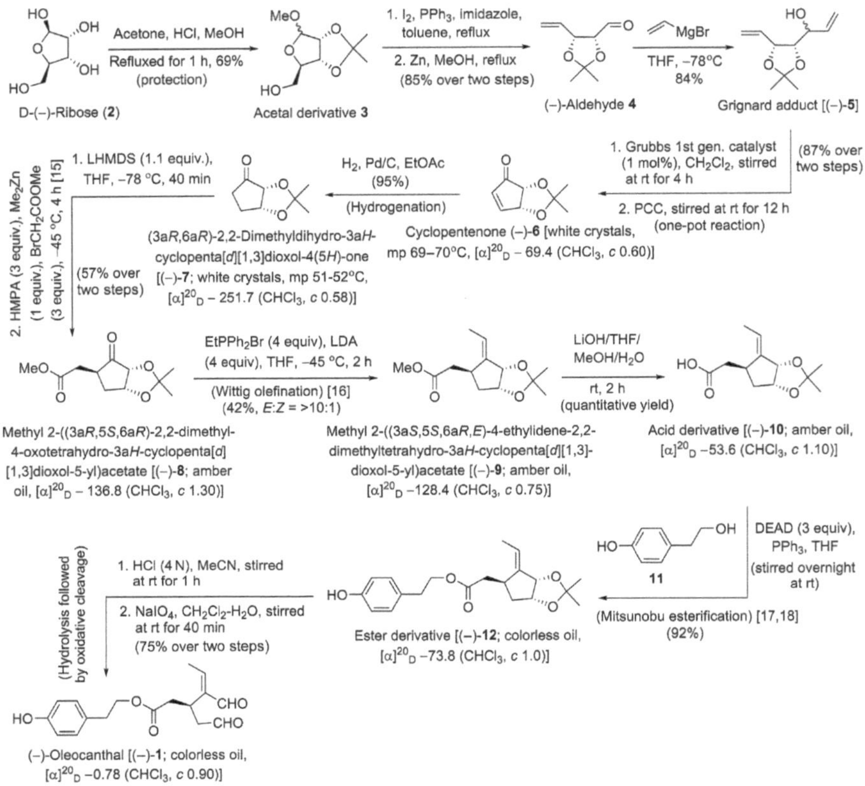

Scheme 1 Total synthesis of (−)-oleocanthal (**1**) [4].

from commercially available D-(−)-ribose in 12 steps with an overall yield of 7% (Scheme 1) [4]. The investigators first prepared the enantiomeric key intermediate (3a*R*,6a*R*)-2,2-dimethyldihydro-3a*H*-cyclopenta[*d*][1,3] dioxol-4(5*H*)-one [(−)-**7**] from D-(−)-ribose (**2**) using a hybrid of known reaction sequences [10–14] as shown in Scheme 1. Alkylation of (−)-cyclopentanone **7** with excess methyl bromoacetate took place in a stereoselective manner to furnish (−)-**8** in 55%–60% yield over the two steps as a single diastereomer [15]. In the next step, Wittig ethylenation of (−)-**8** was achieved with ethyltriphenylphosphonium bromide in the presence of LDA in THF at −45°C with excellent stereoselectivity favoring the *E*-isomer (−)-**9** (c. 10:1 *E*/*Z*) [16]. The caroboxylic acid derivative (−)-**10**, obtained on hydrolysis of the ester (−)-**9** with LiOH/THF/ MeOH/H_2O, was then subjected to Mitsunobu esterification with 4-hydroxyphenethyl alcohol (**11**) to afford ester (−)-**12** in 92% yield with complete chemoselectivity at the primary hydroxyl function [17, 18]. Finally, deprotection of the vicinal diol moiety in (−)-**12** with 4 N HCl/MeCN followed by oxidative cleavage with sodium periodate furnished (−)-oleocanthal (**1**) in 75% yield over the two steps; the synthetic compound was found to have physical and spectral properties almost identical to those of the natural product [1].

Besides the pioneering work of Smith III and collaborators [4], a few other reports appeared later on. In 2009, English and Williams [19] succeeded in synthesizing racemic (±)-oleocanthal (**1**) through a tandem intramolecular Michael cyclization/Horner-Wadsworth-Emmons olefination. In addition, two more formal syntheses were reported in the same year, involving the use of a Johnson-Claisen (orthoester) rearrangement strategy [20] and an SmI_2-promoted intramolecular coupling of bromoalkyne with α,β-unsaturated ester [21], respectively. A concise and efficient total synthesis of (±)-oleocanthal (**1**) has been achieved starting from cyclopentane-lactone **13** in just eight steps with 9% overall yield by Valli and coworkers (Scheme 2) [22]. Lactone **13** [23, 24] was first converted into the corresponding hydroxy carboxylate salt **14** in 95% yield upon its exposure to aqueous KOH in EtOH at 60°C overnight. The carboxylate salt was then treated with a large excess (20 equiv.) of finely powdered MnO_2 in absolute ethanol to obtain potassium (*S*)-2-(2-oxocyclopent-3-en-1-yl)acetate (**15**) through chemoselective oxidation of the allylic hydroxyl group [25]. Subsequently, crude salt **15** was exposed to readily

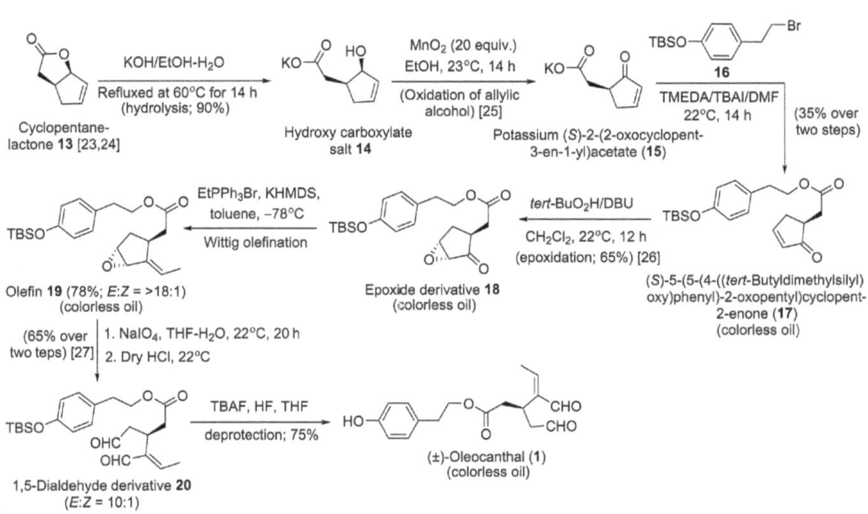

Scheme 2 Total synthesis of (±)-oleocanthal (**1**) [22].

prepared O-TBS-protected tyrosol bromide **16** in dry DMF containing tetramethylethylenediamine (TMEDA; 1.1% v/v) and a catalytic amount of tetrabutylammonium iodide (TBAI); the reaction proceeded smoothly at 22°C giving rise to ester **17** in 35% overall yield.

Cyclopentenone-ester **17** was then subjected to epoxidation with excess ^{t}BuOOH and a catalytic amount of DBU in dichloromethane at 22°C [26] to furnish epoxide **18**, which subsequently underwent Wittig ethylenation in the presence of $EtPh_3PBr$ and KHMDS in toluene at −78°C resulting in the formation of olefin **19** in a stereoselective manner favoring the *E*-isomer (*E*:*Z*> 18:1). Oxidative cleavage of **19** with sodium periodate followed by treatment with anhydrous HCl yielded 1,5-dialdehyde derivative **20** in a stereoselective manner (*E*:Z = 10:1) [27]. Finally, deprotection of phenolic hydroxyl within **20** with tetra-*n*-butylammonium fluoride (TBAF) afforded (±)-oleocanthal **1** in 75% yield, the spectral data of which were found to be completely identical with those of the natural (−)-enantiomer [1].

References

[1] G. Montedoro, M. Servili, M. Baldioli, R. Selvaggini, E. Miniati, A. Macchioni, J. Agric. Food Chem. 41 (1993) 2228.
[2] P. Andrewes, J. Busch, T. de Joode, A. Groenewegen, H. Alexandre, J. Agric. Food Chem. 51 (2003) 1415.
[3] G. Beauchamp, R. Keast, D. Morel, J. Liu, J. Pika, Q. Han, C. Lee, A.B. Smith III, P. Breslin, Nature 437 (2005) 45.
[4] A.B. Smith III, Q. Han, P.A.S. Breslin, G. Beauchamp, Org. Lett. 7 (2005) 5075.
[5] J. Pitt, W. Roth, P. Lacor, A.B. Smith III, M. Blankenship, P. Velasco, F. de Felice, P.A.S. Breslin, W.L. Klein, Toxicol. Appl. Pharmacol. 240 (2009) 189.
[6] W. Li, J.B. Sperry, A. Crowe, J.Q. Trojanowski, A.B. Smith III, V.M.Y. Lee, J. Neurochem. 110 (2009) 1339.
[7] M.C. Monti, L. Margarucci, A. Tosco, R. Riccio, A. Casapullo, Food Funct. 2 (2011) 423.
[8] M.C. Monti, L. Margarucci, R. Riccio, A. Casapullo, J. Nat. Prod. 75 (2012) 1584.
[9] A.Y. Elnagar, P.W. Sylvester, K.A. El Sayed, Planta Med. 77 (2011) 1013.
[10] H. Moon, W. Choi, H. Kim, L. Jeong, Tetrahedron Asymmetry 13 (2002) 1189.
[11] Y. Jin, P. Liu, J. Wang, R. Baker, J. Huggins, C. Chu, J. Org. Chem. 68 (2003) 9012.
[12] M. Yang, Y. Wei, S. Schneller, J. Org. Chem. 69 (2004) 3993.
[13] A. Palmer, V. Jager, Eur. J. Org. Chem. 66 (2001) 1293.
[14] L. Paquette, S. Bailey, J. Org. Chem. 60 (1995) 7849.
[15] Y. Morita, M. Suzuki, R. Noyori, J. Org. Chem. 54 (1989) 1787.
[16] H. El Fakih, F. Pautet, H. Fillion, J. Luche, Tetrahedron Lett. 33 (1992) 4909.
[17] O. Mitsunobu, Synthesis (1981) 1.
[18] G. Appendino, A. Minassi, N. Daddario, F. Bianchi, G. Tron, Org. Lett. 4 (2002) 3839.
[19] B.J. English, R.M. Williams, Tetrahedron Lett. 50 (2009) 2713.
[20] J.T.B. Kuch, P.D. O'Connor, H. Hügel, M.A. Brimble, ARKIVOC (2009) 58.
[21] K. Takahashi, H. Morita, T. Honda, Tetrahedron Lett. 53 (2012) 3342.
[22] M. Valli, E.G. Peviani, A. Porta, A. D'Alfonso, G. Zanoni, G. Vidari, Eur. J. Org. Chem. (2013) 4332.

[23] H. Kapeller, H. Baumgartner, C. Marschner, R. Pucher, H. Griengl, Monatsh. Chem. 128 (1997) 953.
[24] (a) R.C. Larock, T.R. Hightower, J. Org. Chem. 1993 (58) (1993) 5298; (b) G. Zanoni, A. Porta, A. Meriggi, M. Franzini, G. Vidari, J. Org. Chem. 67 (2002) 6064.
[25] G. Tojo, M. Fernández, Oxidation of Alcohols to Aldehydes and Ketones, Springer, New York, 2006.
[26] C. Oger, Y. Brinkmann, S. Bouazzaoui, T. Durand, J.-M. Galano, Org. Lett. 10 (2008) 5087.
[27] C.M. Binder, D.D. Dixon, E. Almaraz, M.A. Tius, B. Singaram, Tetrahedron Lett. 49 (2008) 2764.

CHAPTER FORTY FIVE

(+)-3-Oxo-α-ionol

Abbreviations

CBS catalyst Corey-Bakshi-Shibata catalyst
CH_2Cl_2 dichloromethane
Et_3N triethylamine
MABR methylaluminum bis(4-bromo-2,6-di-*tert*-butyl-4-methylphenoxide)
NMO *N*-methylmorpholine-*N*-oxide
PMB *p*-methoxybenzyl
TBS *tert*-butylsilyl
THF tetrahydrofuran
TPAP tetrapropylammonium perruthenate

Systematic name: (*R*)-4-((*R*,*E*)-3-Hydroxybut-1-en-1-yl)-3,5,5-trimethylcyclohex-2-enone

Compound class: Bisnorsesquiterpene

Structure:

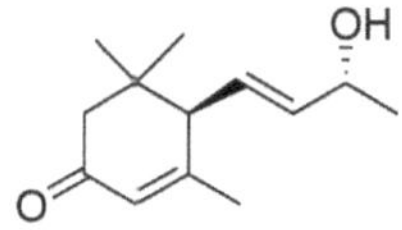

(+)-3-Oxo-α-ionol (**1**)

Natural source: Various tobacco leaves as neutral aroma constituents and Kudzu oil (*Pueraria lobata* Ohwi; family: Fabaceae) [1]; *Vulpia myuros* (family: Poaceae; rattail fescue; aqueous extract) [2]

Pharmaceutical potential: Allelopathic [1–4]

Synthetic route:

Shishido and his group successfully completed the first enantiocontrolled total synthesis of the allelopathic bisnorsesquiterpene (+)-3-oxo-α-ionol (**1**) in 12 steps with an overall yield of 8.5% (Scheme 1) [5]. They started the synthesis with the readily available (*R*)-3-((4-methoxybenzyl)oxy)butanal (**2**) [6]; treatment of **2** with (2-methylprop-1-enyl)magnesium bromide gave the alcohol **3**, as a mixture

https://doi.org/10.1016/B978-0-08-102822-3.00045-6

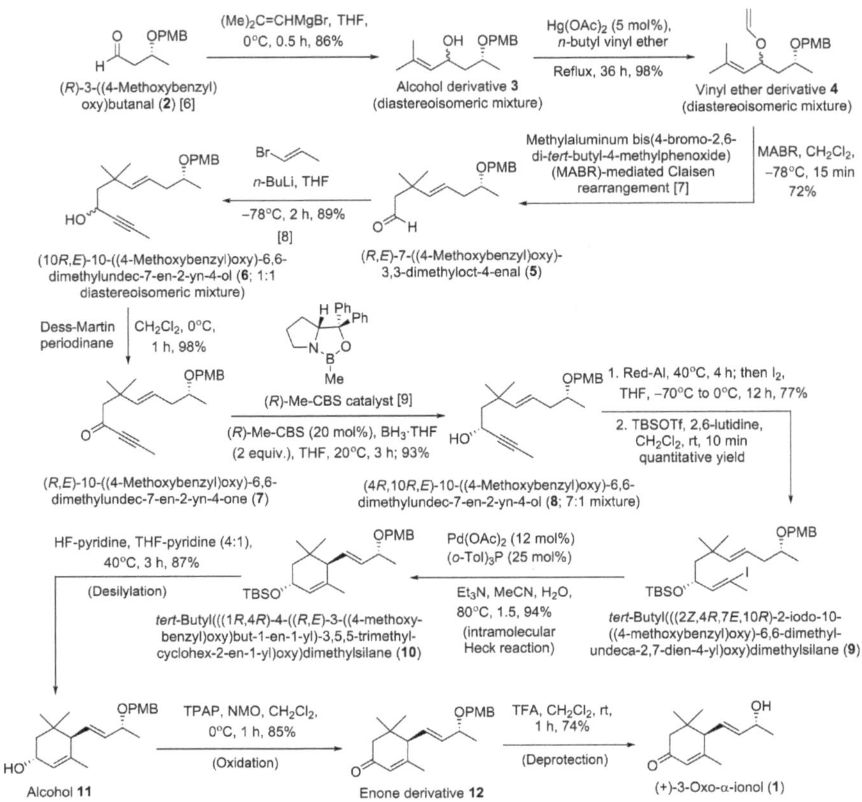

Scheme 1 Total synthesis of (+)-3-oxo-α-ionol (**1**) [5].

of diastereoisomers, which was then converted into the vinyl ether **4** in good yield. Compound **4** underwent methylaluminum bis(4-bromo-2,6-di-*tert*-butyl-4-methylphenoxide) (MABR)-mediated Claisen rearrangement [7] to produce (*R*,*E*)-7-((4-methoxybenzyl)oxy)-3,3-dimethyloct-4-enal (**5**) in 72% yield. The aldehyde **5** reacted in the next step with in situ prepared prop-1-ynyllithium [8] to afford compound **6**, as a 1:1 mixture of diastereoisomers, which was then oxidized with Dess-Martin periodinane to give the ynone derivative (*R*,*E*)-10-((4-methoxybenzyl)oxy)-6,6-dimethylundec-7-en-2-yn-4-one (**7**). Enantioselective reduction of the ynone **7** furnished alcohol **8** in 93% yield as a 7:1 mixture of stereoisomers following the Corey-Bakshi-Shibata protocol [9]. This acetylenic alcohol was then sequentially treated with Red-Al and iodine in one pot to provide the *Z*-vinyl iodide, the secondary alcohol moiety of which was protected as the TBS ether to provide **9** in a 77% yield over the two steps. Compound

9 underwent smooth intramolecular Heck reaction in the presence of Pd $(OAc)_2$/(*o*-tol)$_3$P (25 mol%) and triethylamine in aqueous acetonitrile at 80°C to generate the desired cyclized product **10** in a 94% yield as a single product. Desilylation of **10** using HF/pyridine provided the allyl alcohol **11**, which was oxidized with TPAP/NMO to give enone **12**. Finally, the PMB ether was cleaved with trifluoroacetic acid yielding (+)-3-oxo-α-ionol (**1**) ($[\alpha]_D^{29}$ 280° ($CHCl_3$, *c* 0.70); lit. $[\alpha]_D$ 269° ($CHCl_3$, *c* 01.29) [1e]); the spectral data of the synthetic compound **1** were found to be in good agreement with those reported for the natural product.

References

[1] (a) A.J. Aasen, B. Kimland, C.R. Enzell, Acta Chem. Scand. 25 (1971) 1481; (b) B. Kimland, A.J. Aasen, C.R. Enzell, Acta Chem. Scand. 26 (1972) 2177; (c) A.J. Aasen, B. Kimland, C.R. Enzell, Acta Chem. Scand. 27 (1973) 2107; (d) T. Fujimori, R. Kasuga, H. Matsushita, H. Kaneko, M. Noguchi, Agric. Biol. Chem. 40 (1976) 303; (e) D. Behr, I. Wahlberg, T. Nishida, C.R. Enzell, Acta Chem. Scand. 32 (1978) 391; (f) B. D'Abrosca, M. DellaGreca, A. Fiorentino, P. Monaco, P. Oriano, F. Temussi, Phytochemistry 65 (2004) 497; (g) J.H. Park, D.G. Lee, S.W. Yeon, H.S. Kwon, J.H. Ko, D.J. Shin, H.S. Park, Y.S. Kim, M.H. Bang, N.I. Baek, Arch. Pharm. Res. 34 (2011) 533.

[2] (a) T. Fujimori, R. Kasuga, M. Noguchi, H. Kaneko, Agric. Biol. Chem. 38 (1974) 891; (b) S. Shibata, A. Katsuyama, M. Noguchi, Agric. Biol. Chem. 42 (1978) 195.

[3] H. Kato-Noguchi, M. Yamamoto, K. Tamura, T. Teruya, K. Suenaga, Y. Fujii, Plant Growth Regul. 60 (2010) 127.

[4] F.A. Macías, R.M. Varela, A. Torres, R.M. Oliva, J.M.G. Molinillo, Phytochemistry 48 (1998) 631.

[5] D. Kikuchi, M. Yoshida, K. Shishido, Tetrahedron Lett. 53 (2012) 145.

[6] P.Y. Dakas, R. Jogireddy, G. Valot, D. Barluenga, N. Wissinger, Chem. Eur. J. 15 (2009) 11490.

[7] K. Maruoka, H. Banno, K. Nonoshita, H. Yamamoto, Tetrahedron Lett. 30 (1989) 1265.

[8] J. Suffert, D. Toussaint, J. Org. Chem. 60 (1995) 3550.

[9] E.J. Corey, R.K. Bakshi, S. Shibata, C.-P. Chen, V.K. Singh, J. Am. Chem. Soc. 109 (1987) 7925.

CHAPTER FORTY SIX

Paecilomycin B

Abbreviations

Bn benzyl
CAN ceric ammonium nitrate
CH_2Cl_2 dichloromethane
DIAD *N,N'*-diisopropylcarbodiimide
DMF *N,N*-dimethylformamide
DMP Dess-Martin periodinane
MeOH methanol
$MeSO_3H$ methane sulfonic acid
MOMCl methoxymethyl chloride
***p*-TSA** *p*-toluene sulfonic acid
rt room temperature
TBAI tetrabutylammonium iodide
TBDMSCl/TBSCl *tert*-butyldimethylsilyl chloride
TIPPLi 2,4,6-triisopropylphenyllithium
TFA trifluoroacetic acid
THF tetrahydrofuran

Systematic name: (4*S*,6*E*,8*R*,9*S*,10*S*,12*R*,13*S*)-10-Hydroxy-9,13-epoxy-8,12,18-trihydroxy-16-methoxy-4-methyl-3-oxabicyclo[12.4.0] octadecane-1(14),6,15,17-tetraene-2-one

Compound class: β-Resorcyclic acid lactone (polyketide)

Structure:

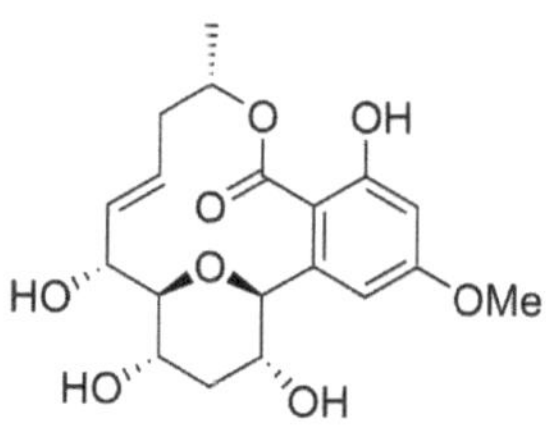

Paecilomycin B (**1**)

Natural source: *Paecilomyces* sp. SC0924 (fungal strain; family: Trichocomaceae; mycelial solid cultures) [1]

Total Synthesis of Bioactive Natural Products
https://doi.org/10.1016/B978-0-08-102822-3.00046-8

Pharmaceutical potential: Mitogen-activated protein kinases inhibitor [2]

Synthetic route:

The first total synthesis of the fungal polyketide paecilomycin B (**1**) was accomplished by Ohba and Nakata [2] via functionalized aryl-β-*C*-glycoside synthesis using 2,4,6-triisopropylphenyllithium under Barbier-type reaction conditions and ring-closing metathesis (RCM) as the key steps. Their retrosynthetic strategy for the target compound **1** is shown in Scheme 1.

Based on their retrosynthetic strategy, the investigators first prepared aryl bromide **6** and δ-lactone **7** starting from known compounds as shown in Schemes 2 and 3, respectively. Once having the starting compounds at hand, they then explored the total synthesis as described in Scheme 4.

Synthesis of aryl bromide 6:

Aryl bromide **6** was prepared starting from the known compound 1-bromo-3,5-dimethoxybenzene (**8**) [3], which was converted to methyl 2-bromo-6-hydroxy-4-methoxybenzoate (**11**) in the next four steps

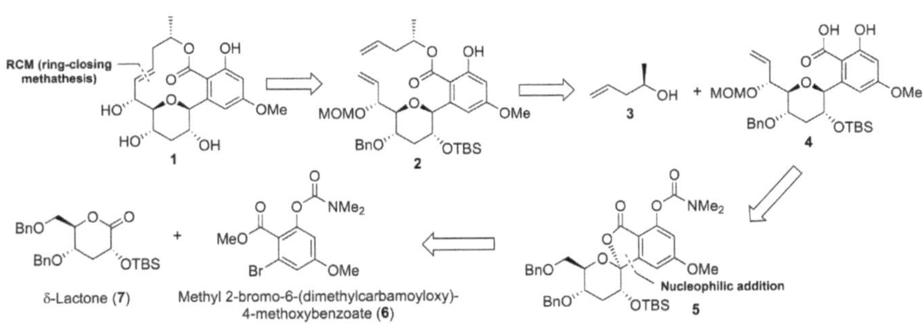

Scheme 1 Retrosynthetic strategy for paecilomycin B (**1**) [2].

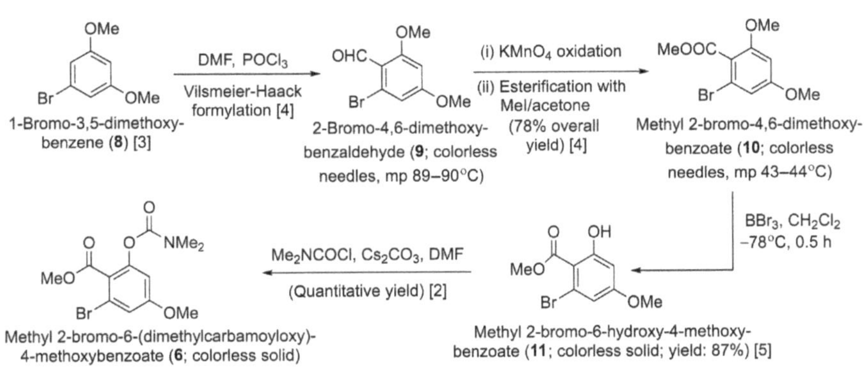

Scheme 2 Synthesis of aryl bromide **6**.

[4, 5]. Aryl bromide **6** was then obtained as colorless solid in quantitative yield upon carbamoylation of the hydroxy group of ester **11** (Scheme 2). The *N,N*-dimethylcarbamoyl group is relatively inert in character under acidic conditions and shows moderate resistant ability for nucleophilic attack [2].

Synthesis of δ-lactone 7:

δ-Lactone **7** was synthesized starting from *p*-methoxyphenyl 4,6-*O*-benzylidene-2-*O*-*tert*-butyldimethylsilyl-3-deoxy-β-D-ribo-hexopyranoside (**15**) [6], which was prepared from *p*-methoxyphenyl β-D-glucopyranoside **12** in five steps following a previously reported procedure [6] (Scheme 2). The benzylidene moiety of 3-deoxy glucoside **15** was then removed by hydrogenation to have the corresponding diol **16**, both the hydroxy groups, which were benzylated to give **17** in 83% yield over two steps. Removal of the *p*-methoxyphenyl group of **17** was achieved using CAN [7] to afford an anomeric mixture of **18**. The lactone **7** was then obtained upon oxidation of the resulting hydroxy group in **18** with Dess-Martin periodinane (DMP) [8] in good yield.

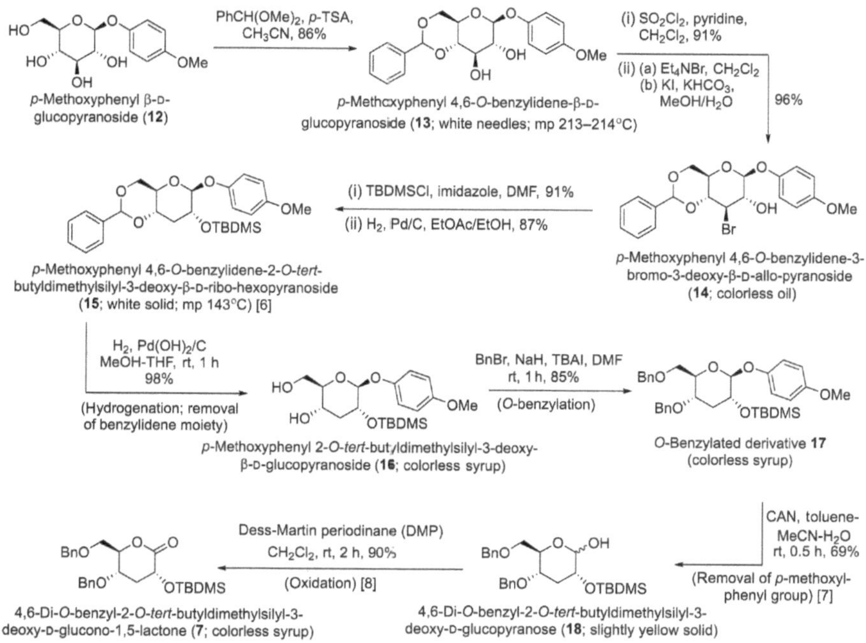

Scheme 3 Synthesis of δ-lactone **7**.

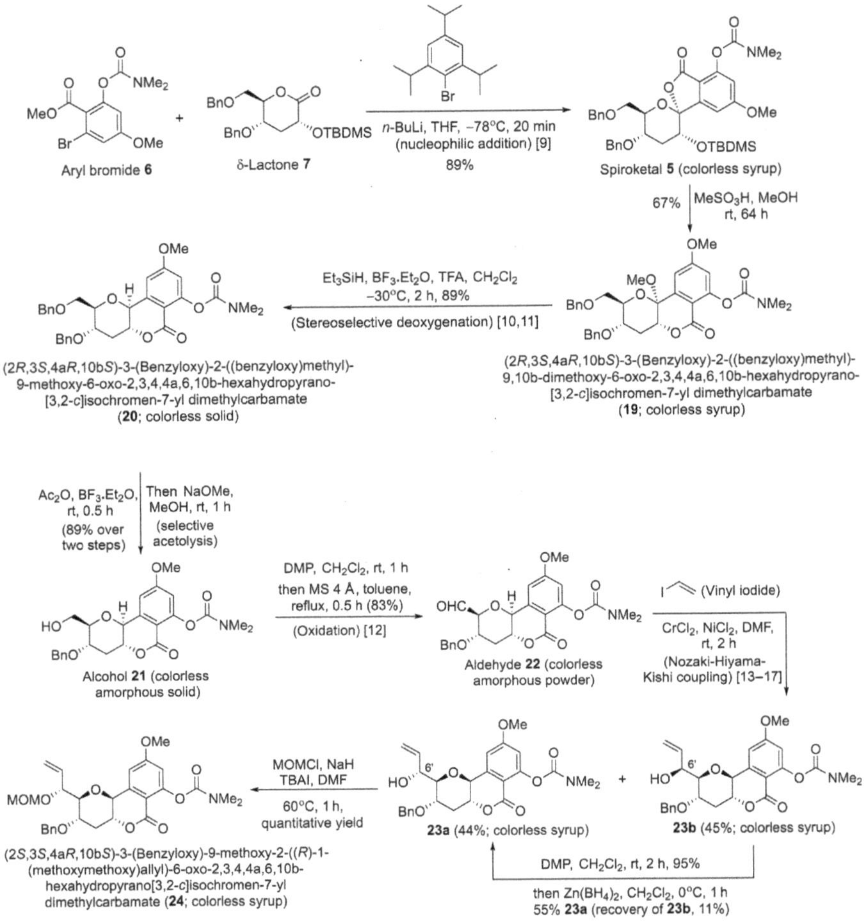

Scheme 4—Cont'd

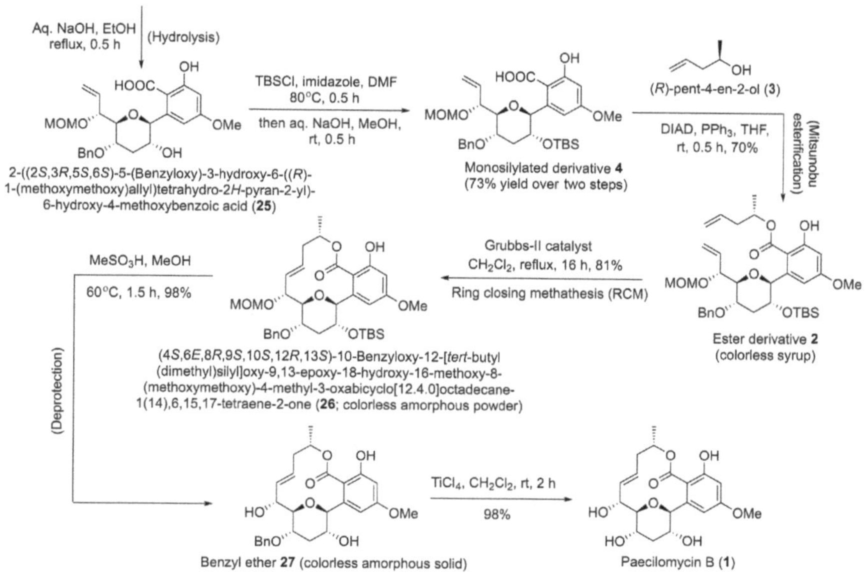

Scheme 4 Synthesis of paecilomycin B (**1**) [2].

Combination of synthons 6 and 7 to build up paecilomycin B (1): With aryl bromide **6** and *δ*-lactone **7** at hand, the investigators then performed the key nucleophilic addition reaction using TIPPLi under Barbier-type reaction conditions [9] to obtain the spiroketal **5** as colorless syrup in an excellent yield of 89%. Treatment of the coupling product **5** with methanesulfonic acid in methanol afforded **19**, deoxygenation of which in the presence of Et_3SiH, $BF_3{\cdot}Et_2O$, and trifluoroacetic acid proceeded smoothly in a stereoselective manner [10, 11] to produce the aryl-β-*C*-glycoside **20** as colorless solid. The selective acetolysis of its primary benzyl ether and subsequent deacetylation afforded the alcohol (2*R*,3*S*,4a*R*,10b*S*)-3-(benzyloxy)-2-(hydroxymethyl)-9-methoxy-6-oxo-2,3,4,4a,6,10b-hexahydropyrano[3,2-*c*]isochromen-7-yl dimethylcarbamate (**21**) in 89% yield over the two steps. Dess-Martin oxidation of the primary hydroxy group of alcohol **21** gave the corresponding aldehyde **22** [12]. Aldehyde **22** upon Nozaki-Hiyama-Kishi coupling [13–17] with vinyl iodide produced a coupling product **23a** in 44% yield along with its C-6′ epimer **23b** in 45% yield; however, the epimer **23b** was converted into **23a** by oxidation and subsequent stereoselective reduction. The allyl alcohol of **23a** was protected as its MOM ether, and the resulting derivative **24** was hydrolyzed with aqueous sodium hydroxide, which was accompanied by the removal of an *N*,*N*-dimethylcarbamoyl group to afford carboxylic acid, 2-((2*S*,3*R*,5*S*,6*S*)-5-(benzyloxy)-3-hydroxy-6-((*R*)-1-(methoxymethoxy)allyl)tetrahydro-2*H*-pyran-2-yl)-6-hydroxy-4-methoxybenzoic acid (**25**), in 96% yield from **23a**. The subsequent treatment of diol **25** with excess TBSCl/imidazole furnished the persilylated compound, which was then hydrolyzed with aqueous sodium hydroxide to produce monosilylated product **4** in 73% yield over two steps. The precursor **2** for RCM was obtained by Mitsunobu esterification between benzoic acid **4** and chiral alcohol **3** in 70% yield. The macrocyclization by RCM as the second key step using the Grubbs-II catalyst successfully yielded (4*S*,6*E*,8*R*,9*S*,10*S*,12*R*,13*S*)-10-benzyloxy-12-[*tert*-butyl(dimethyl)silyl]oxy-9,13-epoxy-18-hydroxy-16-methoxy-8-(methoxymethoxy)-4-methyl-3-oxabicyclo[12.4.0]-octadecane-1(14),6,15,17-tetraene-2-one (**26**) as the sole cyclization product. The deprotection of **26** with methanesulfonic acid gave benzyl ether **27**, and the subsequent exposure of **27** to $TiCl_4$ furnished paecilomycin B (**1**) in excellent yield. The ^{1}H NMR, ^{13}C-NNR, and MS spectral data of this synthetic compound were found to be identical to the reported data for natural paecilomycin B [1]. The overall reactions are shown in Scheme 4.

The same group of investigators [18] explored two more routes for the convergent total synthesis of the natural product paecilomycin B and its 6′-epimer (6′-epi-paecilomycin B) by a Barbier-type reaction using 2,4,6-triisopropylphenyllithium based on a functionalized aryl-β-*C*-glycoside synthetic method for the construction of the 2-aryltetrahydropyran skeleton.

References

[1] L. Xu, Z. He, J. Xue, X. Chen, X. Wei, J. Nat. Prod. 73 (2010) 885. *ibid*, 75 (2012) 1006.
[2] K. Ohba, M. Nakata, Org. Lett. 17 (2015) 2890.
[3] N.B. Dean, W.B. Whalley, J. Chem. Soc. (1954) 4638. R.A. Benkeser, R.A. Hickner, D.I. Hoke, O.H. Thomas, J. Am. Chem. Soc. 80 (1985) 5289.
[4] M.V. Sargent, J. Chem. Soc. Perkin Trans. I (1982) 403.
[5] P. Sellès, R. Lett, Tetrahedron Lett. 43 (2002) 4621.
[6] M. Nitz, D.R. Bundle, J. Org. Chem. 65 (2000) 3064.
[7] A. Kimura, A. Imamura, H. Ando, H. Ishida, M. Kiso, Synlett (2006) 2379.
[8] D.B. Dess, J.C. Martin, J. Org. Chem. 48 (1983) 4155.
[9] K. Ohba, Y. Koga, S. Nomura, M. Nakata, Tetrahedron Lett. 56 (2015) 1007.
[10] P.P. Deshpande, B.A. Ellsworth, F.G. Buono, A. Pullockaran, J. Singh, T.P. Kissick, M.-H. Huang, H. Lobinger, T. Denzel, R.H. Mueller, J. Org. Chem. 72 (2007) 9746.
[11] M. Terauchi, H. Abe, A. Matsuda, S. Shuto, Org. Lett. 6 (2004) 3751.
[12] A. Dondoni, A. Marra, Tetrahedron Lett. 44 (2003) 13.

[13] Y. Okude, S. Hirano, T. Hiyama, H. Nozaki, J. Am. Chem. Soc. 99 (1977) 3179.
[14] H. Jin, J. Uenishi, W.J. Christ, Y. Kishi, J. Am. Chem. Soc. 108 (1986) 5644.
[15] K. Takai, M. Tagashira, T. Kuroda, K. Oshima, K. Utimoto, H. Nozaki, J. Am. Chem. Soc. 108 (1986) 6048.
[16] G.C. Hargaden, P.J. Guiry, Adv. Synth. Catal. 349 (2007) 2407.
[17] X. Liu, X. Li, Y. Chen, Y. Hu, Y. Kishi, J. Am. Chem. Soc. 134 (2012) 6136.
[18] K. Ohba, M. Nakata, J. Org. Chem. 83 (2018) 7019.

CHAPTER FORTY SEVEN

Paecilomycine A

Abbreviations

CH_2Cl_2 dichloromethane
DMF *N,N*-dimethylformamide
imid imidazole
MeOH methanol
$NaBH_4$ sodium borohydride
NMO *N*-methylmorpholine *N*-oxide
TBS *tert*-butyldimethylsilyl
TBSCl *tert*-butyldimethylsilyl chloride
THF tetrahydrofuran
TMP 2,2,6,6-tetramethylpiperidine
TMS trimethylsilyl
TPAP tetra-*n*-propylammonium perruthenate

Systematic name: (3*R*,4a*R*,5*S*,7a*S*,11a*R*)-4a,9-Dimethyl-3,4,4a,5,6,7a,10,11-octahydro-1*H*-3,5-methanopyrano[4,3-*d*]chromene-3,5-diol

Compound class: Trichothecene-type sesquiterpenoid

Structure:

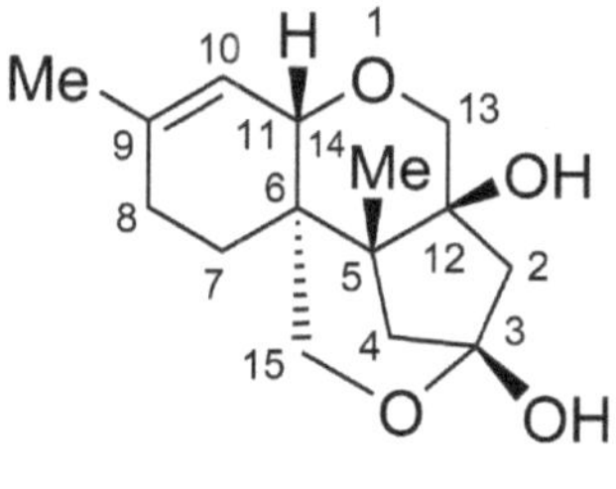

(±)-Paecilomycine A (**1**)

Natural source: *Paecilomyces tenuipes* (*Isaria japonica*) (entomopathogenic fungal strain; family: Trichocomaceae) [1]

Pharmaceutical potential: Neuroprotective activity [1]

Total Synthesis of Bioactive Natural Products
https://doi.org/10.1016/B978-0-08-102822-3.00047-X

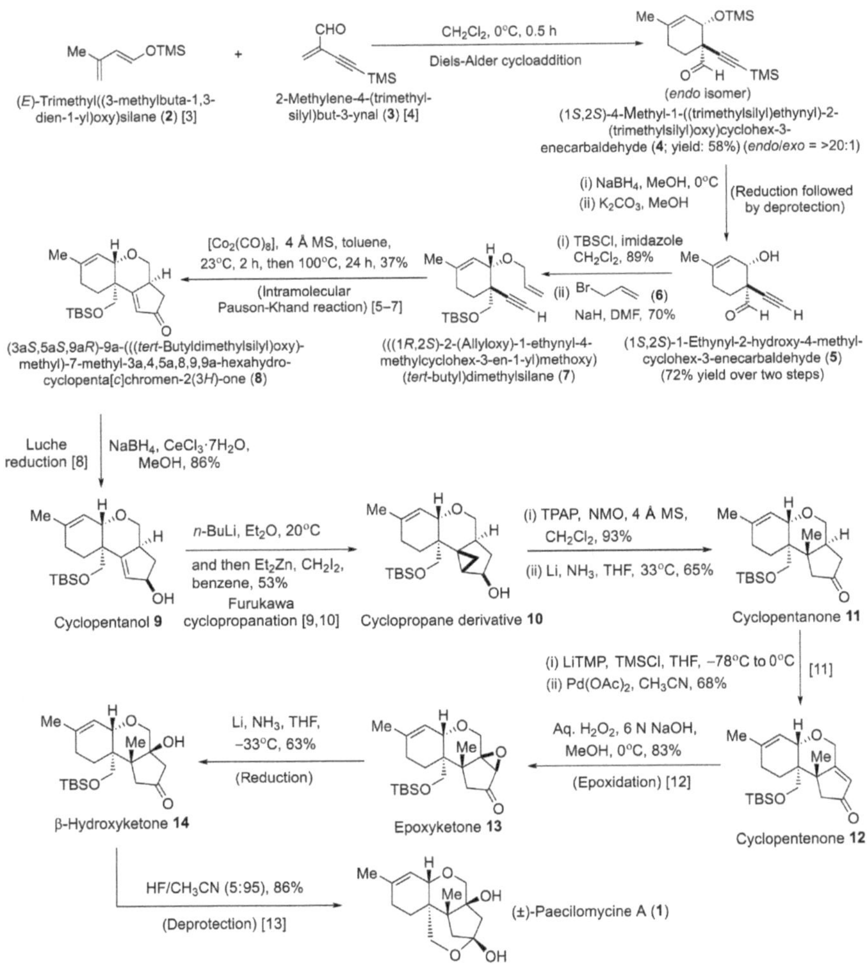

Scheme 1 Total synthesis of (±)-paecilomycine A (**1**) [2].

Synthetic route:

The first total synthesis of the trichothecene-type sesquiterpenoid (±)-paecilomycine A (**1**) was reported by Min and Danishefsky [2] (Scheme 1). The investigators accomplished the total synthesis of the target molecule in 15 steps starting from a Diels-Alder cycloaddition between (*E*)-trimethyl((3-methylbuta-1,3-dien-1-yl)oxy)silane (**2**) [3] and 2-methylene-4-(trimethylsilyl)but-3-ynal (**3**) [4] to form the cycloadduct **4** with high *endo*-selectivity (*endo*:*exo* ≥20:1). Compound **4** upon reduction with sodium borohydride in methanol, followed by deprotection with potassium carbonate, gave (1*S*,2*S*)-1-ethynyl-2-hydroxy-4-methylcyclohex-3-enecarbaldehyde (**5**) in 72% yield over the two steps. Compound **5** then yielded the key intermediate **7** over two steps on treating with TBSCl/imidazole/DCM, followed by *O*-alkylation with allyl bromide (**6**) under strong basic conditions (NaH/DMF).

Intermediate **7** was then subjected to intramolecular Pauson-Khand reaction [5–7] to obtain a single stereoisomer, **8**, in 37% yield. The Pauson-Khand enone **8** was reduced under Luche conditions [5–7] to provide (2*R*,3a*S*,5a*S*,9a*R*)-9a-(((*tert*-butyldimethylsilyl)oxy)methyl)-7-methyl-2,3,3a,4,5a,8,9,9a-octahydrocyclopenta[*c*]chromen-2-ol (**9**) that underwent cyclopropanation with the in situ generated Furukawa reagent prepared by addition of diiodomethane to diethylzinc [9, 10]. The resulting cyclopropane derivative **10** on oxidation of its hydroxy group, followed by dissolving metal reduction of the resultant activated cyclopropane, afforded the cyclopentanone **11** in good yield. Ketone **11** was converted into cyclopentenone **12** by dehydrogenation of the silyl enol ether following the protocol of Itoh et al. [11]. Treatment of the resulting enone **12** with alkaline hydrogen peroxide imposed an epoxy ring within the molecule in an orientation *syn* to the angular methyl group [12] to give the expoxyketone **13**, which on dissolving metal reduction furnished β-hydroxyketone **14** in good yield. In the final step, the TBS protecting group was successfully removed through the use of hydrofluoric acid in acetonitrile [13] whereupon the resulting free hydroxyketone spontaneously cyclized to afford (±)-paecilomycine A (**1**) in 86% yield. The spectral data for the synthetic compound were found to be identical to those reported for natural product [1]. The overall synthetic steps are shown in Scheme 1. Later on (2012), Mehta et al. [14] reported an alternative route for attaining the key intermediate **8**.

References

[1] H. Kikuchi, Y. Miyagawa, Y. Sahashi, S. Inatomi, A. Haganuma, N. Nakahata, Y. Oshima, Tetrahedron Lett. 45 (2004) 6225.
[2] S.-J. Min, S.J. Danishefsky, Org. Lett. 46 (2007) 2199.
[3] E.W. Colvin, I.G. Thom, Tetrahedron 42 (1986) 3137.
[4] C. Thongsornkleeb, R.L. Danheiser, J. Org. Chem. 70 (2005) 2364.
[5] L. PMrez-Serrano, L. Casarrubios, G. Dominguez, J. PMrez-Castells, Org. Lett. 1 (1999) 1187.
[6] M. Ishizaki, K. Iwahara, K. Kyoumura, O. Hoshino, Synlett (1999) 587.
[7] J. Castro, A. Moyano, M.A. PericNs, A. Riera, A.E. Greene, Tetrahedron Asymmetry 5 (1994) 307.
[8] L. Luche, J. Am. Chem. Soc. 100 (1978) 2226.
[9] J. Furukawa, N. Kawabata, J. Nishimura, Tetrahedron 24 (1968) 53.

[10] J. Furukawa, N. Kawabata, J. Nishimura, Tetrahedron Lett. 7 (1966) 3353.
[11] Y. Itoh, T. Hirao, T. Saegusa, J. Org. Chem. 43 (1978) 1011.
[12] M. Miyashita, T. Suzuki, A. Yoshikoshi, J. Am. Chem. Soc. 111 (1989) 3728.
[13] R.F. Newton, D.P. Reynolds, M.A.W. Finch, D.R. Kelly, S.M. Roberts, Tetrahedron Lett. 20 (1979) 3981.
[14] G. Mehta, R. Samineni, P. Srihari, Tetrahedron Lett. 53 (2012) 829.

CHAPTER FORTY EIGHT

Pochonin A

Abbreviations

DEAD diethylazodicarboxylate
EDTA ethylenediaminetetraacetic acid
P(*m*-ClPh)$_3$ tris-(3-chlorophenyl)phosphine
SEM-Cl 2-(trimethylsilyl)ethoxymethyl chloride
THF tetrahydrofuran

Systematic name: (1a*R*,14*R*,15a*R*,*E*)-8-Chloro-9,11-dihydroxy-14-methyl-2,3,15,15a-tetrahydro-1a*H*-benzo[*c*]oxireno[2,3-*k*][1]oxacyclotetradecine-6,12(7*H*,14*H*)-dione

Compound class: Macrocyclic antibiotic (resorcylic macrolide)

Structure:

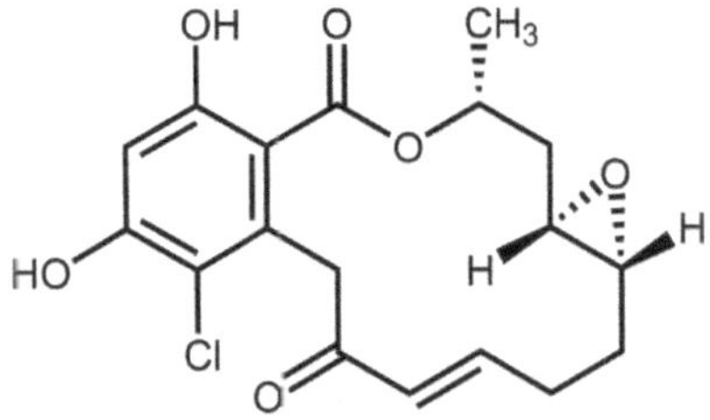

Pochonin A (**1**)

Natural source: *Pochonia chlamydosporia* var. *catenulata* strain P 0297 (fungus; fermentation broth) [1]

Pharmaceutical potentials: Herpes simplex virus 1 inhibitor [1]; heat-shock protein 90 inhibitor [2]

Synthetic route:

A concise and expedient synthesis of pochonin A (**1**), a member of resorcylic macrolides (resorcylic acid lactones) originating from the fermentation broth of *Pochonia chlamydosporia* var. *catenulata* [1], was developed by Winssinger and coworkers [2] starting from 3-chloro-4,6-dihydroxy-2-methylbenzoic acid (**2**). Among the two synthetic outlines for pochonin

Total Synthesis of Bioactive Natural Products
https://doi.org/10.1016/B978-0-08-102822-3.00048-1

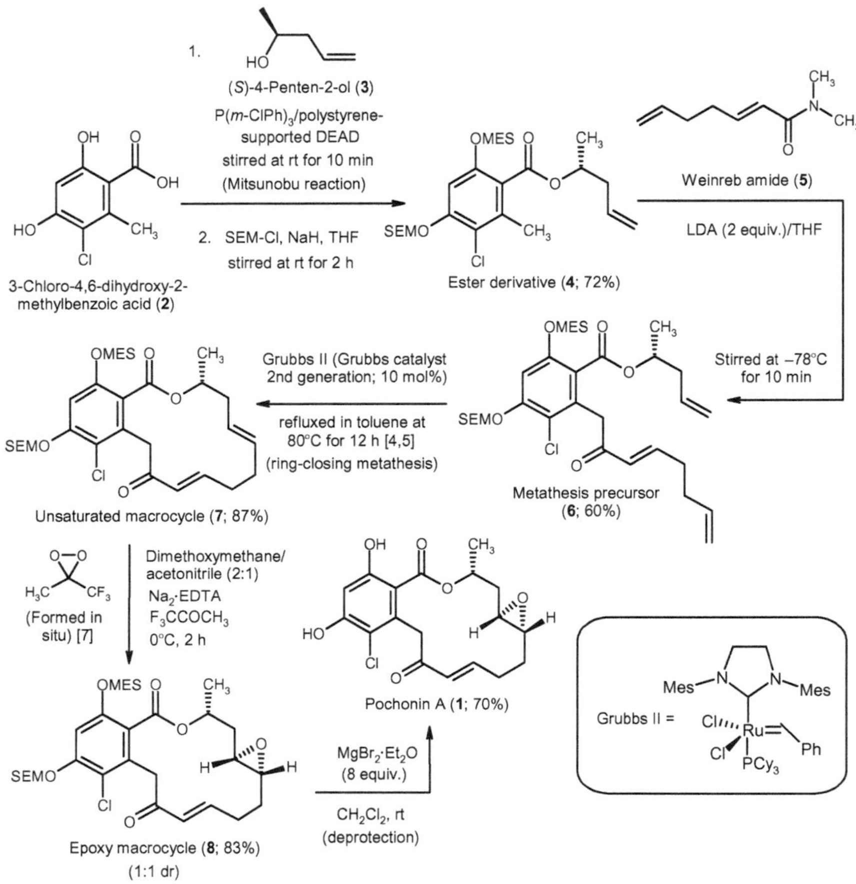

Scheme 1 Total synthesis of pochonin A (**1**).

A described by the investigators in their report, only the modified method is presented herein (Scheme 1). The benzoic acid **2** underwent selective Mitsunobu reaction with (*S*)-4-penten-2-ol (**3**) in the presence of tris-(3-chlorophenyl)phosphine and polymer-bound DEAD followed by subsequent protection with 2-(trimethylsilyl)ethoxymethyl chloride (SEM-Cl)/sodium hydride in THF at 0°C to afford the ester derivative **4** in 72% yield [3]. Deprotonation from the aromatic-methyl group within **4** with LDA in THF at −78°C followed by addition of Weinreb amide **5** furnished the metathesis precursor **6** in 60% yield. Ring-closing metathesis of compound **6** using second-generation Grubbs catalyst [4,5] under thermodynamic conditions [6] gave the unsaturated macrocycle **7** in excellent yield of 87%. Epoxidation of the unconjugated olefin **7** was carried out with methyl(trifluoromethyl)-dioxirane generated in situ [7] to yield the epoxy macrocycle **8** in 83% yield albeit in 1:1 diastereomeric ratio (inseparable); deprotection of the SEM groups in the resulting epoxy macrocycle **8** using $MgBr_2 \cdot Et_2O$ in dichloromethane at room temperature eventually produced the desired pochonin A (and its diastereoisomer as a separable mixture) in 70% yield.

The same group of investigators [8] accomplished a diversity-oriented synthesis of a library of compounds based on the pochonin scaffold by using solid-supported reagents, and evaluated their inhibitory activity against a panel of 24 kinases—the results of testing the library against the kinase enzymes at 10 μM afforded a >14% hit rate, thereby demonstrating the potential of the resorcylides toward the inhibition of therapeutically relevant kinases [8].

References

[1] V. Hellwig, A. Mayer-Bartschmid, H. Mueller, G. Greif, G. Kleymann, W. Zitzmann, H.-V. Tichy, M. Stadler, J. Nat. Prod. 66 (2003) 829.
[2] E. Moulin, S. Barluenga, N. Winssinger, Org. Lett. 7 (2005) 6537.
[3] E. Moulin, V. Zoete, S. Barluenga, M. Karplus, N. Winssinger, J. Am. Chem. Soc. 127 (2005) 6999.
[4] M. Scholl, S. Ding, C.W. Lee, R.H. Grubbs, Org. Lett. 1 (1999) 953.
[5] A.K. Chatterjee, J.P. Morgan, M. Scholl, R.H. Grubbs, J. Am. Chem. Soc. 122 (2000) 3783.
[6] C.W. Lee, R.H. Grubbs, Org. Lett. 2 (2000) 2145.
[7] D. Yang, M.-K. Wong, Y.-C. Yip, J. Org. Chem. 60 (1995) 3887.
[8] E. Moulin, S. Barluenga, F. Totzke, N. Winssinger, Chem. Eur. J. 12 (2006) 8819.

CHAPTER FORTY NINE

Rhinacanthin A

Abbreviations

Ar argon gas
$B(OCH_3)_3$ trimethyl borate
CAN ceric(IV) ammonium nitrate
CH_2Cl_2 dichloromethane
CH_3CN acetonitrile
DMD dimethyldioxirane
DMF *N,N*-dimethylformamide
DMM dimethoxymethane
MCPBA *meta*-chloroperbenzoic acid
MOMCl methoxymethyl chloride/methyl chloromethyl ether
Na_2(EDTA) ethylenediamine tetraacetic acid disodium salt
$Na_2B_4O_7{\cdot}10H_2O$ borax (sodium tetraborate decahydrate)
$NaBH_3CN$ sodium cyanoborohydride
rt room temperature
sec-BuLi sec-butyl lithium
TBFA tetra-*n*-butylammonium bromide
TBSCl *tert*-butylchlorodimethylsilane
TFA trifluoroacetic acid
THF tetrahydrofuran
TMEDA *N,N,N′,N′*-tetramethylethylenediamine

Systematic name: 3-Hydroxy-2,2-dimethyl-3,4-dihydro-2*H*-benzo[*g*]chromene-5,10-dione

Compound class: Pyranonaphthoquinone

Structure:

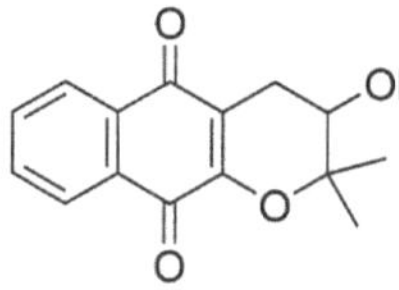

Rhinacanthin A (**1**)

Natural source: *Rhinacanthus nasutus* (L.) Kurz (roots; family: Acanthaceae) [1]

Total Synthesis of Bioactive Natural Products
https://doi.org/10.1016/B978-0-08-102822-3.00049-3

Pharmaceutical potentials: Antiproliferative [1, 2], inhibitor against mosquito cytochrome P450 enzymes [3]

Synthetic routes:

Kimachi and his group accomplished the total synthesis of rhinacanthin A in both racemic (*rac*-**1**) and enantioenriched [(*R*)-**1**)] forms starting from a reduced lapachol compound (**3**), in eight steps in overall yields of 47% and 49%, respectively (Schemes 1 and 3) [2].

Total synthesis of racemic rhinacanthin A (*rac*-1):

The reduced lapachol (**3**) was first prepared from 1,4-dimethoxynaphthalene (**2**) followed by protection of the hydroxy group with MOMCl affording 1,4-dimethoxy-2-(methoxymethoxy)naphthalene (**4**) as pale

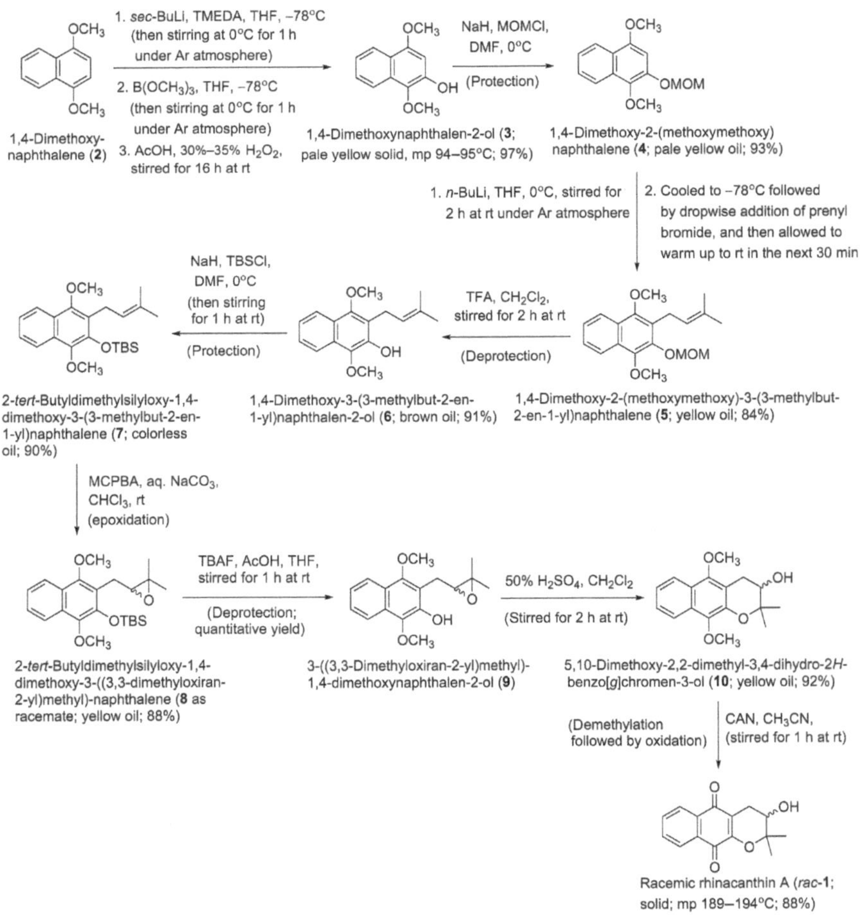

Scheme 1 Kimachi's total synthesis of (±)-rhinacanthin A (*rac*-**1**) [2].

yellow oil in 93% yield. The MOM ether **4** was then converted to 1,4-dimethoxy-2-(methoxymethoxy)-3-(3-methylbut-2-enyl)naphthalene (**5**; yellow oil; 84%) using a metalation–substitution reaction. In the next steps, the exchange of a protecting group from MOM in **5** to TBS in **7** was carried out smoothly. Epoxidation of the olefinic double bond in 2-*tert*-butyldimethylsilyloxy-1,4-dimethoxy-3-(3-methylbut-2-enyl) naphthalene (**7**) was effected by reacting with MCPBA to afford **8** as a racemate (yellow oil; 88%). The deprotection of a TBS group in **8** with acetic acid in the presence of TBAF produced the epoxynaphthol **9**, the ensuing acidic treatment of which at room temperature promoted the intramolecular ring opening of an epoxide function to afford 5,10-dimethoxy-2,2-dimethyl-3,4-dihydro-2*H*-benzo[*g*]chromen-3-ol (**10**, yellow oil; 92%). Finally, conversion of this dimethoxynaphthalene **10** to (±)-rhinacanthin A (*rac*-**1**, 88%) was achieved by CAN oxidation. The synthetic material (Scheme 1) had analytical and spectroscopic data consistent with those in the literature [1]. Later on, Lee et al. [4] outlined another route for the total synthesis of (±)-rhinacanthin A (*rac*-**1**) starting from dehydro-α-lapachone (**11**) (Scheme 2). Reaction of **11** with dimethyldioxirane (DMD) in acetone afforded 2,2-dimethyl-3,4-epoxy-2*H*-naphtho[2,3-*b*]pyran-5,10-dione (**12**), which was then chemo- and regioselectively reduced with $NaBH_3CN$ in the presence of $BF_3{\cdot}OEt_2$ in THF to furnish (±)-**1** in 72% yield, the spectral data of which were also found to be in good agreement with those reported in the literature [2].

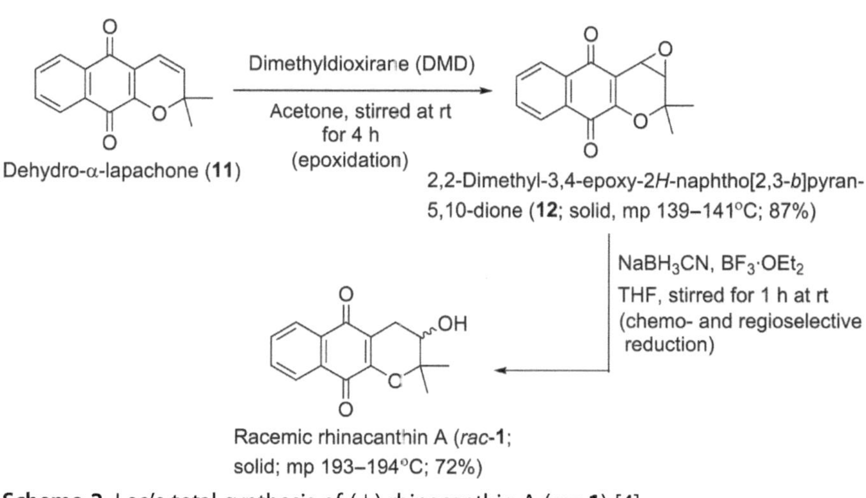

Scheme 2 Lee's total synthesis of (±)-rhinacanthin A (*rac*-**1**) [4].

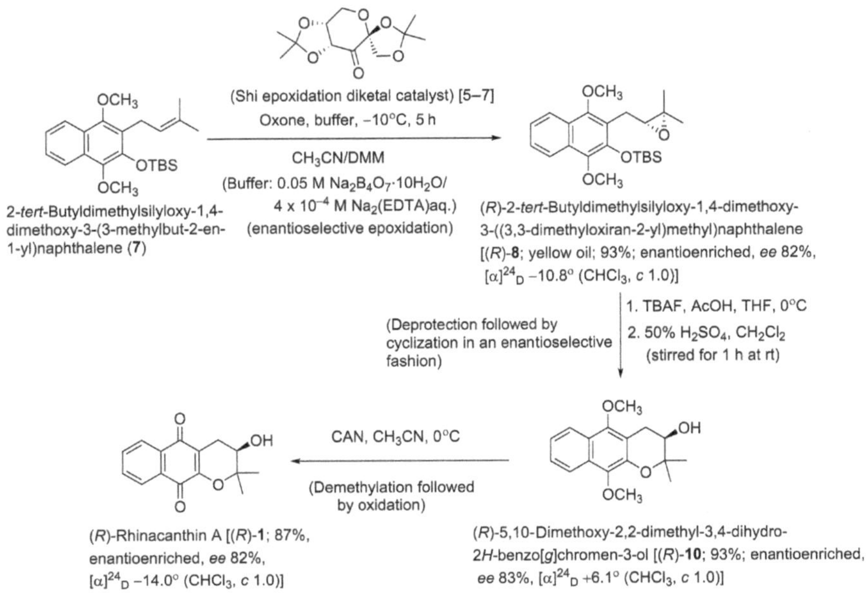

Scheme 3 Kimachi's total synthesis of (*R*)-rhinacanthin A [(*R*)-**1**] [2].

Total synthesis of enantioenriched (*R*)-rhinacanthin A:

Kimachi et al. [2] also accomplished the asymmetric total synthesis of (*R*)-rhinacanthin A using the same partway for the racemic route (Scheme 1) in eight steps from **3** in overall 49% yield with high enantiomeric purity (*ee* 82%). The chiral dioxirane species prepared in situ from the Shi epoxidation diketal catalyst [5–7] and Oxone in buffered media promoted the epoxidation reaction of **7** to afford optically active (*R*)-**9** in good chemical in 93% yield with high enantioselectivity (*ee* 82%). Eventually, (*R*)-**9** was converted to (*R*)-**1** via (*R*)-**10** in 87% yield in an enantioenriched fashion (*ee* 82%) as shown in Scheme 3. The enantiomeric excess of synthetic rhinacanthin A (82% *ee*) was found to be increased on recrystallization from hexane/AcOEt up to *ee*>99% [2].

References

[1] T.-S. Wu, H.-J. Tien, M.-Y. Yeh, K.-H. Lee, Phytochemistry 27 (1988) 3787.
[2] T. Kimachi, E. Torii, R. Ishimoto, A. Sakue, M. Ju-ichi, Tetrahedron Asymmetry 20 (2009) 1683.
[3] S. Pethuan, P. Duangkaew, S. Sarapusit, E. Srisook, P. Rongnoparut, J. Med. Entomol. 49 (2012) 993.
[4] X. Wang, Y. Chen, Y.R. Lee, Bull. Kor. Chem. Soc. 32 (2011) 153.
[5] Y. Tu, Z.-X. Wnag, Y. Shi, J. Am. Chem. Soc. 118 (1996) 9806.
[6] H. Tian, X. She, Y. Shi, Org. Lett. 3 (2001) 715.
[7] O.A. Wong, Y. Shi, Chem. Rev. 108 (2008) 3958.

CHAPTER FIFTY

Rubriflordilactone A

Abbreviations

CH_2Cl_2 dichloromethane
DAST diethylaminosulfur trifluoride
DMSO dimethylsulfoxide
Et_3N triethylamine
NIS *N*-iodosuccinamide
NMP *N*-methyl-2-pyrrolidone
rt room temperature
TBHP *tert*-butylhydroperoxide
TES triethylsilyl
THF tetrahydrofuran
TMS trimethylsilyl

Systematic name: (3a*R*,5a*S*,9a*R*,10*S*,11*S*,14a*R*)-5,5,10-Trimethyl-11-(*S*)-4-methyl-5-oxo-2,5-dihydrofuran-2-yl)-3,3a,5,5a,6,7,8,9,9a,10,11,14-dodecahydro-2*H*-cyclopenta[*de*]furo[3″,2″:2′,3′]furo[3′,4′:4,5]cyclohepta[1,2-*g*]chromen-2-one

Compound class: Bisnortriterpenoid

Structure:

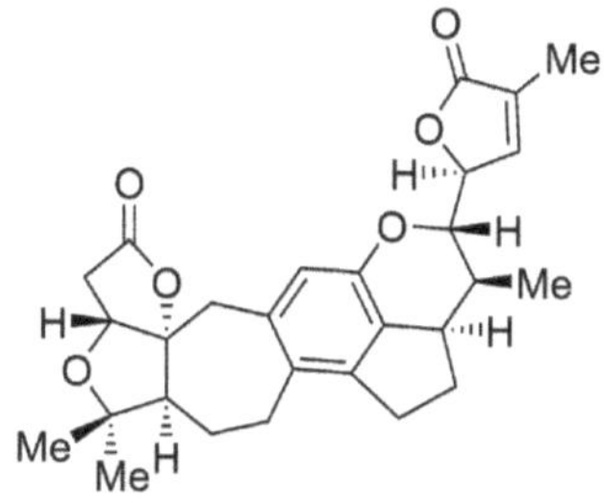

Rubriflordilactone A (**1**)

Natural source: *Schisandra rubriflora* (leaves and stems; family: Schisandraceae) [1]

Total Synthesis of Bioactive Natural Products
https://doi.org/10.1016/B978-0-08-102822-3.00050-X

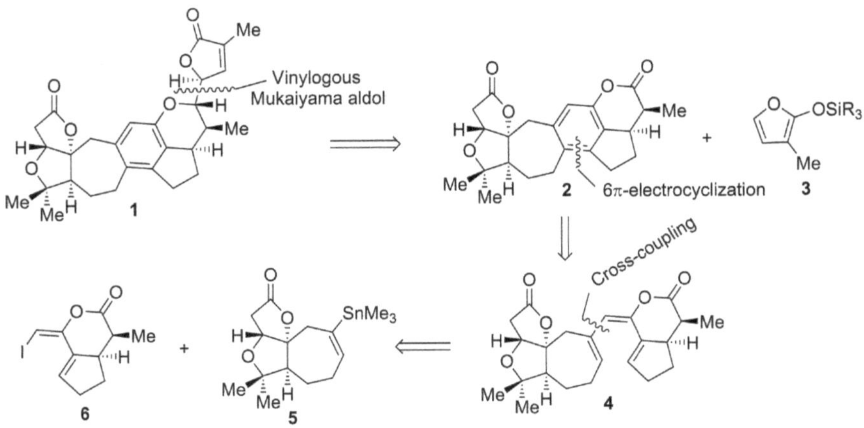

Scheme 1 Retrosynthetic approach for rubriflordilactone A (**1**) [2].

Pharmaceutical potential: Anti-HIV [1]

Synthetic route:

Li and his group [2] accomplished the first and asymmetric total synthesis of rubriflordilactone A, a bisnortriterpenoid isolated from *Schisandra rubriflora*. They envisioned a retrosynthetic approach for the molecule as shown in Scheme 1.

The investigators synthesized the key fragments and then assembled them in the final step to obtain the target compound. Their total synthetic approach is discussed step-wise.

Synthesis of fragment 6:

They started with 2-iodo-2-cyclopentenone (**7**), which on Corey-Bakshi-Shibata reduction [3] followed by Sonogashira coupling [4] of the resulting (*R*)-2-iodocyclopent-2-enol (**9**) with TMS-acetylene gave compound (*R*)-2-((trimethylsilyl)ethynyl)cyclopent-2-enol (**10**) in overall 81% yield as yellow oil. This alcohol **10** was then subjected to the conditions of Johnson-Claisen rearrangement [5, 6] to convert it into its methyl ester **11**. On treatment with aqueous LiOH/THF (1:1), the methyl ester underwent saponification along with desilylation to afford carboxylic acid derivative **12** in 93% yield. Then compound **12** was exposed to *N*-iodosuccinamide/$NaHCO_3$ to form (4*S*,4a*R*,*E*)-1-(iodomethylene)-4-methyl-4,4a,5,6-tetrahydrocyclopenta[*c*]pyran-3(1*H*)-one (**6**, colorless oil, 52%) via a stereospecific iodolactonization process [7]. The overall reaction sequences are shown in Scheme 2.

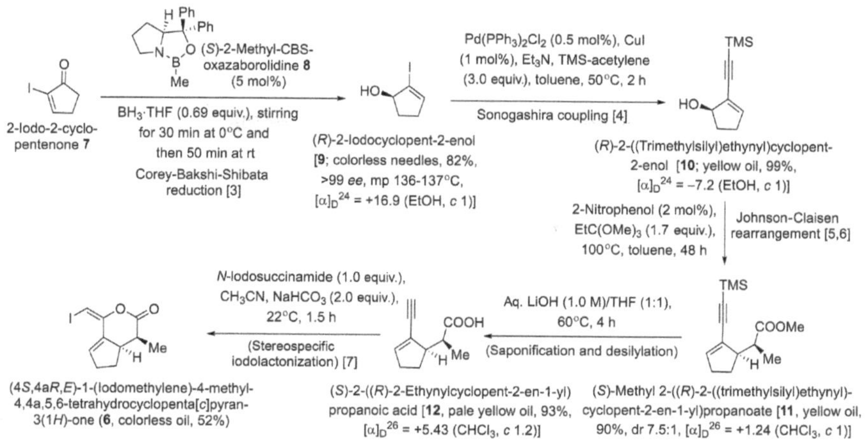

Scheme 2 Synthesis of fragment **6** [2].

Synthesis of fragment 5:

Synthesis of fragment **5** is depicted in Scheme 3. First, the diene **13** and dienophile **14** underwent the Cu-catalyzed asymmetric Diels-Alder reaction [8, 9] in the presence of copper triflate and bis-oxazoline ligand **15** to give the cycloadduct **16** as yellow oil in 80% yield (76% *ee*). The thioester **17** obtained on replacing the oxazolidinone moiety with EtSLi was subjected to Pd-catalyzed methyl coupling with methyl zinc iodide in the presence of $Pd_2(dba)_3$/S-Phos to prepare (1*R*,6*R*)-methyl 6-acetyl-3-((*tert*-butyldimethylsilyl)oxy)cyclohex-3-enecarboxylate (**18**) with good overall efficiency. The bromoenone **20** was then prepared with excellent level of enantiopurity (>99% *ee*) from the ester **18** in five steps via the intermediacy of lactone **19** following the procedure of Yang and his group [12–15]. Bromoenone **20** was converted into triflate **22** in two steps involving hydrogenation and debromination with Pd/C-H_2 followed by treating with the resulting cycloheptanone **21** with LiHMDS/$PhNTf_2$ in a regioselective fashion. Removal of trimethylsilyl moiety within molecule **22** with acetyl group was successfully achieved by reacting with $Sc(OTf)_3/Ac_2O$ to afford acetate derivative **23** in excellent yield (96%). Intramolecular Dieckmann condensation provided a hemiketal intermediate **24**, which underwent subsequent cationic deoxygenation [16] on treating with $Et_3SiH/BF_3{\cdot}OEt_2$ to furnish tricycle **25** in 65% yield along with 29% of the recovered starting material. Stannylation with $Pd(PPh_3)_4/Me_3SnSnMe_3$/LiCl [17] of **25** afforded (3a*R*,5a*S*,10a*R*)-5,5-dimethyl-9-(trimethylstannyl)-3,3a,5,5a,6,7-hexahydrocyclohepta[*c*]furo[3,2-*b*]furan-2(10*H*)-one (fragment **5**) as a yellow oil in 75% yield (Scheme 3).

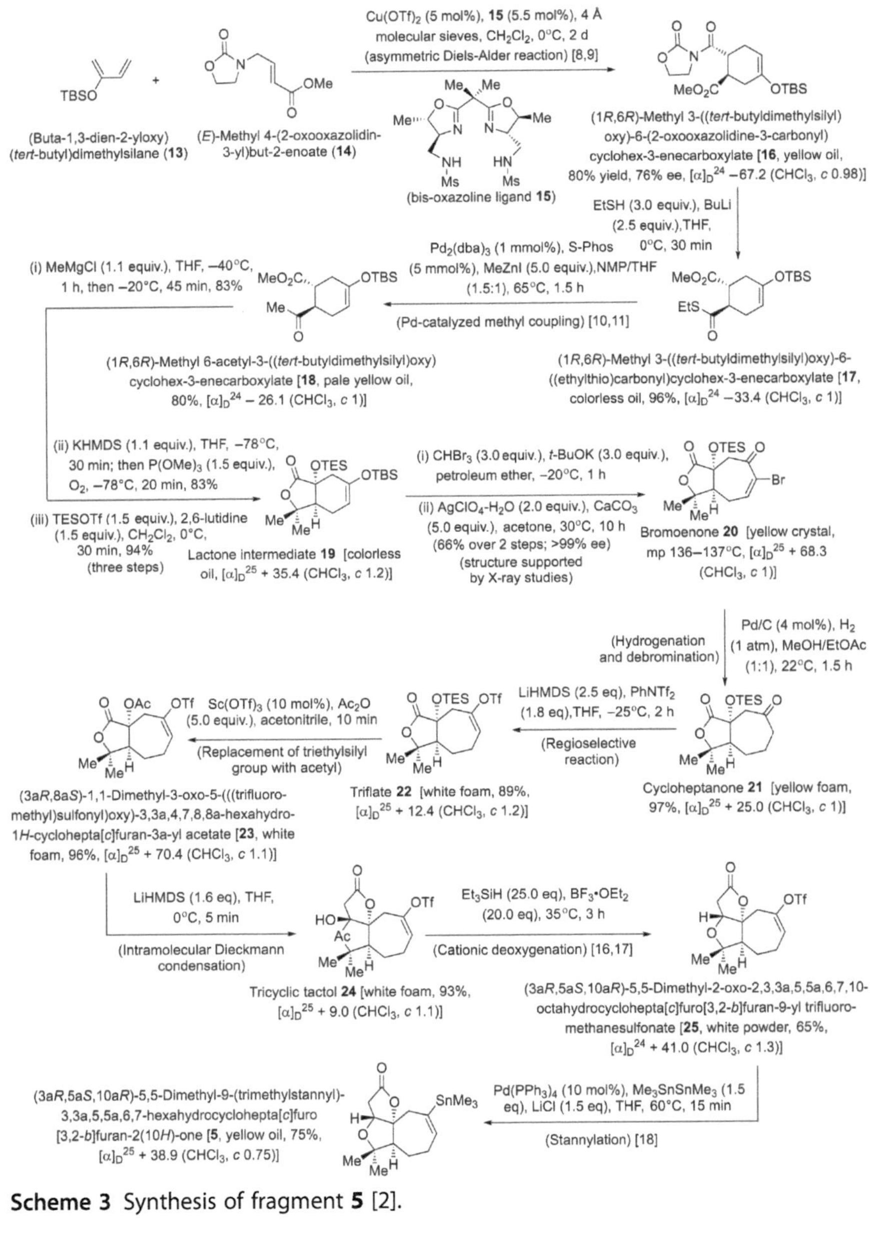

Scheme 3 Synthesis of fragment **5** [2].

Assembly of fragments 5 and 6 and synthesis of rubriflordilactone A (1):

With both the fragments **5** and **6** at hand, the investigators then proceeded to assemble them with a motto to reach the goal (Scheme 4).

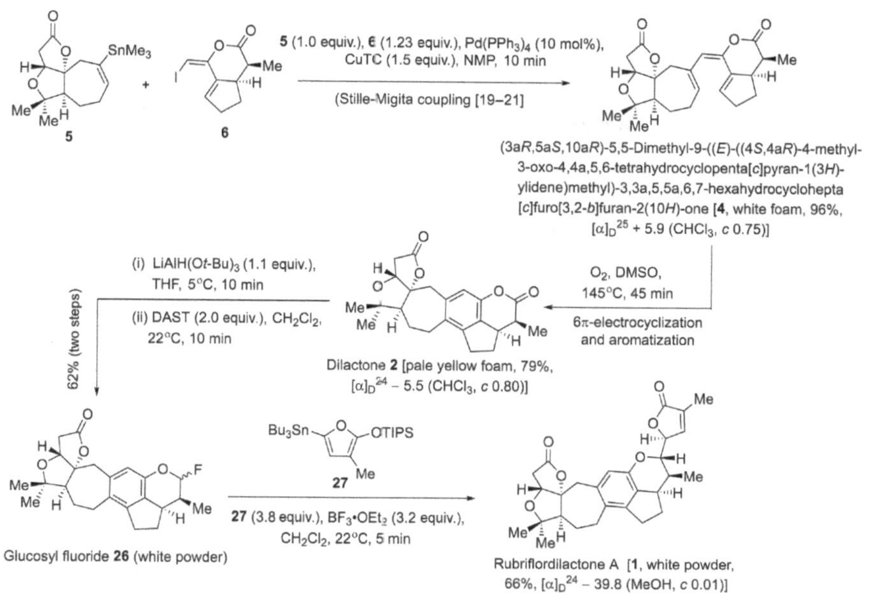

Scheme 4 Assembly of fragments **5** and **6**, and synthesis of rubriflordilactone A (**1**) [2].

They performed a Stille-Migita reaction using Pd(0)/CuTC catalytic systems [19–21] that promoted the coupling between fragments **5** and **6** efficiently to afford triene **4** in 96% yield. Heating **4** in DMSO at 145°C under an air atmosphere accomplished the 6π-electrocyclization and aromatization in one pot, giving arene dilactone **2** in 73% yield. The glucosyl fluoride **26** was prepared from dilactone **2** in two successive steps on treating with bulky $LiAlH(Ot\text{-}Bu)_3$ (selective reduction of carbonyl) [22, 23] and the resulting intermediates (lactol/aldehyde mixture) with DAST [24, 25]. In the final step, the stannane **27** (triisopropyl((3-methyl-5-(tributylstannyl)furan-2-yl)oxy)silane) reacted smoothly with **26** under the activation offered by $BF_3{\cdot}OEt_2$ to furnish the target compound rubriflordilactone A (**1**) in 66% yield as a single detectable diastereomer. The spectral and physical properties of the synthetic sample were found to be identical to those reported for the natural product [1].

References

[1] W.-L. Xiao, L.-M. Yang, N.-B. Gong, L. Wu, R.-R. Wang, J.-X. Pu, X.-L. Li, S.-X. Huang, Y.-T. Zheng, R.-T. Li, Y. Lu, Q.-T. Zheng, H.-D. Sun, Org. Lett. 8 (2006) 991.

[2] J. Li, P. Yang, M. Yao, J. Deng, A. Li, J. Am. Chem. Soc. 136 (2014) 16477.

[3] E.J. Corey, C.J. Helal, Angew. Chem. Int. Ed. 37 (1998) 1986.

[4] M.M. Kabat, J. Kiegiel, N. Cohen, K. Toth, P.M. Wovkulich, M.R. Uskoković, J. Org. Chem. 61 (1996) 118.

[5] M. Toyota, T. Asoh, M. Matsuura, K. Fukumoto, J. Org. Chem. 61 (1996) 8687.

[6] R.A. Fernandes, A.K. Chowdhury, P. Kattanguru, Eur. J. Org. Chem. (2014) 2833.
[7] M.J. Sofia, J.A. Katzenellenbogen, J. Org. Chem. 50 (1985) 2331.
[8] A. Sakakura, R. Kondo, Y. Matsumura, M. Akakura, K. Ishihara, J. Am. Chem. Soc. 131 (2009) 17762.
[9] I.B. Seiple, S. Su, I.S. Young, A. Nakamura, J. Yamaguchi, L. Jorgensen, R.A. Rodriguez, D.P. O'Malley, T. Gaich, M. Kock, P.S. Baran, J. Am. Chem. Soc. 133 (2011) 14710.
[10] H. Tokuyama, S. Yokoshima, T. Fukuyama, Tetrahedron Lett. 39 (1998) 3189.
[11] S.D. Walker, T.E. Barder, J.R. Martinelli, S.L. Buchwald, Angew. Chem. Int. Ed. 43 (2004) 1871.
[12] Q. Xiao, W.-W. Ren, Z.-X. Chen, T.-W. Sun, Y. Li, Q.-D. Ye, J.-X. Gong, F.-K. Meng, L. You, Y.-F. Liu, M.-Z. Zhao, L.-M. Xu, Z.-H. Shan, Y.-F. Tang, J.-H. Chen, Z. Yang, Angew. Chem. Int. Ed. 50 (2011) 7373.
[13] W.-W. Ren, Z.-X. Chen, Q. Xiao, Y. Li, T.-W. Sun, Z.-Y. Zhang, Q.-D. Ye, F.-K. Meng, L. Yon, M.-Z. Zhao, L.-M. Xu, Y.-F. Tang, J.-H. Chen, Z. Yang, Chem. Asian J. 7 (2012) 2341.
[14] T.-W. Sun, W.-W. Ren, Q. Xiao, Y.-F. Tang, Y.-D. Zhang, Y. Li, F.-K. Meng, Y.-F. Liu, M.-Z. Zhao, L.-M. Xu, J.-H. Chen, Z. Yang, Chem. Asian J. 7 (2012) 2321.
[15] Y. Li, Z.-X. Chen, Q. Xiao, Q.-D. Ye, T.-W. Sun, F.-K. Meng, W.-W. Ren, L. You, L.-M. Xu, Y.-F. Wang, J.-H. Chen, Z. Yang, Chem. Asian J. 7 (2012) 2334.
[16] G.A. Kraus, K.A. Frazier, B.D. Roth, M.J. Taschner, K. Neuenschwander, J. Org. Chem. 46 (1981) 2417.
[17] M.D. Lewis, K.C. Cha, Y. Kishi, J. Am. Chem. Soc. 104 (1982) 4976.
[18] R.C. Winstead, T.H. Simpson, G.A. Lock, M.D. Schiavelli, D.W. Thompson, J. Org. Chem. 51 (1986) 277.
[19] G.D. Allred, L.S. Liebeskind, J. Am. Chem. Soc. 118 (1996) 2748.
[20] A. Furstner, J.A. Funel, M. Tremblay, L.C. Bouchez, C. Nevado, M. Waser, J. Ackerstaff, C.C. Stimson, Chem. Commun. (2008) 2873.
[21] H.N. Lim, K.A. Parker, J. Am. Chem. Soc. 133 (2011) 20149.
[22] W.E. Parham, L.D. Huestis, J. Am. Chem. Soc. 84 (1962) 813.
[23] X. Xiong, Y. Li, Z. Lu, M. Wan, J. Deng, S. Wu, H. Shao, A. Li, Chem. Commun. (2014) 5294.
[24] G.H. Posner, S.R. Haines, Tetrahedron Lett. 26 (1985) 5.
[25] K. Toshima, Carbohydr. Res. 327 (2000) 15.

CHAPTER FIFTY ONE

(+)-Sattazolin

Abbreviations

$(Boc)_2O$ di-*tert*-butyl dicarbonate
DCE 1,2-dichloroethane
DMAP 4-dimethylaminopyridine
DMF *N,N*-dimethylformamide
MeCN acetonitrile
***n*-BuLi** *n*-butyllithium
rt room temperature
TBAF tetrabutylammonium fluoride
TBS-Cl *tert*-butylchlorodimethylsilane
THF tetrahydrofuran
$Yb(OTf)_3$ ytterbium triflate

Systematic name: (*S*)-2-Hydroxy-1-(1*H*-indol-3-yl)-5-methylhexan-3-one

Compound class: Indole acyloin

Structure:

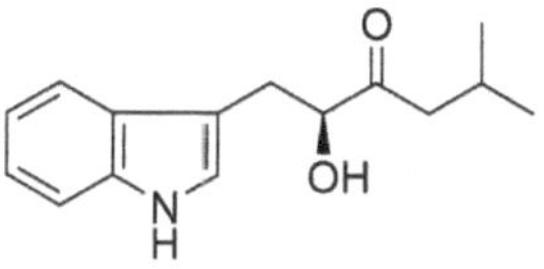

(+)-Sattazolin (**1**)

Natural source: *Bacillus* sp. strain B-60 (fermentation broth of the soil bacteria; family: Bacillaceae) [1].

Pharmaceutical potential: Antiviral [1].

Synthetic route:

The first asymmetric total synthesis of the antiviral natural product (+)-sattazolin (**1**) was accomplished by Miller and his group in seven overall steps and five chromatographic purifications from commercially available starting materials (Scheme 1) [2]. Commercially available (2*S*)-methyl glycidate (**3**; *ee* 99%) was treated with an excess of indole (**2**) under refluxing in

https://doi.org/10.1016/B978-0-08-102822-3.00051-1

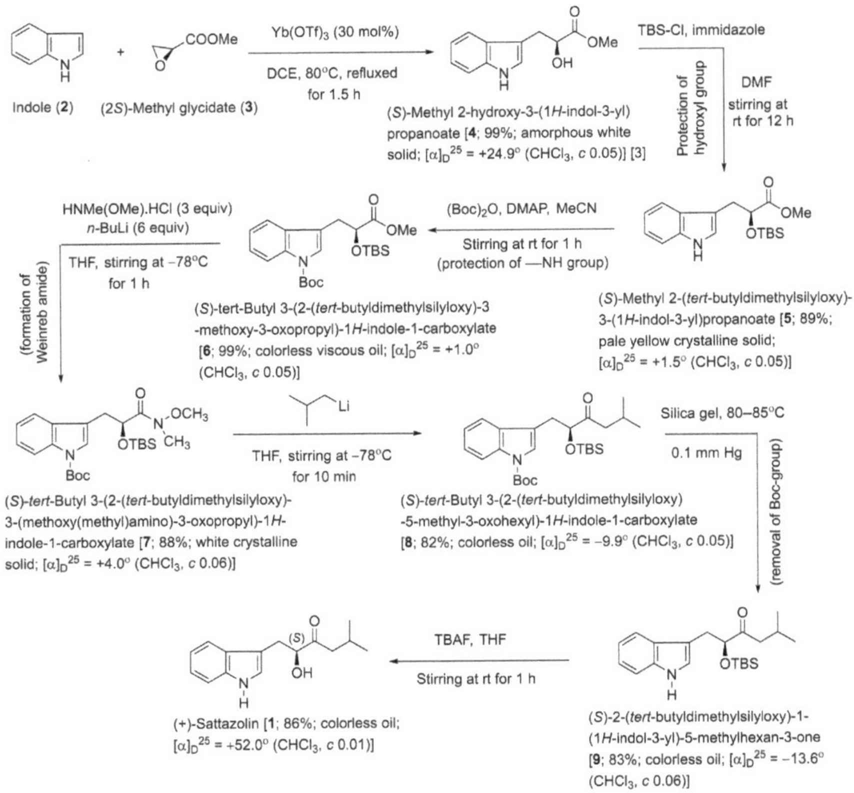

Scheme 1 Miller's total synthesis of (+)-sattazolin (**1**) [2].

DCE in the presence of 30 mol% Yb(OTf)$_3$ when (*S*)-methyl 2-hydroxy-3-(1*H*-indol-3-yl)propanoate (**4**) was obtained as amorphous white solid (99% yield) [3]. Treatment of **4** with TBSCl and imidazole resulted in a straightforward protection of the hydroxyl group as the silyl ether **5** in excellent yield. In the next step, the indole nitrogen was protected as the corresponding *tert*-butyl carbamate **6** (colorless viscous oil) in quantitative yield. Treatment of this doubly protected ester **6** with the anion of *N*,*O*-dimethylhydroxylamine in THF at −78°C efficiently afforded the key Weinreb amide (*S*)-*tert*-butyl 3-(2-(*tert*-butyldimethylsilyloxy)-3-(methoxy(methyl)amino)-3-oxopropyl)-1*H*-indole-1-carboxylate (**7**; 88%) as white crystalline solid. The Weinreb amide **7** on treatment with a slight excess of isobutyllithium gave the corresponding ketone **8**, a protected version of the natural product. Afterward, the Boc-group was removed first by gently heating **8** at 80–85°C in the presence of silica gel under vacuum.

Finally, treatment of the silyl ether **9** with excess TBAF smoothly afforded (+)-sattazolin as a colorless solid. The physical properties, optical rotation, and spectral data for the synthetic compound were found to be comparable with those reported for the natural product [1].

References

[1] G. Lampis, D. Deidda, C. Maullu, M.A. Madeddu, R. Pompei, F.D. Monache, G. Satta, J. Antibiot. 48 (1995) 967.
[2] K.M. Snyder, T.S. Doty, S.P. Heins, A.L. DeSouchet, K.A. Miller, Tetrahedron Lett. 54 (2013) 192.
[3] S. Tsuchiya, T. Sunazuka, T. Shirahata, T. Hirose, E. Kaji, S. Omura, Heterocycles 72 (2007) 91–94.

CHAPTER FIFTY TWO

Somocystinamide A

Abbreviations

CH_2Cl_2 dichloromethane
DMAP 4-dimethylaminopyridine
DMF *N,N*-dimethylformamide
EDC *N'*-ethylcarbodiimide
MeOH methanol
rt room temperature
TBAF tetrabutylammonium fluoride
TEA triethylamine
TFA trifluoroacetic acid
THF tetrahydrofuran
TsOH *p*-toluene sulfonic acid

Systematic name: (10*E*,10′*E*,12*R*,12′*R*)-13,13′-Disulfanediylbis(12-acetamido-*N*-methyl-*N*-((*E*)-penta-1,4-dien-1-yl)tridec-10-enamide)

Compound class: Lipopeptide (disulfide dimer-type enamide)

Structure:

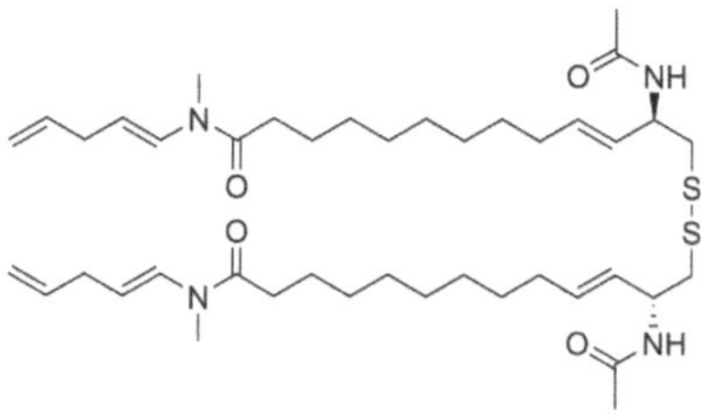

Somocystinamide A (**1**)

Natural source: *Lyngbya majuscula* (cyanobacteria; family: Oscillatoriaceae) [1,2]

Pharmaceutical potential: Anticancer [1,3]

Synthetic route:

Suyama and Gerwick [4] accomplished total synthesis of the marine cyanobacteria-derived lipopeptide **1** in a stereospecific and concise manner as depicted in Scheme 1. The synthesis was started with a known compound, (2*R*,4*R*)-*tert*-butyl 4-formyl-2-phenylthiazolidine-3-carboxylate (**5**), prepared

https://doi.org/10.1016/B978-0-08-102822-3.00052-3

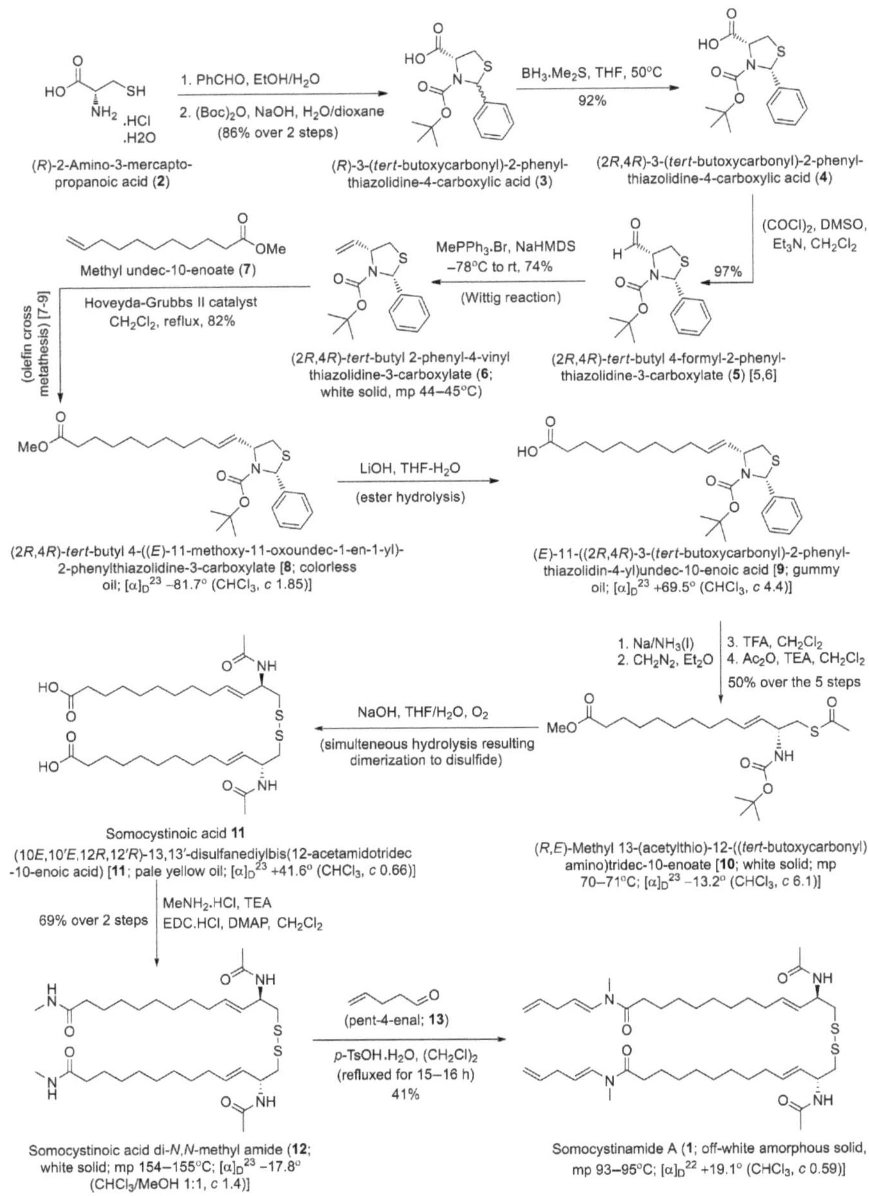

Scheme 1 Total synthesis of somocystinamide A (**1**) [4].

from (*R*)-2-amino-3-mercaptopropanoic acid hydrochloride salt (**2**) in four steps [5,6]. The investigators envisioned that the key carbon-carbon connection at the internal olefin could be feasible by olefin cross-metathesis using a ruthenium-based Grubbs catalyst [7–9]. Accordingly, the terminal olefin **6** was prepared from aldehyde **5** via a Wittig reaction. The olefin cross-coupling product **8** was obtained upon such a coupling reaction between the terminal olefinic compounds **6** and methyl undec-10-enoate (**7**) under the influence of a second-generation Hoveyda-Grubbs catalyst [8] in good yield and with high stereoselectivity (*trans*:*cis* = 18:1). The methyl ester **8** was then hydrolyzed to obtain carboxylic acid derivative **9** to avoid undesired reduction to the primary alcohol in the next step. The thiol and the carbamate of compound **9** were then reductively deprotected by sodium in liquid ammonia [6]; reprotection of the carboxylic acid as a methyl ester, deprotection of the amine, and acetylation yielded **10** in good yield. The simultaneous basic hydrolysis of the methyl ester and the thioacetate with sodium hydroxide in the presence of oxygen led to dimerization to the disulfide in one pot to produce (10*E*,10′*E*,12*R*,12′*R*)-13,13′-disulfanediylbis(12-acetamidotridec-10-enoic acid) (**11**; somocystinoic acid) as a pale yellow oil, followed by its conversion to the di-*N*,*N*-methyl amide derivative **12** in two steps with overall yield of 69%. Compound **12**, the final-stage precursor of the target molecule **1**, was condensed with pent-4-enal (**13**) [10] under acidic conditions to afford somocystinamide A (**1**) in good yield as off-white amorphous solid [mp 93–95°C; $[\alpha]_D^{22}$ +19.1° ($CHCl_3$, *c* 0.59)]. The analytical data (^1H- and ^{13}C-NMR, MS, UV, IR, and optical rotation) for synthetic compound **1** were found to be fully identical to those for natural somocystinamide A [1,2]. Scheme 1 depicts the overall reactions involved in the total synthetic protocol for molecule **1**.

References

[1] L.M. Nogle, W.H. Gerwick, Org. Lett. 4 (2002) 1095.
[2] R. Van Wagoner, A.K. Drummond, J.L.C. Wright, Adv. Appl. Microbiol. 61 (2007) 89.
[3] W. Wrasidlo, A. Mielgo, V.A. Torres, S. Barbero, K. Stoletov, T.L. Suyama, R.L. Klemke, W.H. Gerwick, D.A. Carson, D.G. Stupack, Proc. Nat. Acad. Sci. U.S.A., 105 (2008) 2313.
[4] T.L. Suyama, W.H. Gerwick, Org. Lett. 10 (2008) 4449.
[5] S.P. Chavan, A.G. Chittiboyina, G. Ramakrishna, R.B. Tejwani, T. Ravindranathan, S.K. Kamat, B. Rai, L. Sivadasan, K. Balakrishnan, S. Ramalingam, V.H. Deshpande, Tetrahedron 61 (2005) 9273.
[6] F.D. Deroose, P.J. De Clercq, J. Org. Chem. 60 (1995) 321.
[7] K.C. Nicolaou, P.G. Bulger, D. Sarlah, Angew. Chem. Int. Ed. 44 (2005) 4490.
[8] S.B. Garber, J.S. Kingsbury, B.L. Gray, A.H. Hoveyda, J. Am. Chem. Soc. 122 (2000) 8168.
[9] M. Scholl, S. Ding, C.W. Lee, R.H. Grubbs, Org. Lett. 1 (1999) 953.
[10] G.A. Griffith, J.M. Percy, S. Pintat, C.A. Smith, N. Spencer, E. Uneyama, Org. Biomol. Chem. 3 (2005) 2701.

CHAPTER FIFTY THREE

(−)-Stagonolide A

Abbreviations

CH_2Cl_2 dichloromethane
DCC 1,3-dicyclohexylcarbodiimide
(−)-DIPT D-(−)-diisopropyl tartarate
2,2-DMP 2,2-dimethoxypropane
DMAP 4-dimethylaminopyridine
ICH_3Ph_3P iodomethyltriphenylphosphine
$LiAlH_4$ lithium aluminum hydride
NMO *N*-methylmorpholine *N*-oxide
PCC pyridinium chlorochromate
PPTS pyridinium *p*-toluenesulfonate
rt room temperature
TBAF tetrabutylammonium fluoride
TBDMS *tert*-butyldimethylsilyl
TBHP *tert*-butylhydroperoxide
TFA trifluoroacetic acid
THF tetrahydrofuran

Systematic name: (9*R*,10*R*,*E*)-9-Hydroxy-10-propyl-4,5,9,10-tetrahydro-2*H*-oxecine-2,8(3*H*)-dione

Compound class: Macrolide

Structure:

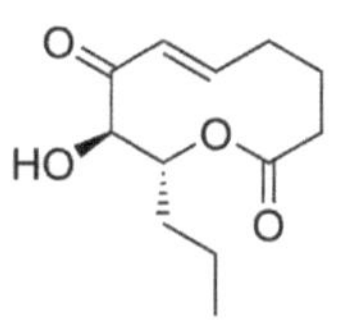

(−)-Stagonolide A (**1**)

Natural source: Fungus *Stagonospora cirsii*, a pathogen of *Cirsium arvense* causing necrotic lesions on leaves [1].

Pharmaceutical potential: Toxic metabolite [1]

Synthetic route:

Srihari et al. [2] accomplished the first stereoselective total synthesis of (−)-stagonolide A (**1**) with an overall yield of 24% (Scheme 1). The

https://doi.org/10.1016/B978-0-08-102822-3.00053-5

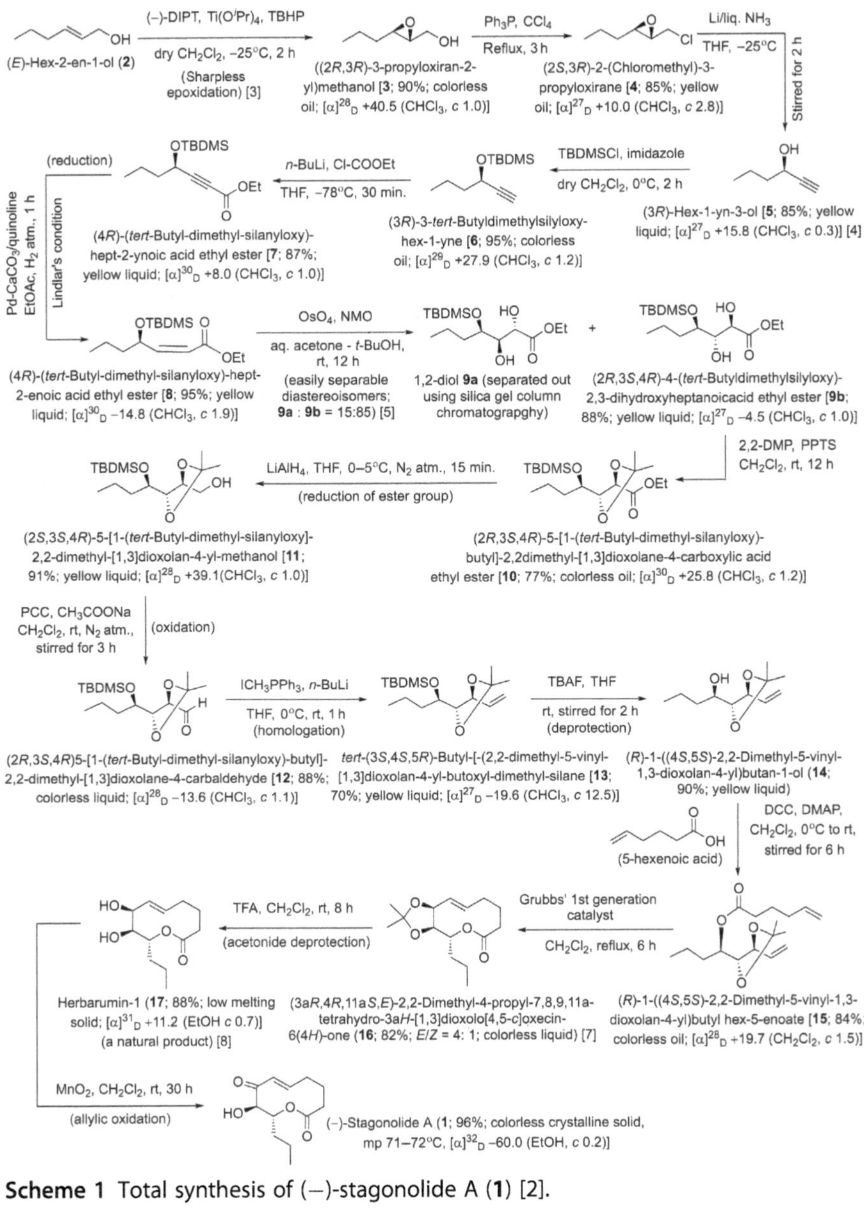

Scheme 1 Total synthesis of (−)-stagonolide A (**1**) [2].

investigators started with the commercially available *trans*-2-hexenol (**2**), which was epoxidized under Sharpless conditions [3] to obtain the chiral epoxy alcohol **3**, followed by its conversion to epoxy chloride **4** on treating with triphenylphosphine and carbon tetrachloride under reflux conditions. On reaction with Li/liq. NH_3 at −25°C following a standard procedure [4], compound **4** yielded chiral propargyl alcohol (3*R*)-hex-1-yn-3-ol (**5**; 85%) as a yellow liquid, the alcoholic group of which was then protected as TBDMS-ether **6**. The terminal acetylene within **6** on treatment with *n*-BuLi/ethyl chloroformate in THF at −78°C produced the α,β-unsaturated ester (4*R*)-(*tert*-butyl-dimethyl-silanyloxy)-hept-2-ynoic acid ethyl ester (**7**) in good yield (87%). The *cis*-reduction of acetylenic triple bond under Lindlar's condition gave *Z*-olefin **8**, which was subjected to dihydroxylation with osmium tetraoxide to give the polyhydroxylated ester **9b** along with undesired **9a** as an easily separable diastereomeric mixture (85:15) [5]. Isopropylidination of the desired major diol **9b** with 2, 2-DMP in CH_2Cl_2 yielded ester **10**, which was reduced with $LiAlH_4$ to furnish alcohol **11**. On oxidation with PCC, alcohol **11** gave aldehyde **12**, which was subsequently treated with ICH_3Ph_3P in the presence of *n*-BuLi resulting in the 1-C homologated product **13**. Silyl deprotection with TBAF yielded the key fragment with the required stereogenic centers (*R*)-1-((4*S*,5*S*)-2,2-dimethyl-5-vinyl-1,3-dioxolan-4-yl)butan-1-ol (**14**) in 90% yield as a yellow liquid. Alternatively, the investigators also synthesized this key intermediate **14** starting from D-ribose utilizing a chiral pool synthetic strategy in a concise manner [6].

After then, alcohol **14** underwent esterification with 5-hexenoic acid under standard DCC-DMAP conditions to give ester **15**, which in turn was subjected to metathesis reaction in the presence of first-generation Grubbs catalyst to produce the desired lactone ring **16** as a diastereomeric mixture *E*/*Z* (4:1) [7]. On acetonide deprotection of the crude *E*/*Z*-mixture of ester **15** with TFA in CH_2Cl_2, the easily separable diastereomers were obtained and eventually herbarumin-I (**17**) was separated out, whose analytical data matched those of the natural product [8] and the synthetic product from an earlier report by Furstner et al. [9]. Finally, oxidation of herbarumin-I (**17**) with MnO_2 afforded nonenolide (−)-stagonolide A (**1**) in a 96% yield as colorless crystalline solid, mp 71–72°C, $[\alpha]_D^{32}$ -60.0 (EtOH, *c* 0.2); the spectroscopic data were found to be identical to those of the natural product [1].

References

[1] Y. Oleg, M. Galin, B. Alexander, J. Agric. Food Chem. 55 (2007) 7707 E. Antonio, C. Alessio, B. Alexander, M. Galina, A. Anna, M. Andera, J. Nat. Prod., 71 (2008) 31.
[2] P. Srihari, B. Kumaraswamy, G. Maheswara Rao, J.S. Yadav, Tetrahedron: Asymmetry 21 (2010) 106.

[3] Y. Gao, R.M. Hanson, J.M. Klunder, S.Y. Ko, H. Masamune, K.B. Sharpless, J. Am. Chem. Soc. 109 (1987) 5765.
[4] S. Takano, K. Samizu, T. Sugihara, K. Ogasawara, J. Chem. Soc. Chem. Commun. (1989) 1344 J.S. Yadav, P.K. Deshpande, G.V.M. Sharma, Tetrahedron 46 (1990) 7033; S. Takano, T. Sugihara, K. Ogasawara, Synlett (1991) 279.
[5] J.K. Cha, W.J. Christ, Y. Kishi, Tetrahedron Lett. 24 (1983) 3943 J.K. Cha, W.J. Christ, Y. Kishi, Tetrahedron 40 (1984) 2247; S. Seiki, M. Yasuhisa, M. Toshio, J. Org. Chem. 55 (1990) 5424; E. Ludmila, N.S. Andre, J. Org. Chem. 71 (2006) 693.
[6] P. Srihari, B. Kumaraswamy, J.S. Yadav, Tetrahedron 65 (2009) 6304.
[7] H.R. Grubbs, S. Chang, Tetrahedron 54 (1998) 4413 A. Furstner, Angew. Chem. Int. Ed. 39 (2000) 3012.
[8] J.F. Rivero-Cruz, G. Garcia-Aguieee, C.M. Cerda-Garcia-Rojas, R. Mata, Tetrahedron 56 (2000) 5337 J.F. Rivero-Cruz, M. Martha, C.M. Cerda-Garcia-Rojas, R. Mata, J. Nat. Prod. 66 (2003) 511.
[9] A. Furstner, K. Radkowski, C. Wirtz, R. Goddard, W.C. Lehmann, R. Mynott, J. Am. Chem. Soc. 124 (2002) 7061.

CHAPTER FIFTY FOUR

(±)-Steenkrotin A

Abbreviations

Bz benzoyl
DBU 1,8-diazabicyclo[5.4.0]undec-7-ene
DMAP *N,N'*-dimethylaminopyridine
HMPA hexamethylphosphoric triamide
NMO 4-methylmorpholine *N*-oxide
PCC pyridinium chlorochromate
py pyridine
TBAF tetrabutylammonium fluoride
TBSOTf *tert*-butyldimethylsilyl trifluoromethanesulfonate
TMSOTf trimethylsilyl trifluoromethanesulfonate
THF tetrahydrofuran
TPAP tetrapropylammonium perruthenate
***p*-TsOH** *p*-toluenesulfonic acid

Systematic name: (1b*S*,2*R*,3*R*,6*S*,7a*R*,9*R*,10a*S*)-2-Hydroxy-1,1,3,6,9-pentamethyl-2,3,6,7,10,10a-hexahydro-1*H*-1a,3-epoxycyclopropa[3,4]cyclohepta[1,2-*d*]indene-5,8(1b*H*,9*H*)-dione

Compound class: Diterpenoid

Structure:

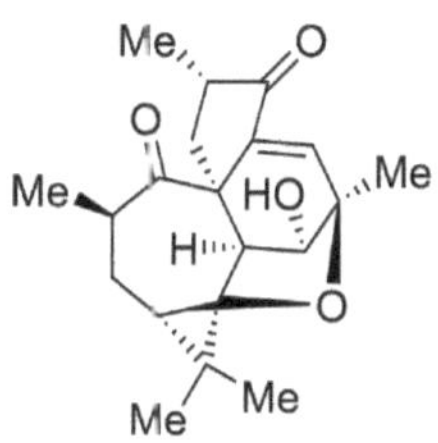

(±)-Steenkrotin A (**1**)

Natural source: *Croton steenkampianus* Gerstner (commonly known as "Marsh Fever-berry" and "Tonga Croton2; Euphorbiaceae) [1]

Pharmaceutical potential: Antimalarial [2]

Total Synthesis of Bioactive Natural Products
https://doi.org/10.1016/B978-0-08-102822-3.00054-7

Synthetic route:

Ding and his group [3], for the first time, achieved the total synthesis of (±)-steenkrotin A (**1**) involving key steps such as a rhodium carbenoid O—H insertion followed by a carbonylene cyclization to construct the sterically congested tetrahydrofuran subunit, two sequential SmI_2-mediated Ueno-Stork and ketyl-olefin cyclizations to install the [5,7]-spirobicyclic skeleton, and an intramolecular aldol condensation/vinylogous retro-aldol/aldol reaction sequence to form the final six-membered ring with the concomitant inversion of the relative configuration at C7. The total synthetic route as adopted by these investigators is outlined in Scheme 1.

The starting enone **2** [4] underwent 1,2-addition with allyllithium followed by Dauben-Michno oxidative rearrangement [5] to afford (4*S*,5*S*)-3-allyl-4-((*tert*-butyldimethylsilyl)oxy)-5-methylcyclohept-2-enone (**3**) as a colorless oil with 72% yield in two steps. The preferred boat transition state **5** added exclusively to the less hindered convex face with in situ generated dimethylcarbene from 2,2-dibromopropane (**4**) under the reaction conditions to produce the bicyclic alcohol **6** in 70% yield as a single diastereomer [6–8], which was then coupled with dimethyl 2-diazomalonate (**7**) through a carbenoid O—H insertion under rhodium acetate catalysis [9, 10] to furnish

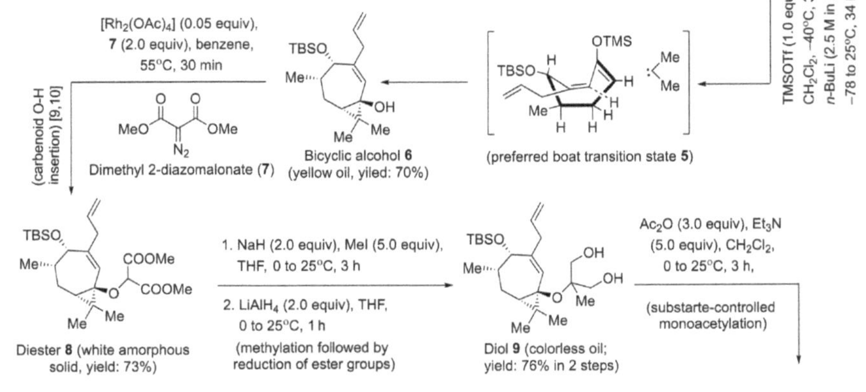

Scheme 1—Cont'd

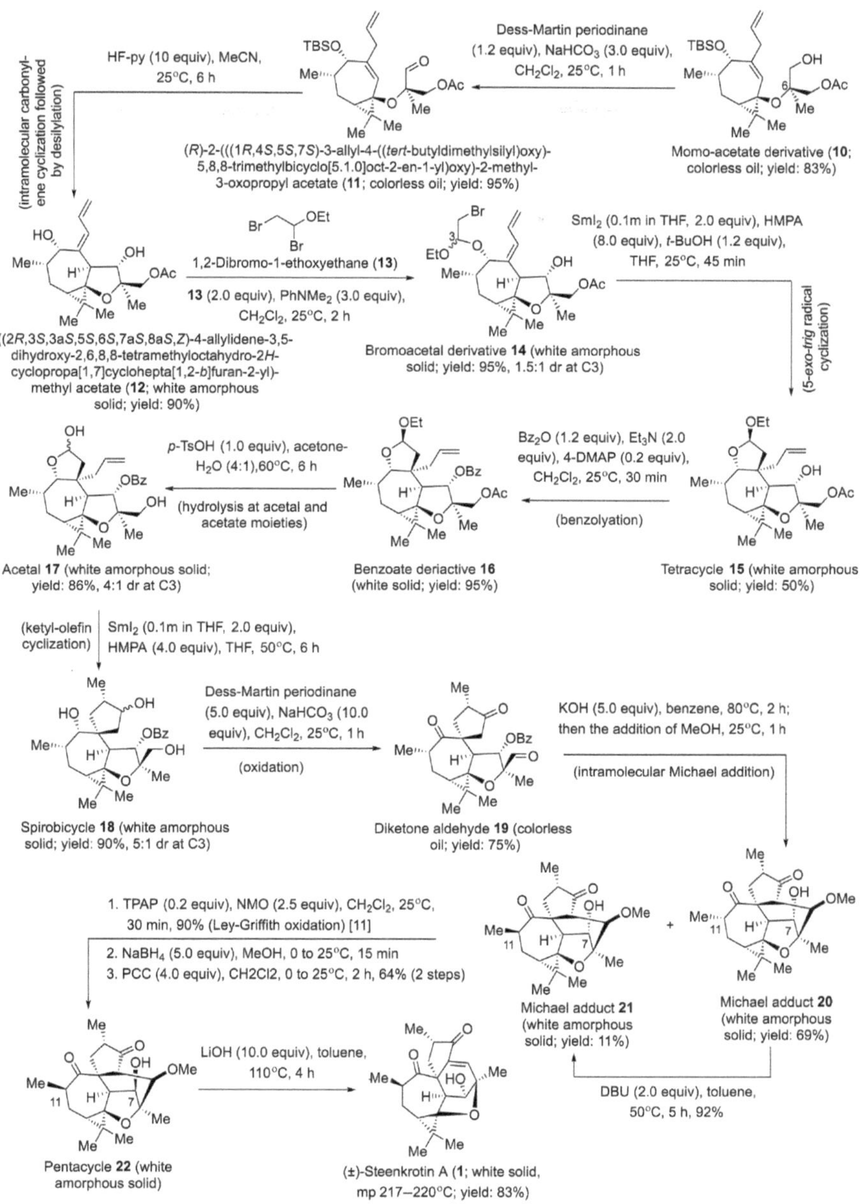

Scheme 1 Total synthesis of (±)-steenkrotin A (**1**) [3].

the diester **8** as white amorphous solid. After converting **8** into the diol **9** through methylation and reduction, the two hydroxy groups were successfully differentiated by a substrate-controlled acylation, thus leading to the monoacetate **10** in 83% yield as colorless oil. Dess-Martin oxidation of **10** resulted in the formation of 3-allyl-4-((*tert*-butyldimethylsilyl)oxy)-5,8,8-trimethylbicyclo[5.1.0]oct-2-en-1-yl)oxy)-2-methyl-3-oxopropyl acetate (**11**) in excellent yield of 95%. This compound **11** then underwent intramolecular carbonyl-ene cyclization on treatment with HF-pyridine in acetonitrile for 6 h to afford the [3,5,7]-tricycle derivative **12** in 90% yield.

After that the investigators carried out a regioselective alkylation of **12** with dibromoacetal **13** using *N,N*-dimethylaniline to generate the derivative **14** as a 1.5:1 mixture of diastereomers in 95% yield. The bromoacetal derivative **14** then proceeded through a 5-*exo-trig* radical cyclization of its C3β-isomer on slow addition of SmI_2 and HMPA in the presence of *t*-BuOH to afford the tetracycle **15** as a single diastereomer in 50% yield, the hydroxy group of which was next benzoylated with benzoylanhydride. Hydrolysis of both the acetal and acetate moieties of **16** delivered **17** in 86% yield (4:1 dr), and was immediately transformed into the [5,7]-spirobicycle **18** (90% yield, 5:1 dr at C3) through an SmI_2-mediated ketyl-olefin cyclization. The diketone aldehyde **19** obtained on Dess-Martin oxidation took part in intramolecular aldol condensation in the presence of potassium hydroxide in benzene, and subsequent removal of the benzoyl group in methanol gave rise to a 6.3:1 mixture of the Michael adducts **20** and **21** in 80% combined yield. The adduct **20** was converted to its C11 epimer **21** on treating with DBU base, and hence the relative configuration at the C7 position needed to be inverted. To achieve this conversion, the investigators carried out Ley-Griffith oxidation [11] of **21** that led to the corresponding triketone, which was then converted into **22** by global 1,2-reduction and selective oxidation in 58% yield over the three steps. Finally, treatment of **22** with LiOH in refluxing toluene afforded (±)-steenkrotin A (**1**) in 83% yield. The spectral properties of the synthetic diterpene **1** were found to be identical to those reported for the natural product [2].

References

[1] E. Pooley, The Complete Field Guide to Trees of Natal, Natal Flora Publication Trust, Durban, Zululand and Transkei, 1993, 222.
[2] A.M. Adelekan, E.A. Prozesky, A.A. Hussein, L.D. Ureña, P.H. van Rooyen, D.C. Liles, J.J.M. Meyer, B. Rodríguez, J. Nat. Prod. 71 (2008) 1919.
[3] S. Pan, J. Xuan, B. Gao, A. Zhu, H. Diang, Angew. Chem. Int. Ed. 54 (2015) 6905.

[4] G.M. Rubottom, J.M. Gruber, J. Org. Chem. 43 (1978) 1599.
[5] W.G. Dauben, D.M. Michno, J. Org. Chem. 42 (1977) 682.
[6] M.G.B. Drew, L.M. Harwood, A.J. Macías-Sánchez, R. Scott, R.M. Thomas, D. Uguen, Angew. Chem. Int. Ed. 40 (2001) 2311.
[7] I. Hayakawa, Y. Asuma, T. Ohyoshi, K. Aoki, H. Kigoshi, Tetrahedron Lett. 48 (2007) 6221.
[8] T. Ohyoshi, S. Funakubo, Y. Miyazawa, K. Niida, I. Hayakawa, H. Kigoshi, Angew. Chem. Int. Ed. 51 (2012) 4972.
[9] A. Padwa, M.M. Sá, J. Braz. Chem. Soc. 10 (1999) 231.
[10] J. Gong, G. Lin, W. Sun, C.-C. Li, Z. Yang, J. Am. Chem. Soc. 132 (2010) 16745.
[11] W.P. Griffith, S.V. Ley, G.P. Whitcombe, A.D. White, J. Chem. Soc. Chem. Commun. 1625 (1987).

CHAPTER FIFTY FIVE

Thiaplakortone A and Its Semisynthetic Derivatives

Abbreviations

Bn benzyl
Boc *tert*-butoxycarbonyl
DMF *N,N*-dimethylformamide
$LiAlH_4$ lithium aluminum hydride
LiHMDS lithium hexamethyl disilazide
rt room temperature
TEA triethylamine
THF tetrahydrofuran

Systematic name: 8-(2-Aminoethyl)-[1,4]thiazino[2,3-*f*]indole-5,9(4*H*,6*H*)-dione 1,1-dioxide (thiaplakortone A)

Compound class: Thiazine alkaloid

Structures:

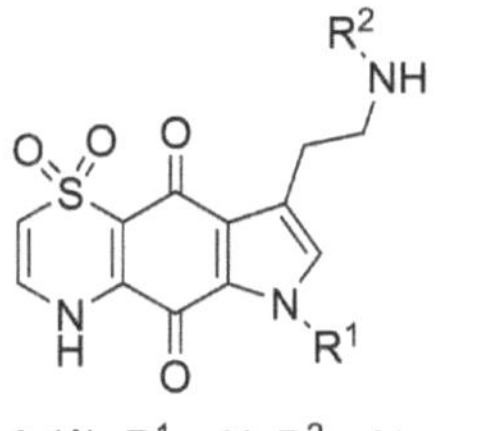

Thiaplakortone A (**1**): R^1 = H, R^2 = H
1a: R^1 = CH_3, R^2 = H
1b: R^1 = CH_3, R^2 = CH_3
1c: R^1 = H, R^2 = CH_3

Natural source: *Plakortis lita* (Australian marine sponge) [1]

Pharmaceutical potential: Antimalarial (displayed significant growth inhibition against chloroquine-sensitive [3D7] and chloroquine-resistant [Dd2] *Plasmodium falciparum* with IC_{50} values of 51 and 6.6 nM, respectively) [1].

https://doi.org/10.1016/B978-0-08-102822-3.00055-9

Synthetic route:

Quinn and his group [3] reported the first total synthesis of the potent antimalarial thiazine-alkaloid thiaplakortone A (**1**) along with its semisynthetic derivatives (**1a–1c**) as metabolically stable leads for antimalarial drugs. Scheme 1 summarizes the synthetic routes for the key intermediates (**5–7** and **10**) starting from commercially available 4-hydroxyindole (**2**). The benzyl ether derivative **3** [2] on Vilsmeier-Haack formylation gave 4-(benzyloxy)-1*H*-indole-3-carbaldehyde (**4**) as an off-white powder, which was then converted into a common intermediate, *tert*-butylcarbamate (**5**), in 80% overall yield. Selective methylation of **5** yielded **6**, which upon subsequent treatment with LiHMDS and MeI resulted in intermediate **7**. On the other hand, the common intermediate **5** was converted into *tert*-butyl(2-(4-(benzyloxy)-1*H*-indol-3-yl)ethyl)(methyl)carbamate (**10**; white solid) following subsequent tosylation, methylation, and deprotection (Scheme 1) [3].

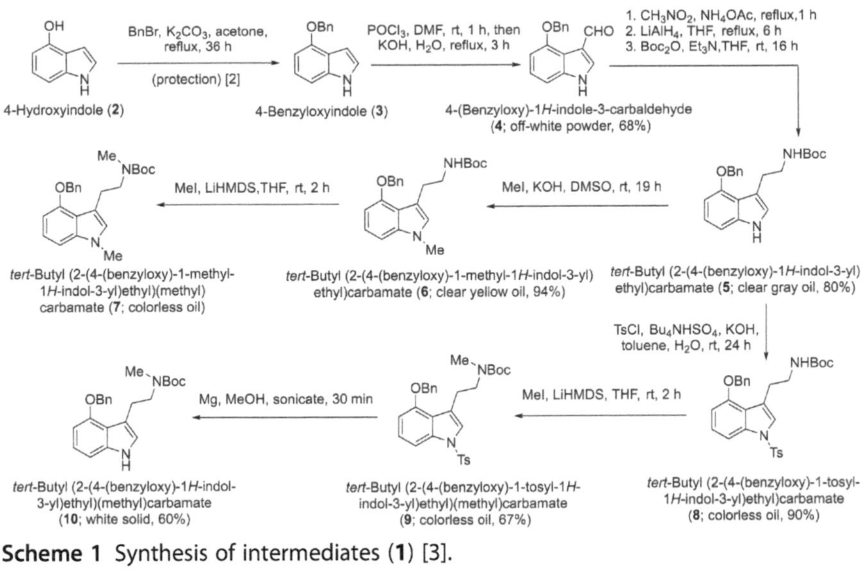

Scheme 1 Synthesis of intermediates (**1**) [3].

The investigators then synthesized the thiazine dioxide pyrroloquinones from these tryptamine derivatives **5–7** and **10** through a general procedure (Scheme 2). Initial hydrogenolysis of the benzyl protecting group was followed by immediate oxidation of the unstable intermediate with Frémy's salt; the resulting quinones participated in a double conjugate addition/oxidation sequence upon treatment with 2-aminoethanesulfinic acid to yield dihydrothiazine dioxides **11a–d** as separable mixtures with their corresponding regioisomers. Dihydrothiazine dioxides **11a–d** were then oxidized and deprotected under acidic conditions to produce the natural product **1** and thiazine dioxides **1a–c** as their hydrochloride salts.

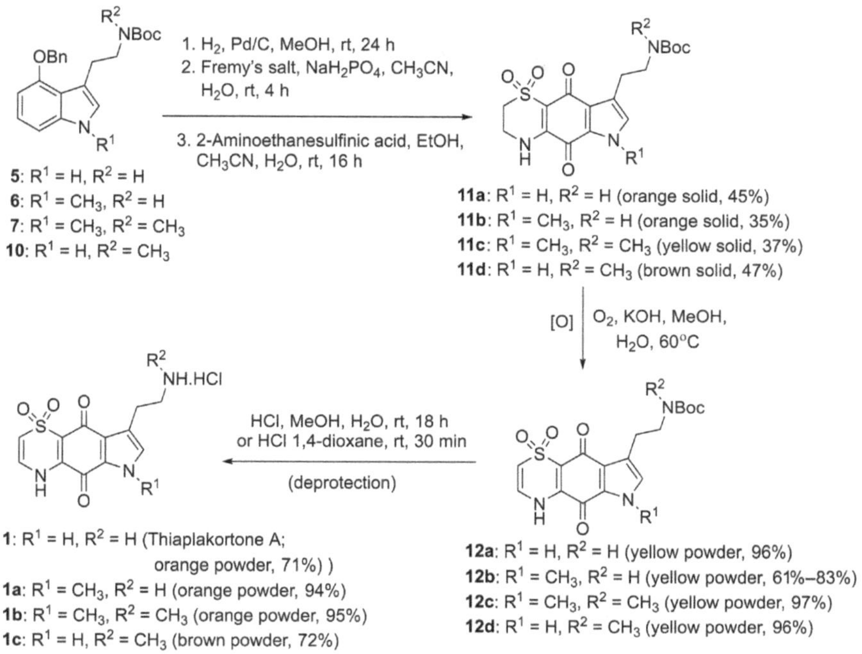

Scheme 2 Synthesis of thiaplakortone A (**1**) and its semisynthetic derivatives [3].

The investigators observed that the synthetic natural product **1** shows inhibitory activity against the chloroquine-sensitive *Plasmodium falciparum* line 3D7 with an IC_{50} of 104 nM; it was found to be 4.3-fold more potent against the multidrug resistant *P. falciparum* line Dd2 (IC_{50}=24 nM). Interestingly, the analogs **1a–c** also showed similar potential, with selectivity for the *P. falciparum* line Dd2 ranging between 1.7- and 3.4-fold.

References

[1] R.A. Davis, S. Duffy, S. Fletcher, V.M. Avery, R.J. Quinn, J. Org. Chem. 78 (2013) 9608.
[2] B.P. Smart, R.C. Oslund, L.A. Walsh, M.H. Gelb, J. Med. Chem. 49 (2006) 2858.
[3] R.H. Pouwer, S.M. Deydier, P.V. Le, B.D. Schwartz, N.C. Franken, R.A. Davis, M.J. Coster, S.A. Charman, M.D. Edstein, T.S. Skinner-Adams, K.T. Andrews, I.D. Jenkins, R.J. Quinn, ACS Med. Chem. Lett. 5 (2014) 178.

CHAPTER FIFTY SIX

Vinigrol

Abbreviations

$(COCl)_2$ oxalyl chloride
AIBN 2,2′-azobis(2-methylpropionitrile)
Bn benzyl
brsm based on recovered starting material
CDMT 2-chloro-4,6-dimethoxy-1,3,5-triazine
CS_2 carbon disulfide
DCE dichloroethane
DCM dichloromethane
DIAD diisopropyl azodicarboxylate
DIBAL diisobutylaluminum hydride
DMAP 4-dimethylaminopyridine
DMF *N*,*N*-dimethylformamide
DMP Dess-Martin periodinane
DMSO dimethyl sulfoxide
DPS *tert*-butyldiphenylsilyl
KHMDS potassium hexamethyldisilazide
LAH lithium aluminum hydride
LDA lithium diisopropylamide
MeOH methanol
MsCl methane sulfonyl chloride
NMM *N*-methylmorpholine
NMO *N*-methylmorpholine-*N*-oxide
***o*-DCB** *ortho*-dichlorobenzene
OsO_4 osmium tetraoxide
$PhNTf_2$ *N*-phenyltrifluorosulfonimide
Piv trimethyl acetyl
PivCl pivaloyl chloride
PTSA *para*-toluenesulfonic acid
rt room temperature
TBAF tetrabutylammonium fluoride
TBS *tert*-butyldimethylsilyl
TEA triethylamine
TEMPO 2,2,6,6-tetramethylpiperidine-1-oxyl
TMEDA tetramethylethylenediamine
$TrisNHNH_2$ Triisopropylbenzenesulfonyl hydrazide

Total Synthesis of Bioactive Natural Products
https://doi.org/10.1016/B978-0-08-102822-3.00056-0

Systematic name: (1*R*,4*S*,5*S*,8*R*,8a*S*,9*R*,12*R*)-3-(Hydroxymethyl)-12-isopropyl-8,9-dimethyl-1,4,4a,5,6,7,8,8a-octahydro-1,5-butanonaphthalene-4,8a-diol

Compound class: Diterpene

Structure:

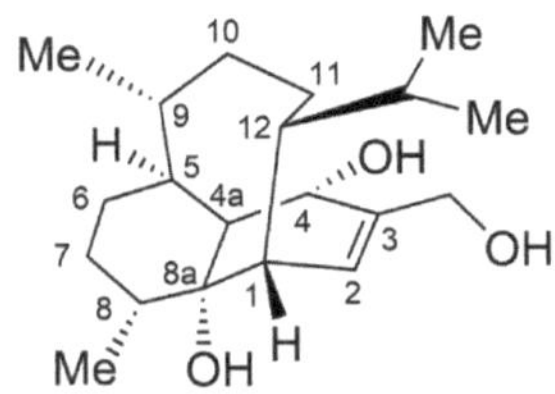

Vinigrol (**1**)

Natural source: *Virgaria nigra* F-5408 (fungal strain) [1]

Pharmaceutical potential: Antiplatelet aggregation [2,3]; antitumor necrosis factor antagonism [4]; antiinflammatory [5,6]

Synthetic routes:

Vinigrol (**1**) has a unique tricyclic core comprising a *cis*-fused [4.4.0] system with a four-carbon bridge between C-1 and C-5 and features eight contiguous stereocenters. The rare boat half chair conformation of the eight-membered ring makes the molecule quite special among the naturally occurring diterpenoids, and extreme difficulty in preparing this molecule originates from an unprecedented and highly congested decahydro-1, 5-butanonaphthalene ring system! This unique structural feature as well as the significant biological activities of vinigrol (**1**) has created considerable attention in the community of synthetic chemists; as a result a number of initiatives for the total synthesis of this compound have been made over the past two decades [7–30]. In 1993, Hanna and coworkers unearthed the first synthesis of the tricyclic core of **1** using an anionic oxy-Cope ring expansion [7]. Eventually, Baran and his group for the first time succeeded in solving this longstanding challenge of the total synthesis of vinigrol (**1**) in 2009 (Scheme 1) [31].

Baran and coworkers accomplished total synthesis of the target molecule in 23 steps with 3% overall yield from commercially available materials (Scheme 1). First, the starting intermediate (1*S*,3a*R*,3a^1*R*,6a*S*,10*R*)-3a-((*tert*-butyldimethylsilyl)oxy)-10-isopropyl-2,3,3a,4,6,6a,7,8-octahydro-1*H*-1,4-methanophenalen-5(3a1H)-one [(±) **8**] was prepared in seven steps from the commercially available cyclohexa-1,3-diene-1,3-diylbis(oxy))bis

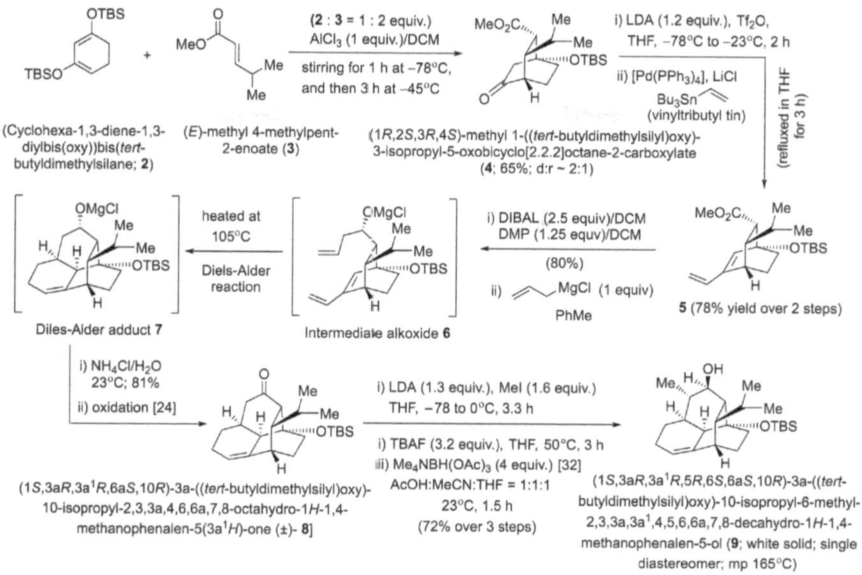

Scheme 1—Cont'd

(Continued)

(*tert*-butyldimethylsilane) (**2**) and (*E*)-methyl 4-methylpent-2-enoate (**3**) following the procedure as they reported earlier [24]. Compound **9** was obtained as a single diastereomer (white solid) in 72% yield over a three-step sequence from **8** via alkylation at C-9 with LDA/MeI, silyl group removal with TBAF, and successive reaction with Evans' $Me_4NBH(OAc)_3$-mediated 4 hydroxyl-directed reduction. Diol **9** on treatment with MsCl followed by KHMDS produced an olefinic ketone that reacted with in situ generated bromonitrile oxide (from the reaction of dibromoformaldoxime and $KHCO_3$) [33] through dipolar cycloaddition, leading to the formation of bromoisoxazole **10** as a single diastereomer in 88% yield on a gram scale. This cycloaddition proceeded with complete control over regio- and positional selectivity to produce the single diastereomer of **10** as evidenced from X-ray crystallographic studies. On ketonic reduction with DIBAL followed by directed olefin hydrogenation (20% Crabtree's catalyst, H_2, B(O-*i*Pr)$_3$) [34], bromoisoxazole **10** afforded alcohol derivative **11** in 83% yield.

Subsequent xanthate formation (NaH, CS_2, MeI) followed by Chugaev elimination resulted in the formation of (3a*R*,6*S*,6a*R*,10*R*,10a*S*,11*R*,14*R*)-3-bromo-11-isopropyl-14-methyl-4,5,6,6a,9,10-hexahydro-3a*H*-6,10-butanonaphtho[1,8a-*d*]isoxazole (**12**; 85% overall yield), which was then

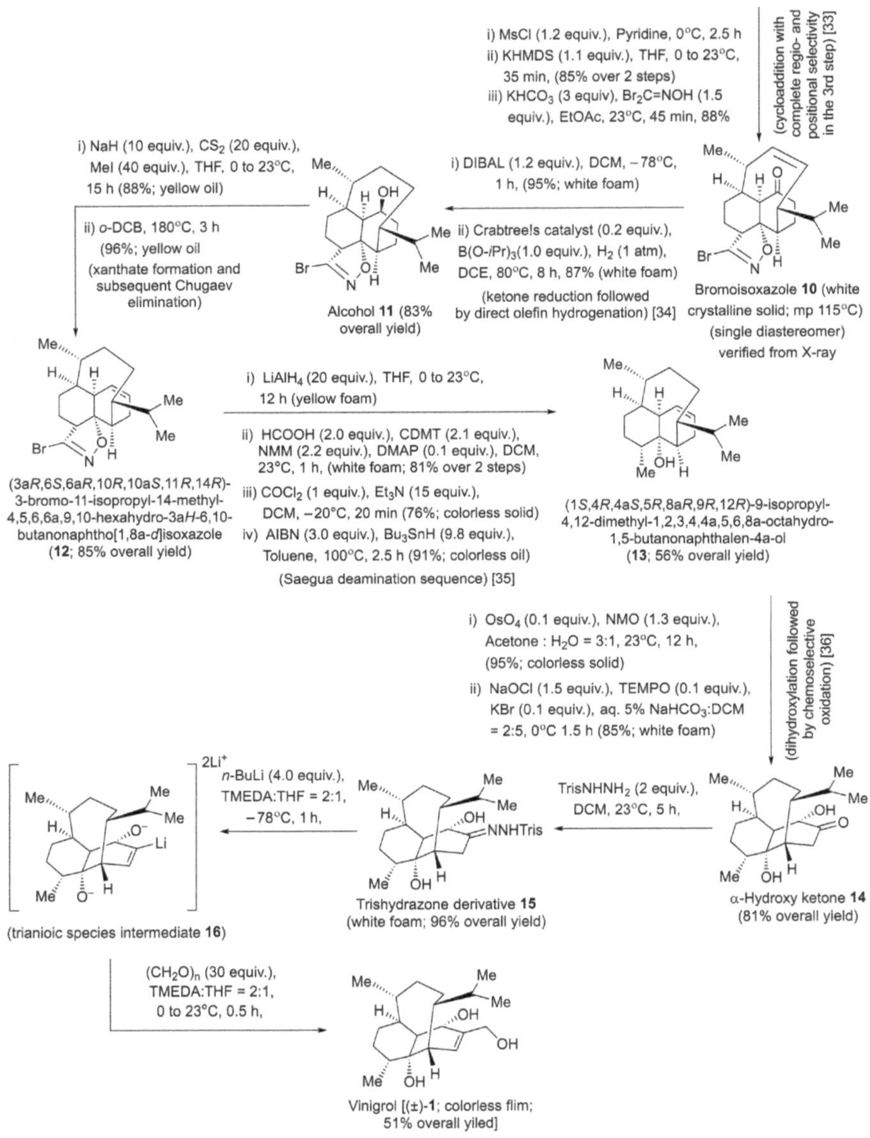

Scheme 1 Baran's total synthesis of vinigrol (**1**) [31].

converted to the desired tertiary alcohol **13** by the Saegusa deamination sequence (Scheme 1) in 56% overall yield [35]. α-Hydroxy ketone **14** was generated by dihydroxylation of **13** with OsO_4 followed by chemoselective oxidation of the resulting diol with NaOCl/TEMPO in 81% overall yield [36]. (±)-Vinigrol (**1**) was then obtained from the trisylhydrazone derivative

15 by the application of the Shapiro reaction [37] presumably via the trianionic species **16**. The synthetic product was found to be spectroscopically identical in all respects to a natural sample of vinigrol, with the exception of optical rotation [1,31].

Barriault and his group reported another synthetic strategy from commercially available starting materials for the formal 24-step total synthesis of vinigrol (**1**) (Scheme 2) that features the construction of the vinigrol carbocyclic core through a sequence involving a stereocontrolled Claisen rearrangement and an intramolecular Diels-Alder reaction as key steps. The starting compounds, (*E*)-6-(*tert*-butyl-diphenyl-silanyloxy)-2-methyl-hex-2-en-1-ol (**17**) and (±)-2,2-dimethyl-propionic acid 4,4-dimethoxy-cyclohexyl ester (**18**), were respectively from butane-1,4-diol and cyclohexane-1,4-diol. A thermal Claisen rearrangement [39,40] of alcohol **17** and ketal **18** in the presence of propionic acid produced ketone **19** (yellow oil; 62% yield), which was then converted into the enol triflate **20** as a 50:50 epimeric mixture at C-3. A Stille reaction between enol triflate intermediate **20** and vinyltributylstannane furnished a mixture of epimeric dienes **21** and **22** in 80% yield; the α-isomer was transformed into the β-isomer **22** via four steps utilizing Mitsunobu reaction as a key step. Removal of the DPS moiety by using TBAF/THF gave the alcohol (±)-(1*R*,5*S*)-5-((*R*)-6-hydroxy-2-methylhexan-3-yl)-4-vinylcyclohex-3-en-1-yl-pivalate (**23**; colorless oil; 84%). The tricycle **24** (±)-(1*S*,3*S*,5*R*,4a*R*,12*R*)-12-isopropyl-9-oxo-1,2,3,4,4a,5,6,7-octahydro-1,5-butanonaphthalen-3-yl pivalate was then prepared from **23** in four steps

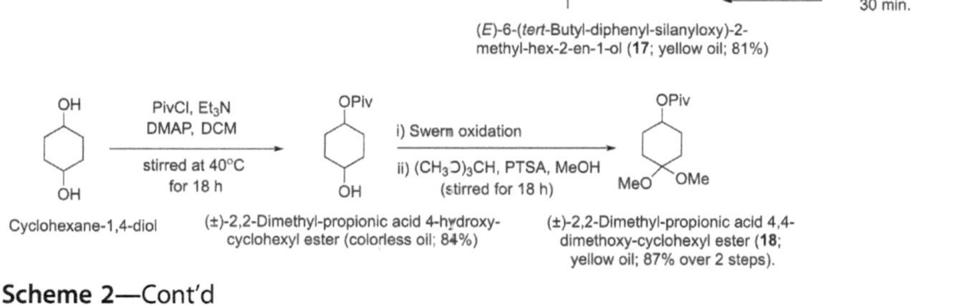

Scheme 2—Cont'd

(Continued)

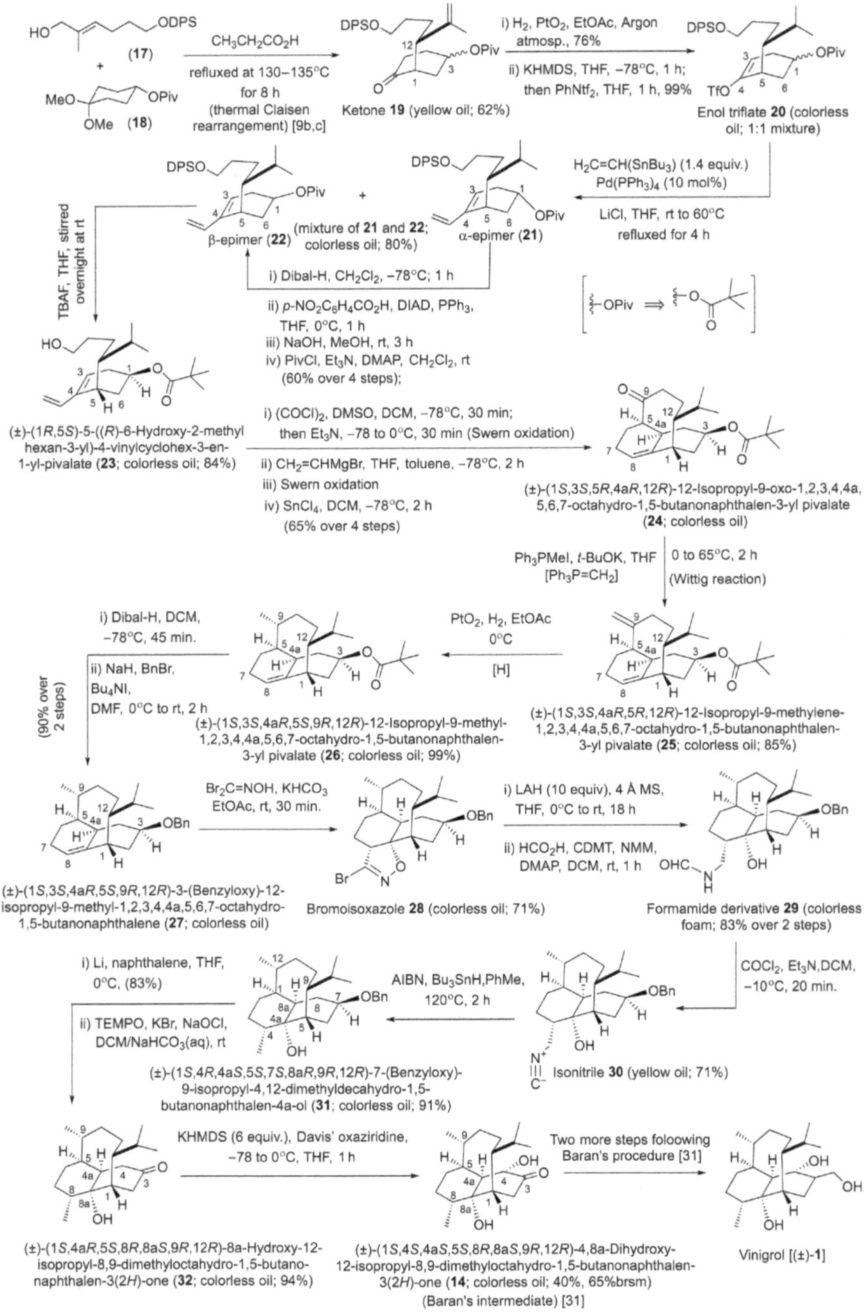

Scheme 2 Barriault's total synthesis of vinigrol (**1**) [38].

in an overall yield of 65%. This, in turn, was transformed into the benzyl ether **27** following the next four steps. This tricyclic alkene **27** underwent a [3 + 2] cycloaddition with bromonitrile oxide to give bromoisoxazole **28** as colorless oil in 71% yield, which was then converted into the isonitrile derivative **30** via formamide **29** [33]. The ketone 8a-hydroxy-12-isopropyl-8,9-dimethyloctahydro-1,5-butano-naphthalen-3(2*H*)-one (**32**; colorless oil) was obtained from isonitrile **30** through a Saegusa deamination [41] followed by removal of the benzyl group and TEMPO oxidation of the secondary alcohol intermediate. In the next step, the investigators carried out the regioselective α-oxygenation at C-4 to obtain the diol (±)-(1*S*,4*S*,4a*S*,5*S*,8*R*,8a*S*,9*R*,12*R*)-4,8a-dihydroxy-12-isopropyl-8,9-dimethyloctahydro-1,5-butanonaphthalen-3(2*H*)-one (**14**; called Baran's intermediate) in 40% yield (65% based on recovered starting material). They converted this intermediate **14** into (±)-vinigrol (**1**) in two more steps following the procedure of Baran and coworkers (Scheme 1).

References

[1] I. Uchida, T. Ando, N. Fukami, K. Yoshida, M. Hashimoto, T. Tada, S. Koda, Y. Morimoto, J. Org. Chem. 52 (1987) 5292.
[2] T. Ando, Y. Tsurumi, N. Ohata, I. Uchida, K. Yoshida, M. Okuhara, J. Antibiot. 41 (1988) 25.
[3] T. Ando, K. Yoshida, M. Okuhara, J. Antibiot. 41 (1988) 31.
[4] D.B. Norris, P. Depledge, A.P. Jackson, PCT Int. Appl. WO 91 07 953 Chem. Abstr. 115 (1991) 64776.
[5] J.T. Keane, PCT Int. Appl. WO 01 00 229 Chem. Abstr. 134 (2001) 80816.
[6] H. Nakajima, N. Yamamoto, T. Kaizu, T. Kino, Jpn. Kokai Tokkyo Koho JP 07 – 206668 Chem. Abstr. 123 (1995) 246812.
[7] J.F. Devaux, I. Hanna, J.Y. Lallemand, J. Org. Chem. 58 (1993) 2349.
[8] J.F. Devaux, I. Hanna, J.Y. Lallemand, T. Prange, J. Chem. Res. Synth. (1996) 32.
[9] G. Mehta, K.S. Reddy, Synlett (1996) 625.
[10] M. Kito, T. Sakai, N. Haruta, H. Shirahama, F. Matsuda, Synlett (1996) 1057.
[11] M. Kito, T. Sakai, H. Shirahama, M. Miyashita, F. Matsuda, Synlett (1997) 219.
[12] J.F. Devaux, I. Hanna, J.Y. Lallemand, J. Org. Chem. 62 (1997) 5062.
[13] F. Matsuda, T. Sakai, N. Okada, M. Miyashita, Tetrahedron Lett. 39 (1998) 863.
[14] F. Matsuda, M. Kito, T. Sakai, N. Okada, M. Miyashita, H. Shirahama, Tetrahedron 55 (1999) 14369.
[15] L.A. Paquette, R. Guevel, S. Sakamoto, I.H. Kim, J. Crawford, J. Org. Chem. 68 (2003) 6096.
[16] L. Gentric, I. Hanna, L. Ricard, Org. Lett. 5 (2003) 1139.
[17] L. Morency, L. Barriault, Tetrahedron Lett. 45 (2004) 6105.
[18] L.A. Paquette, I. Efremov, Z.S. Liu, J. Org. Chem. 70 (2005) 505.
[19] L.A. Paquette, I. Efremov, J. Org. Chem. 70 (2005) 510.
[20] L.A. Paquette, Z.S. Liu, I. Efremov, J. Org. Chem. 70 (2005) 514.
[21] L. Morency, L. Barriault, J. Org. Chem. 70 (2005) 8841.
[21] C.M. Grise, G. Tessier, L. Barriault, Org. Lett. 9 (2007) 1545.
[22] G. Tessier, L. Barriault, Org. Prep. Proc. Int. 39 (2007) 311 (Review).

[23] M.S. Souweha, G.D. Enright, A.G. Fallis, Org. Lett. 9 (2007) 5163.
[24] T.J. Maimone, A.-F. Voica, P.S. Baran, Angew. Chem. Int. Ed. 47 (2008) 3054.
[25] J.G.M. Morton, L.D. Kwon, J.D. Freeman, J.T. Njardarson, Tetrahedron Lett. 50 (2009) 1684.
[26] J.G.M. Morton, L.D. Kwon, J.D. Freeman, J.T. Njardarson, Synlett. (2009) 23.
[27] J.G.M. Morton, C. Draghici, L. Kwon, J.T. Njardarson, Org. Lett. 11 (2009) 4492.
[28] M. Harmata, N.L. Calkins, Chemtracts 22 (2009) 205 (Review).
[29] J.-Y. Lu, D.G. Hall, Angew. Chem. 122 (2010) 2336 Angew. Chem. Int. Ed. 49 (2010) 2286 (Review).
[30] A.D. Huters, N.K. Garg, Chem. Eur. J. 16 (2010) 8586 (Review).
[31] T. Maimone, J. Shi, S. Ashida, P.S. Baran, J. Am. Chem. Soc. 131 (2009) 17066.
[32] D.A. Evans, K.T. Chapman, Tetrahedron Lett. 49 (1986) 5939.
[33] D.M. Vyas, Y. Chiang, T.W. Doyle, Tetrahedron Lett. 25 (1984) 487.
[34] R.H. Crabtree, M.W. Davis, J. Org. Chem. 51 (1986) 2655.
[35] T. Saegusa, S. Kobayashi, Y. Ito, N. Yasuda, J. Am. Chem. Soc. 90 (1968) 4182.
[36] N.Z. Burns, P.S. Baran, Angew. Chem. Int. Ed. 46 (2007) 205.
[37] R.H. Shapiro, M.J. Heath, J. Am. Chem. Soc. 89 (1967) 5734.
[38] J. Poulin, C.M. Grisé-Bard, L. Barriault, Angew. Chem. Int. Ed. 51 (2012) 2111.
[39] P.F. Shuda, S.J. Potlock, H. Ziffer, Tetrahedron 43 (1987) 463.
[40] M. Komada, K. Fukuzumi, M. Kumano, Chem. Pharm. Bull. 37 (1989) 1691.
[41] T. Saegusa, S. Kobayashi, Y. Ito, N. Yasuda, J. Am. Chem. Soc. 90 (1968) 4182.

CHAPTER FIFTY SEVEN

Withaferin A

Abbreviations

DDQ 2,3-dichloro-5,6-dicyano-1,4-benzoquinone
DMF *N,N'*-dimethylformamide
***i*Pr** isopropyl
LAH lithium aluminum hydride
LICHA lithium isopropylcyclohexylamide
***m*-CPBA** *meta*-chloroperbenzoic acid
MOM methoxymethyl
NMO *N*-methylmorpholine *N*-oxide
PCC pyridinium chlorochromate
PDC pyridinium dichromate
***p*-TsOH** *para*-toluene sulfonic acid
py pyridine
TBDMSCl *tert*-butyldimethylsilylchloride
THF tetrahydrofuran

Systematic name: (4*S*,4a*R*,5a*R*,6a*S*,6b*S*,9*R*,9a*S*,11a*S*,11b*R*)-4-Hydroxy-9-((*S*)-1-((*R*)-5-(hydroxymethyl)-4-methyl-6-oxo-3,6-dihydro-2H-pyran-2-yl) ethyl)-9a,11b-dimethyl-5a,6,6a,6b,7,8,9,9a,10,11,11a,11b-dodecahydro-cyclopenta[1,2]phenanthro[8a,9-*b*]oxiren-1(4*H*)-one

Alternatively

4β,27-Dihydroxy-5β,6β-epoxy-1-oxo-(20*R*,22*R*)witha-2,24-dienolide/ (4β,5β,6β,20*R*,22*R*)-4,27-dihydroxy-5,6:22,26-diepoxyergosta-2,24-diene-1,26-dione/ (4β,5β,6β,20*R*,22*R*)-5,6-epoxy-4,22,27-trihydroxy-1-oxoergosta-2,24-dien-26-oic acid δ-lactone

Compound class: Steroidal lactone

Total Synthesis of Bioactive Natural Products
https://doi.org/10.1016/B978-0-08-102822-3.00057-2

Structure:

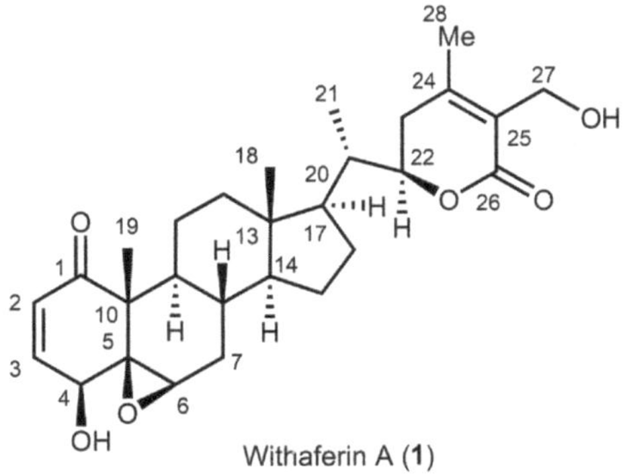

Withaferin A (**1**)

Natural sources: *Withania somnifera* Dunal (leaves; family: Solanaceae) ("Ashwagandha" in Ayurveda, Indian ginseng) [1]; *Acnistus arborescens* (L.) Schlecht (leaves; family: Solanaceae) [2, 3].

Pharmaceutical potentials: Antitumor and anticancerous [4–19]; antiangiogenic [20–23]; immunomodulatory [24–27]; antiinflammatory [28–33]; antiherpes simplex viral activity [34].

Synthetic route:

The first stereoselective total synthesis of withaferin A (**1**) was reported by Hirayama et al. [35]; they synthesized the steroidal lactone from an α, β-unsaturated δ-lactone intermediate (**14**) that was also prepared earlier by them from commercially available 3β-hydroxy-22,23-bisnorchol-5-enoic acid (**2**) via a number of synthetic steps as depicted in Scheme 1 [36]. Compound **2** on reduction with LAH followed by oxidation with an excess amount of DDQ in refluxing dioxane gave the enone-alcohol **3**, which was converted into the triol diacetate **4** as per the procedure of Fuerst et al. [37]. The 1,3-bis(MOM) ether derivative **5** with a newly formed unsaturation at C_{22} was obtained in good yield (81%) from **4** in five steps. A 5:1 mixture of tosylates **6a** (syrup) and **6b** (solid; mp 83–85°C) was formed on osmium tetroxide oxidation of **5** followed by tosylation with *p*-TsCl/py—the major isomer **6a** was separated out in 62% yield and on treatment with K_2CO_3 in methanol it underwent epoxidation to afford chiral 22(*S*),23-epoxide derivative **7** in quantitative yield. A regiospecific alkylation at C_{23} of **7** was carried out with 2-methyl-1,3-dithiane anion followed by dethioketalization with mercuric oxide-boron trifluoride etherate to yield the 22-hydroxy-24-one derivative **8** (76%), which was subsequently transformed into the

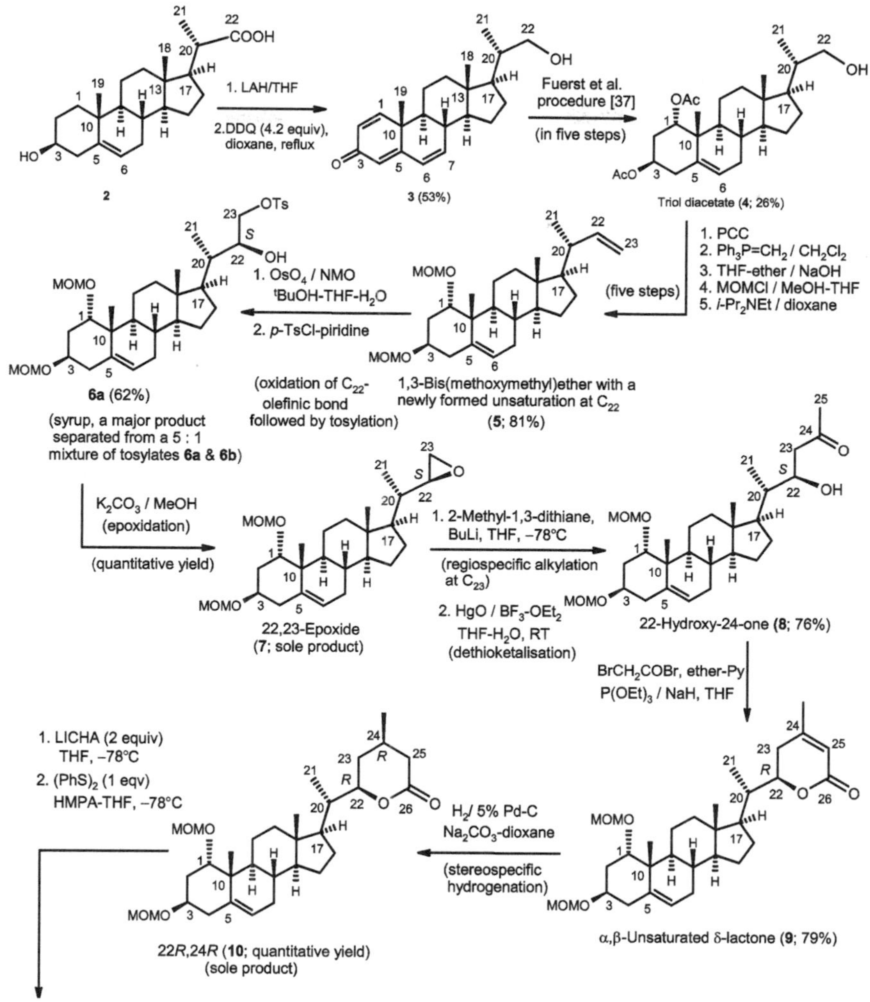

Scheme 1—Cont'd

(Continued)

unsaturated δ-lactone **9** (79%) following the methodology developed by Weihe and McMorris [38]. Compound **9** in turn was converted into the saturated lactone **10** with 24(*R*) configuration in quantitative yield through stereospecific hydrogenation with H_2/Pd-C. This saturated lactone **10** afforded a 3:2 mixture of sulfides **11a** (25*S*; mp 140–141°C) and **11b** (25*R*; mp 90–92°C) on sulfenylation with diphenyl disulfide by

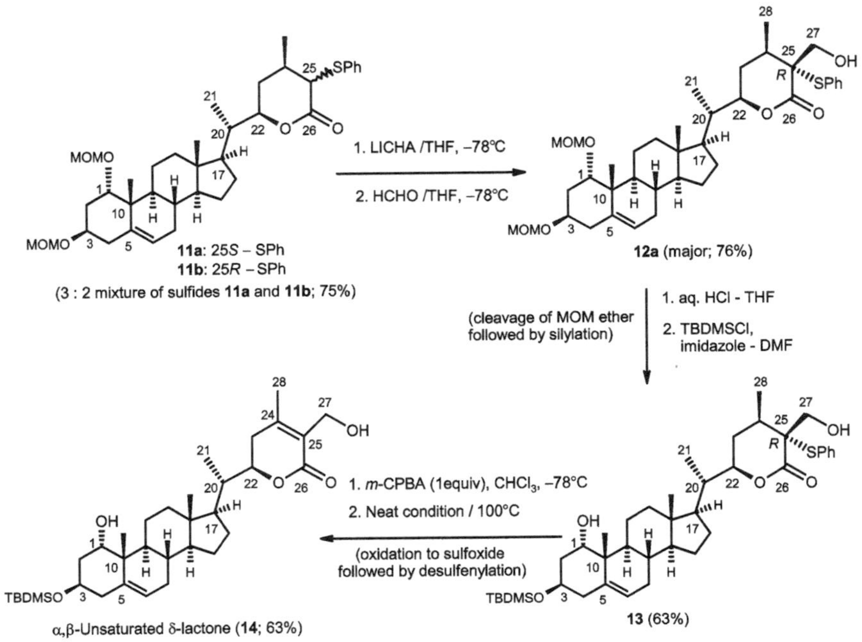

Scheme 1 Preparation of α,β-unsaturated δ-lactone (**14**) from 3β-hydroxy-22, 23-bisnorchol-5-enoic acid (**2**).

the inverse quench method [39]—the epimer **11b** (at C_{25}) was then converted into the major isomer **11a** with excess amount of LICHA. Treatment of **11a** with excess monomeric formaldehyde yielded a 9:1 mixture of stereoisomers **12a** (syrup) and **12b** (solid; mp 179–181°C). The major stereoisomer **12a** (25*R*; 76%) underwent cleavage reaction of 1,3-bis(MOM) ether linkages with acid followed by silylation with TBDMSCl at room temperature to afford the 3-TBDMS ether (**13**; mp 230–233°C) in 63% yield. Oxidation of **13** with *m*-CPBA to the sulfoxide followed by desulfenylation ultimately yielded the α,β-unsaturated δ-lactone derivative (**14**; 63%; mp 197–198°C).

Withaferin A (**1**) was then synthesized from the α,β-unsaturated δ-lactone intermediate (**14**) (Scheme 2) [35]; on epoxidation with *m*-CPBA followed by selective protection of the C_{27}-OH group and subsequent oxidation of C_1-OH with PDC, compound **14** afforded 5α-,6α-epoxy-1-one derivative (**15**) stereospecifically in 49% yield (solid; mp 173–174°C). The expoxy moiety in **15** underwent cleavage in a regio- and stereospecific manner with thiophenol in the presence of aluminum oxide to form 6β-phenylthio-5α-ol (**16**; amorphous solid) in 37% yield. On heating at

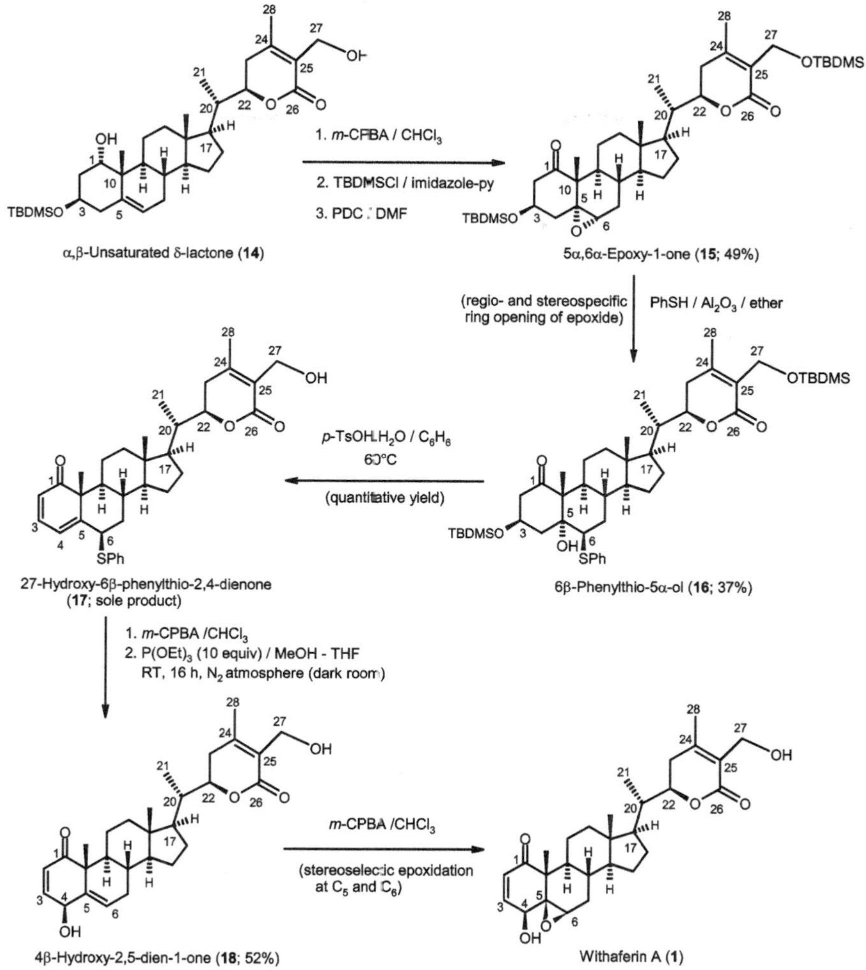

Scheme 2 Synthesis of withaferin A (**1**) from α,β-unsaturated δ-lactone (**14**).

60°C in benzene in the presence of *p*-TsOH.H_2O, the 6β-phenylthio-5α-ol derivative (**16**) underwent simultaneous dienone formation and deprotection of the hydroxy group at C_{27} to give 27-hydroxy-6β-phenylthio-2,4-dienone (**17**) in quantitative yield. The investigators carried out oxidation of derivative **17** with *m*-CPBA followed by quick treatment of the resulting sulfoxide with excess trimethyl phosphite (10 equiv) at room temperature for 16 h under nitrogen atmosphere in a dark room to obtain the desired 4β-hydroxy-2,5-dien-l-one (**18**; mp 198–199°C) in 52% yield. Epoxidation of **18** at the C_5–C_6 double bond with *m*-CPBA ultimately yielded withaferin A stereoselectively (**1**; $C_{28}H_{38}O_6$; mp 243–245°C, $[\alpha]_D$ +114° ($CHCl_3$, *c* 0.56) [1]; mp 252–253°C, $[\alpha]_D^{28}$ +125° ($CHCl_3$, *c* 1.30) [3]).

References

[1] D. Lavie, E. Glotter, Y. Shvo, J. Org. Chem. 30 (1965) 1774.
[2] S.M. Kupchan, R.W. Doskotch, P. Bollinger, A.T. McPhail, R.M. Smith, J.A.S. Renauld, J. Am. Chem. Soc. 87 (1965) 5805.
[3] S.M. Kupchan, W.K. Anderson, P. Bollinger, R.W. Doskotch, R.M. Smith, J.A.S. Renauld, H.K. Schnoes, A.L. Burlingame, D.H. Smith, J. Org. Chem. 34 (1969) 3858.
[4] B. Shohat, S. Gitter, A. Abraham, D. Lavie, Cancer Chemother. Rep. 51 (1967) 271.
[5] B. Jayaprakasam, Y. Zhang, N.P. Seeram, M.G. Nair, Life Sci. 74 (2003) 125.
[6] S.D. Stan, E.-R. Hahm, R. Warin, S.V. Singh, Cancer Res. 68 (2008) 7661.
[7] M.N. Ndlovu, C. Van Lint, K. Van Wesemael, P. Callebert, D. Chalbos, G. Haegeman, W.V. Berghe, Mol. Cell Biol. 29 (2009) 5488.
[8] K. Panjamurthy, S. Manoharan, S. Balakrishnan, K. Suresh, M.R. Nirmal, N. Senthil, L.M. Alias, Afr. J. Tradit. Complement Altern. Med. 6 (2009) 1.
[9] A. Grover, A. Shandilya, A. Punetha, V.S. Bisaria, D. Sundar, BMC Genomics 11 (Suppl 4) (2010) S15, https://doi.org/10.1186/1471-2164-11-S4-S15.
[10] A. Grover, A. Shandilya, A. Punetha, V.S. Bisaria, D. Sundar, BMC Genomics 11 (Suppl 4) (2010) S25, https://doi.org/10.1186/1471-2164-11-S4-S25.
[11] W. Suttana, S. Mankhetkorn, W. Poompimon, A. Palagani, S. Zhokhov, S. Gerlo, G. Haegeman, W.V. Berghe, Mol. Cancer 9 (2010) 99.
[12] S. Koduru, R. Kumar, S. Srinivasan, M.B. Evers, C. Damodaran, Mol. Cancer Ther. 9 (2010) 202.
[13] A.K. Samadi, X. Tong, R. Mukerji, H. Zhang, B.N. Timmermann, M.S. Cohen, J. Nat. Prod. 73 (2010) 1476.
[14] Y. Yu, A. Hamza, T. Zhang, M. Gu, P. Zou, B. Newman, Y. Li, A.A.L. Gunatilaka, L. Whitesell, C.-G. Zhan, D. Sun, Biochem. Pharmacol. 79 (2010) 542.
[15] J. Lee, E.-R. Hahm, S.V. Singh, Carcinogenesis 31 (2010) 1991.
[16] R. Munagala, H. Kausar, C. Munjal, R.C. Gupta, Carcinogenesis 32 (2011) 1697.
[17] A. Grover, A. Shandilya, V. Agrawal, P. Pratik, D. Bhasme, V.S. Bisaria, D. Sundar, BMC Genomics 12 (Suppl 1) (2011) S30, https://doi.org/10.1186/1471-2105-12-S1-S30.
[18] E.-R. Hahm, M.B. Moura, E.E. Kelley, B.V. Houten, S. Shiva, S.V. Singh, PLoS One 6 (2011)e23354https://doi.org/10.1371/journal.pone.0023354.
[19] E.S. Yang, M.J. Choi, J.H. Kim, K.S. Choi, T.K. Kwon, Chem. Biol. Interact. 190 (2011) 9.
[20] R. Mohan, H.J. Hammers, P. Bargagna-Mohan, X.H. Zhan, C.J. Herbstritt, A. Ruiz, L. Zhang, A.D. Hanson, B.P. Conner, J. Rougas, V.S. Pribluda, Angiogenesis 7 (2004) 115.
[21] Y. Yokota, P. Bargagna-Mohan, P.P. Ravindranath, K.B. Kim, R. Mohan, Bioorg. Med. Chem. Lett. 16 (2006) 2603.
[22] P. Bargagna-Mohan, A. Hamza, Y. Kim, Y.K. Ho, N. Mor-Vaknin, N. Wendschlag, J. Liu, R.M. Evans, D.M. Markovitz, C.-G. Zhan, K.B. Kim, R. Mohan, Chem. Biol. 14 (2007) 623.
[23] P. Bargagna-Mohan, P.P. Ravindranath, R. Mohan, Invest. Ophthalmol. Vis. Sci. 47 (2007) 4138.
[24] B. Shohat, I. Kirson, D. Lavie, Biomedicine 28 (1978) 18.
[25] M. Rasool, P. Varalakshmi, Vascular Pharmacol. 44 (2006) 406.
[26] K. Panjamurthy, S. Manoharan, M.R. Nirmal, L. Vellaichamy, Invest. New Drugs 27 (2009) 447.

[27] Q. Uddin, L. Samiulla, V.K. Singh, S.S. Jamil, J. Appl. Pharm. Sci. 2 (2012) 170.
[28] M.K. al-Hindawi, S.H. al-Khafaji, M.H. Abdul-Nabi, J. Ethnopharmacol. 37 (1992) 113.
[29] R. Maitra, M.A. Porter, S. Huang, B.P. Gilmour, J. Inflamm. (Lond) 6 (2009) 15.
[30] M. Kaileh, W. Vanden Berghe, A. Heyeric, J. Horion, J. Piette, C. Libert, D. De Keukeleire, T. Essawi, G. Haegeman, J. Biol. Chem. 282 (2007) 4253.
[31] J.H. Oh, T.K. Kwon, Int. Immunopharmacol. 9 (2009) 614.
[32] J.H. Oh, E.J. Park, J.-W. Park, J. Lee, S.H. Lee, T.K. Kwon, Int. Immunopharmacol. 10 (2010) 572.
[33] K.-J. Min, K. Choi, T.K. Kwon, Int. Immunopharmacol. 11 (2011) 1137.
[34] A. Grover, V. Agrawal, A. Shandilya, V.S. Bisaria, D. Sundar, BMC Bioinformatics 12 (Suppl 13) (2011) S22, https://doi.org/10.1186/1471-2105-12-S13-S22.
[35] M. Hirayama, K. Gamoh, N. Ikekawa, Tetrahedron Lett. 23 (1982) 4725.
[36] M. Hirayama, K. Gamoh, N. Ikekawa, J. Am. Chem. Soc. 104 (1982) 3735.
[37] A. Fuerst, L. Labler, W. Meier, Ger. Offen. 2 (1978) 746,107 Chem. Abstr. 89 (1978) 60008f.
[38] G.R. Weihe, T.C. McMorris, J. Org. Chem. 43 (1978) 3942.
[39] B.M. Trost, T.N. Salzmann, K. Horii, J. Am. Chem. Soc. 98 (1976) 4887.

CHAPTER FIFTY EIGHT

Withanolide A

Abbreviations

DIPEA diisopropylethylamine
DMPU *N*,*N'*-dimethylpropyleneurea
Et_3N triethylamine
IBX 2-iodoxybenzoic acid
LiHMDS lithium hexamethyldisilazide
MOMCl (chloromethyl)methyl ether
MPO 4-methoxypyridine-*N*-oxide
NMO *N*-methylmorpholine *N*-oxide
PDC pyridinium dichromate
TBSCl *tert*-butyldimethylsilyl chloride
TPAP tetrapropylammonium perruthenate
TPP *meso*-tetraphenylporphyrin
Triton B benzyltrimethylammonium hydroxide

Systematic name: 5α,20β(*R*)-Dihydroxy-6α,7α-epoxy-1-oxo-(22*R*) witha-2,24-dienolide

Compound class: Steroidal lactone

Structure:

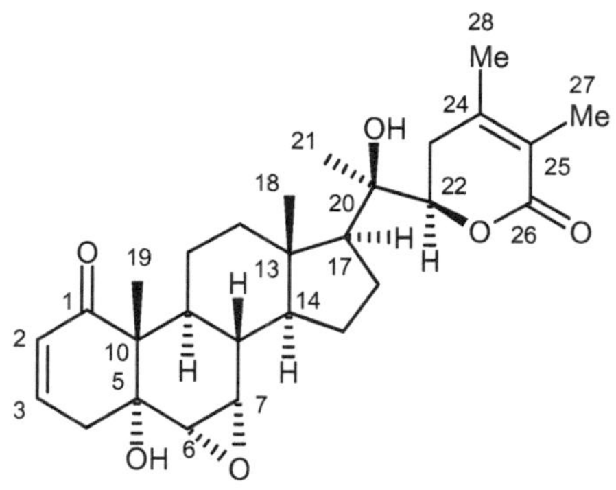

Withanolide A (**1**)

Natural sources: *Withania coagulans* Dunal (roots; family: Solanaceae) [1]; *Withania somnifera* Dunal (roots; "Ashwagandha" in Ayurveda, Indian ginseng) [2]

Total Synthesis of Bioactive Natural Products
https://doi.org/10.1016/B978-0-08-102822-3.00058-4

Pharmaceutical potentials: Neuroprotective [2, 3]; antineurodegenerative diseases like Alzheimer's [4, 5]; immunomodulator [6]

Synthetic route:

Gademann and his group [7] reported the first successful synthesis of withanolide A (**1**) from pregnenolone (**2**) following a synthetic route that involves vinylogous aldol condensation, singlet oxygen ene reaction, and Wharton carbonyl transposition as the key reactions (Scheme 1). The hydroxyl group in

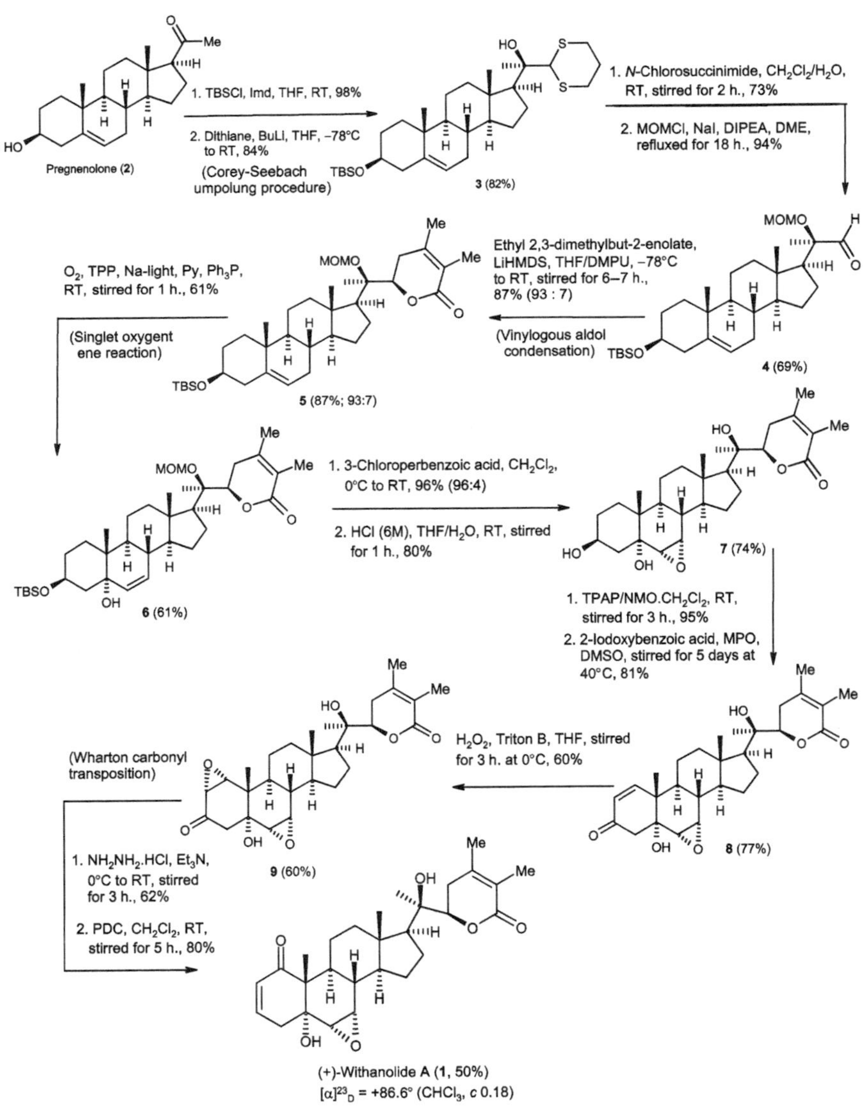

Scheme 1 Total synthesis of withanolide A (**1**).

the starting compound, pregnenolone (**2**), was first protected as TMS-ether and was then converted into a 1,3-dithiane derivative (**3**) as per the Corey-Seabach umpolung method [8]. This dithiane moiety was oxidatively cleaved by *N*-chlorosuccinimide yielding a C_{22}-aldehyde derivative, of which the tertiary C_{20}-OH group was protected as MOM ether (**4**); the aldehyde function in **4** underwent a stereoselective vinylogous aldol reaction with the vinylogous enolate, generated from ethyl 2,3-dimethylbut-2-enolate and LiHMDS, following Ikekawa's procedure [9] to furnish the lactone **5** in excellent yield and stereoselectivity (87%, dr=93:7).

A hydroxyl group was then introduced at the C_5-position in lactone **5** employing the singlet oxygen-mediated photooxygenative olefin migration technique to obtain an allylic tertiary alcohol **6**; the C_6–C_7 double bond in **6** was epoxidated by 3-chloroperbenzoic acid in a stereoselective manner to provide an epoxyalcohol that gave the triol derivative **7** on deprotection of the hydroxyl groups by the treatment of aqueous HCl. The corresponding enone **8** was prepared by treating the triol **7** with TAAP/NMO followed by IBX. The enone **8** was then epoxidated by aqueous hydrogen peroxide in the presence of Triton B, resulting in the epoxy ketone **9**. The investigators utilized Wharton carbonyl transposition [10] as the final key step in their synthetic journey; hence the epoxy ketone **9** was treated with hydrazine hydrochloride in the presence of Et_3N as base, followed by subsequent oxidation of the resulting rearranged alylic alcohol with PDC to obtain withanolide A (**1**) in good yield (50% over the two steps) as white solid, mp 284.2–285°C, R_f=0.5 (EtOAc:hexane: $CHCl_3$=8:1:1), $[\alpha]_D^{23}$ +86.6 ($CHCl_3$, *c* 0.18).

References

[1] S.S. Subramanian, P.D. Sethi, E. Glotter, I. Kirson, D. Lavie, Phytochemistry 10 (1971) 685.

[2] J. Zhao, N. Nakamura, M. Hattori, T. Kuboyama, C. Tohda, K. Komatsu, Chem. Pharm. Bull. 50 (2002) 760.

[3] T. Kuboyama, C. Tohda, K. Komatsu, Br. J. Pharmacol. 144 (2005) 961.

[4] S.P. Patil, S. Maki, S.A. Khedkar, A.C. Rigby, C. Chan, J. Nat. Prod. 73 (2010) 1196.

[5] M.H. Mirjalili, E. Moyano, M. Bonfill, R.M. Cusido, J. Palazón, Molecules 14 (2009) 2373.

[6] V. Bahr, R. Hansel, Planta Medica 44 (1982) 32.

[7] C.K. Jana, J. Hoecker, T.M. Woods, H.J. Jessen, M. Neuburger, K. Gademann, Angew. Chem. Int. Ed. 50 (2011) 8407.

[8] B.B. Shingate, B.G. Hazra, V.S. Pore, R.G. Gonnade, M. Bhadbhade, Tetrahedron 63 (2007) 5622.

[9] K. Gamoh, M. Hirayama, N. Ikekawa, J. Chem. Soc. Perkin Trans. 1 (1984) 449.

[10] C. Dupuy, J.L. Luche, Tetrahedron 45 (1989) 3437.

CHAPTER FIFTY NINE

(+)-Xylarenal A

Abbreviations

$CH(OMe)_3$ trimethylorthoformate
CH_2Cl_2 dichloromethane
DMAP 4-dimethylaminopyridine
DMF *N,N*-dimethylformamide
DMPU 1,3-dimethyl-3,4,5,6-tetrahydro-2-pyrimidinone
HMPA hexamethylphosphoramide
LDA lithium diisopropylamide
$LiBH(C_2H_5)_3$ lithium triethylborohydride
MeOH methanol
MsCl methane sulfonyl chloride
$NaBH_4$ sodium borohydride
$NaIO_4$ sodium periodate
OsO_4 osmium tetroxide
rt room temperature
TBAF tetrabutylammonium fluoride
TBSCl *tert*-butyldimethylsilyl chloride
TEA triethylamine
THF tetrahydrofuran
TMSCl trimethylsilyl chloride
TsOH *p*-toluene sulfonic acid

Systematic name: (1*R*,4*S*,5*R*,7*S*)-4,5-Dimethyl-8-oxo-7-(3-oxoprop-1-en-2-yl)-1,2,3,4,5,6,7,8-octahydronaphthalen-1-yl decanoate

Alternatively

1-Decyloxycarbonyl-8-oxo-9,11(13)-eremophiladien-12-al

Compound class: Sesquiterpenoid (eremophilane type)

Structure:

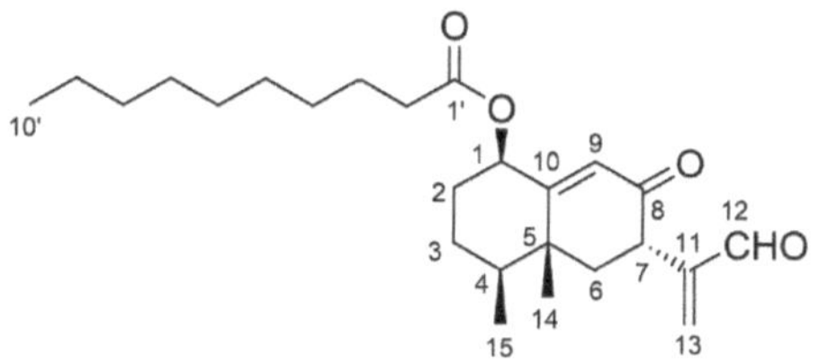

(+)-Xylarenal A (**1**)

https://doi.org/10.1016/B978-0-08-102822-3.00059-6

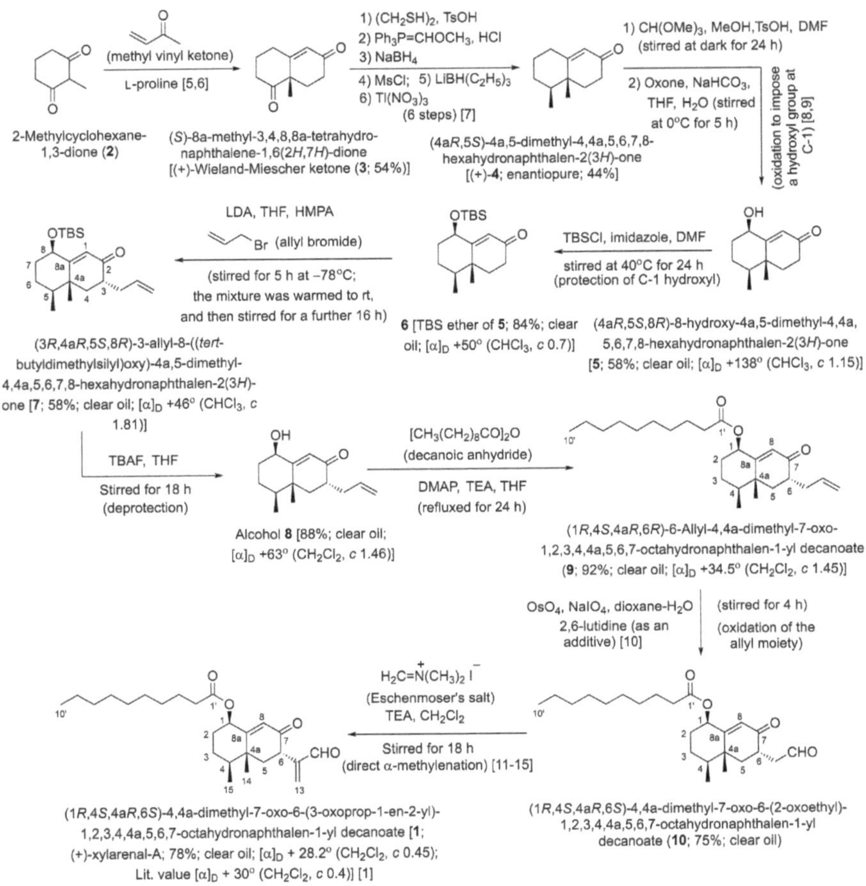

Scheme 1 Synthesis of (+)-xylarenal A (**1**) [4].

Natural source: *Xylaria persicaria* (fungal strain; fermentation broth) [1]

Pharmaceutical potential: Selective ligand for NPY Y5 receptor [2,3]

Synthetic route:

The first total synthesis of the sesquiterpenoid (+)-xylarenal A (**1**) reported by Bonjoch and his group [4] (Scheme 1) is the first example of a synthetic entry to an eremophilane sesquiterpene having a vinyl aldehyde linked to the decalin ring skeleton. The investigators accomplished total synthesis of the target molecule in 14 steps from (+)-Wieland-Miescher ketone [(+)-**3**] with 7% overall yield.

The enantiopure compound **3** was prepared from 2-methyl-1, 3-cyclohexanedione (**2**) through an asymmetric process promoted by L-proline in 54% yield [5,6]. (4a*R*,5*S*)-4a,5-Dimethyl-4,4a,5,6,7, 8-hexahydronaphthalen-2(3*H*)-one [(+)-**4**], the enantiopure building block, was then prepared in 44% yield from the (+)-Wieland-Miescher ketone [(+)-**3**] in six consecutive steps following the protocol developed by Paquette and coworkers [7]. Oxidation of **4** to impose a hydroxyl group at C-1 was carried out through its dienol ether [8,9] yielding (4a*R*,5*S*,8*R*)-8-hydroxy-4a,5-dimethyl-4,4a,5,6,7,8-hexahydronaphthalen-2(3*H*)-one as clear oil (**5**; 58%). Alcohol **5** was then protected with TBSCl to give compound **6**, which on treatment with allyl bromide in the presence of LDA/THF diastereoselectively produced compound **7** as clear oil in 55% yield. After removal of the TBS group from **7** using TBAF/THF, alcohol **8** was isolated in 88% yield. The investigators then carried out esterification of alcohol **8** with decanoic anhydride to incorporate the characteristic side chain of the target molecule at C-1, and obtained the ester (1*R*,4*S*,4a*R*,6*R*)-6-allyl-4,4a-dimethyl-7-oxo-1,2,3,4,4a,5,6,7-octahydronaphthalen-1-yl decanoate as clear oil (**9**) in excellent yield (92%). Oxidation of the allyl moiety within the ester molecule **9** with a catalytic amount of OsO_4 using 2,6-lutidine as an additive [10] furnished aldehyde **11** in 75% yield. In the last step, the direct α-methylenation of the aldehyde **11** at C-13 was achieved smoothly in only one step by a reaction of **11** with dimethyl methyleneammonium iodide (Eschenmoser's salt) in CH_2Cl_2 and triethylamine as a base [11–15], thereby affording (+)-xylarenal A (**1**) in 78% yield (clear oil; $[\alpha]_D$ + 28.2° [CH_2Cl_2, *c* 0.45]; lit. value $[\alpha]_D$ +30° [CH_2Cl_2, *c* 0.4]) [1]. The synthetic compound showed similar physical and spectral properties to those reported for the natural product [1].

References

[1] C.J. Smith, N.R. Morin, G.F. Bills, A.W. Dombrowski, G.M. Salituro, S.K. Smith, A. Zhao, D.J. MacNeil, J. Org. Chem. 67 (2002) 5001.
[2] D.R. Gehlert, Neuropeptides 33 (1999) 329.
[3] C. Blum, X. Zheng, S. de Lombaert, J. Med. Chem. 47 (2004) 2318.
[4] S. Díaz, A. González, B. Bradshaw, J. Cuesta, J. Bonjoch, J. Org. Chem. 70 (2005) 3749.
[5] P. Buchschacher, A. Fürst, Org. Synth. 63 (1985) 37.
[6] N. Harada, T. Sugioka, H. Uda, T. Kuriki, Synthesis (1990) 53.
[7] L.A. Paquette, T.-Z. Wang, C.M.G. Philippo, S. Wang, J. Am. Chem. Soc. 116 (1994) 3367.
[8] S.N. Suryawanshi, P.L. Fuchs, J. Org. Chem. 51 (1986) 902 and references therein.
[9] D.N. Kirk, J.M. Wiles, J. Chem. Soc. Chem. Commun. (1970) 1015.
[10] W. Yu, Y. Mei, Y. Kang, Z. Hua, Z. Jin, Org. Lett. 6 (2004) 3217.
[11] S. Takano, K. Inomata, K. Samizu, S. Tomita, M. Yanase, M. Suzuki, Y. Iwabuchi, T. Sugihara, K. Ogasawara, Chem. Lett. (1989) 1283–1284.
[12] K.C. Nicolaou, K.R. Reddy, G. Skokotas, F. Sato, X.-Y. Xiao, C.-K. Hwang, J. Am. Chem. Soc. 115 (1993) 3558.
[13] H.W. Lam, G. Pattenden, Angew. Chem. Int. Ed. 41 (2002) 508.
[14] H. Mizutani, M. Watanabe, T. Honda, Tetrahedron 58 (2002) 8929.
[15] A.B. Smith III, C. Sfouggatakis, D.B. Gotchev, S. Shirakami, D. Bauer, W. Zhu, V.A. Doughty, Org. Lett. 6 (2004) 3637.

CHAPTER SIXTY

Xyolide

Abbreviations

BAIB [bis(acetoxy)iodo]benzene
BnBr benzyl bromide
CH_2Cl_2 dichloromethane
DBU 1,8-diazabicyclo[5.4.0]undec-7-ene
DCC 1,3-dicyclohexylcarbodiimide
DDQ 2,3-dichloro-5,6-dicyano-1,4-benzoquinone
DIBAL-H diisobutylaluminum hydride
DMAP 4-dimethylaminopyridine
DMF *N,N*-dimethylformamide
DMP 2,2-dimethoxypropane
DMSO dimethylsulfoxide
Et_3N triethylamine
IBX 2-iodoxy benzoic acid
PMB 4-methoxybenzyl
PPTS pyridinium *p*-toluenesulfonate
py pyridine
RCM ring-closing metathesis
rt room temperature
TBS *tert*-butyldimethylsilyl
TEMPO tetramethylpiperdinyloxy free radical
THF tetrahydrofuran
TMSI trimethylsulfonium iodide

Systematic name: 4-(((5*S*,8*S*,9*S*,10*R*,*E*)-10-Heptyl-8,9-dihydroxy-2-oxo-3,4,5,8,9,10-hexahydro-2*H*-oxecin-5-yl)oxy)-4-oxobutanoic acid

Compound class: 10-Membered macrolide (nonenolide)

Structure:

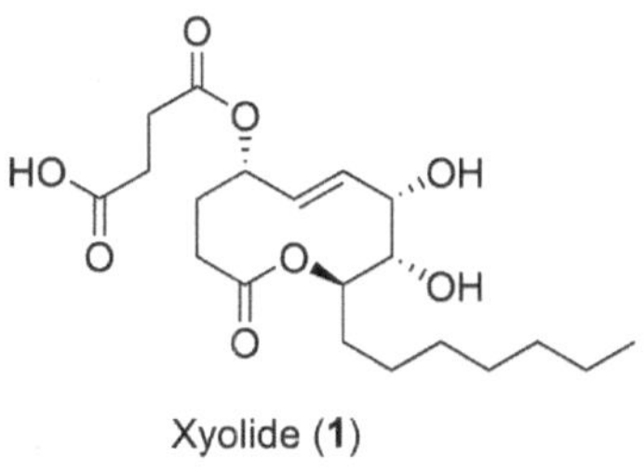

Xyolide (**1**)

Total Synthesis of Bioactive Natural Products
https://doi.org/10.1016/B978-0-08-102822-3.00060-2

Natural source: *Xylaria feejeensis* (an Amazonian endophytic fungus) [1]

Pharmaceutical potential: Antipathogenic (showed minimum inhibitory concentration value of 425 μM against *Pythium ultimum*, an oomycete plant pathogen) [1]

Synthetic route:

Reddy et al. [2] reported a concise and efficient stereoselective total synthesis of xyolide (**1**) employing MacMillan α-hydroxylation, Steglich esterification, and RCM as the key steps. Their approach for the total synthesis of **1** is summarized in Schemes 1–3.

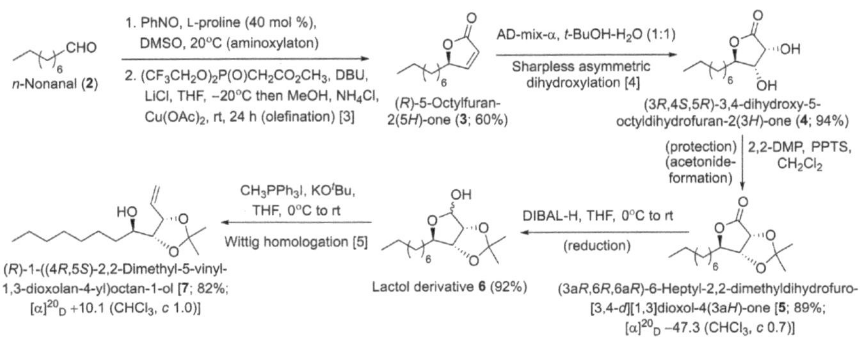

Scheme 1 Synthesis of alkenol fragment **7** [2].

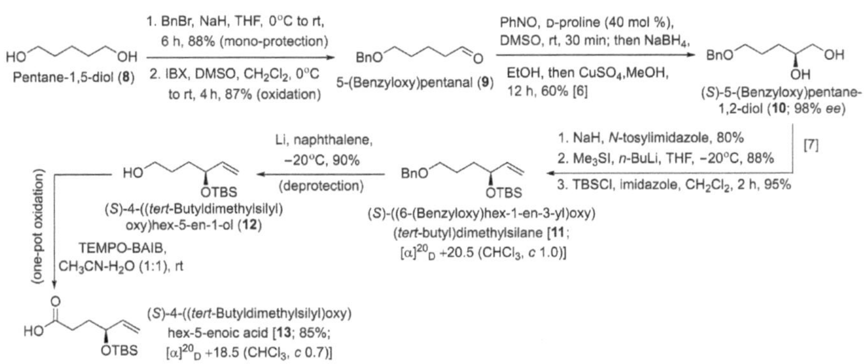

Scheme 2 Synthesis of carboxylic acid fragment **13** [2].

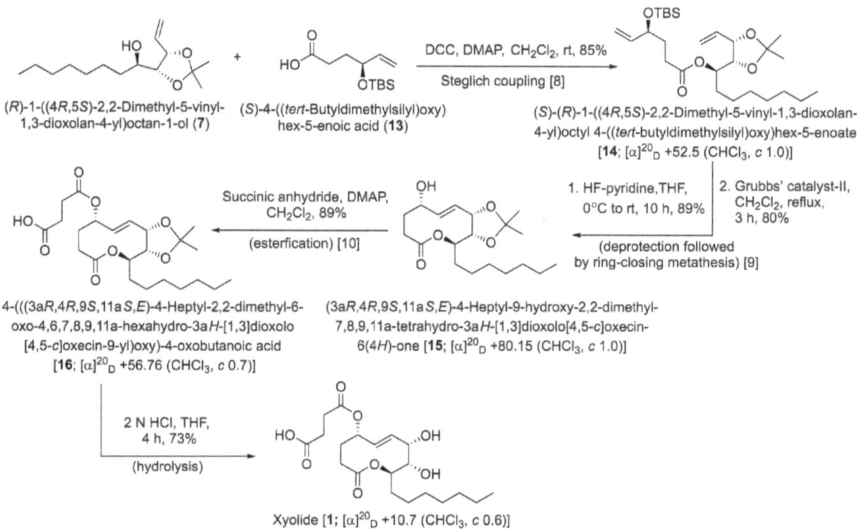

Scheme 3 Coupling of fragments **7** and **13**, and total synthesis of xyolide **1** [2].

At first the investigators synthesized the alkenol fragment (2,2-dimethyl-5-vinyl-1,3-dioxolan-4-yl)octan-1-ol (**7**) starting from *n*-nonanal (**2**) (Scheme 1). Aldehyde **2** was subjected to a sequential L-proline-catalyzed aminoxylation followed by olefination to obtain the γ-butenolide (*R*)-5-octylfuran-2(5*H*)-one (**3**) in 60% yield [3]. Sharpless asymmetric dihydroxylation [4] of compound **3** with AD-mix-α in *tert*-butanol and a water system afforded the diol **4** in 94% yield. Protection of the diol with DMP in the presence of PPTS produced the lactone 6-heptyl-2, 2-dimethyldihydrofuro-[3,4-*d*][1,3]dioxol-4(3a*H*)-one (**5**; 89%), which was then reduced with DIBAL-H to furnish the lactol **6** in 92% yield. On Wittig homologation with methyltriphenylphosphonium iodide in the presence of KO^{t}Bu, lactol **6** afforded the desired alkenol **7** in 82% yield [5] (Scheme 1).

In the next phase, they prepared another key intermediate, (*S*)-4-((*tert*-butyldimethylsilyl)oxy)hex-5-enoic acid (**13**), from pentane-1,5-diol (**8**). Monoprotection of the diol **8** with BnBr in the presence of sodium hydride in THF with subsequent oxidation with IBX afforded the aldehyde **9** in good yield. α-Aminooxylation of compound **9** with nitrasobenzene using D-proline followed by reduction with sodium borohydride and subsequent cleavage of the aminooxy alcohol with copper sulfate produced the desired diol (*S*)-5-(benzyloxy)pentane-1,2-diol (**10**) in 60% yield with 98% *ee* [6]. Treatment of the diol **10** with sodium hydride and *N*-tosylimidazole gave an epoxide in 80% yield, which was then treated with trimethylsulfonium iodide in the presence of *n*-BuLi in THF to obtain the desired allylic alcohol [7] isolated as its TBS ether **11**. On deprotection with Li/naphthalene, compound **11** furnished the alcohol (*S*)-4-((*tert*-butyldimethylsilyl)oxy)hex-5-en-1-ol (**12**) in 90% yield. One-pot oxidation of compound **12** with TEMPO-BAIB eventually produced the key intermediate **13** in 85% yield (Scheme 2).

Finally, the investigators coupled fragment **7** with fragment **13** to construct a 10-membered ring via RCM reaction. The coupling took place under Steglich conditions [8] to afford the corresponding ester **14** in 85% yield. Removal of TBS ether using HF-pyridine followed by RCM of **14** using second-generation Grubbs catalyst [9] in CH_2Cl_2 under reflux conditions for 6 h yielded the 10-membered macrolide **15** exclusively as its *E*-isomer. Esterification of the macrolide **15** with succinic anhydride [10] followed by removal of the acetonide using 2 N HCl eventually resulted in the formation of xyolide **1** in 73% yield (Scheme 3). The physical and spectral data of the synthetic xyolide (**1**) were found to be identical in all respects to those reported for the natural compound [1].

Mohapatra et al. [11] accomplished a nine-step total synthesis for the nonenolide **1** with 30% overall yield following a carbohydrate-based approach starting from D-(−)-ribose; their approach involves epoxide opening with a long-chain aliphatic Grignard reagent, Yamaguchi esterification, and RCM as the key reactions (Schemes 4–6).

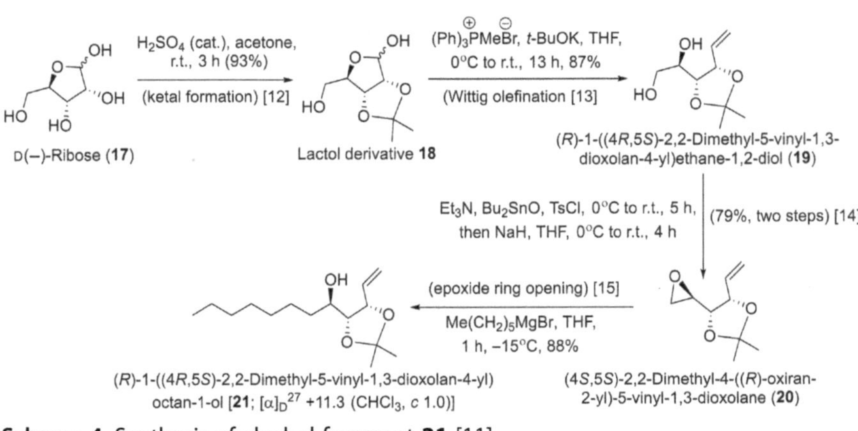

Scheme 4 Synthesis of alcohol fragment **21** [11].

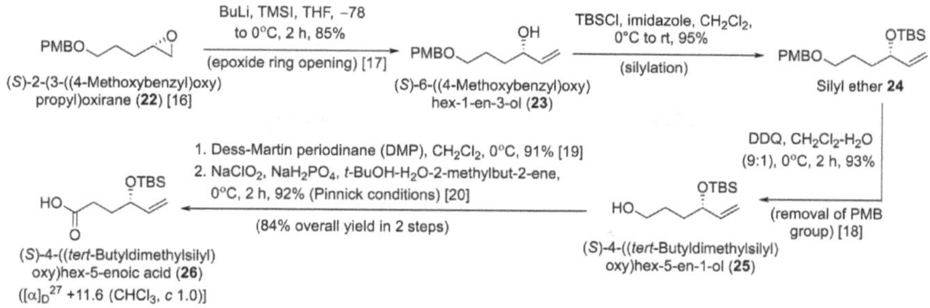

Scheme 5 Synthesis of carboxylic acid fragment **26** [11].

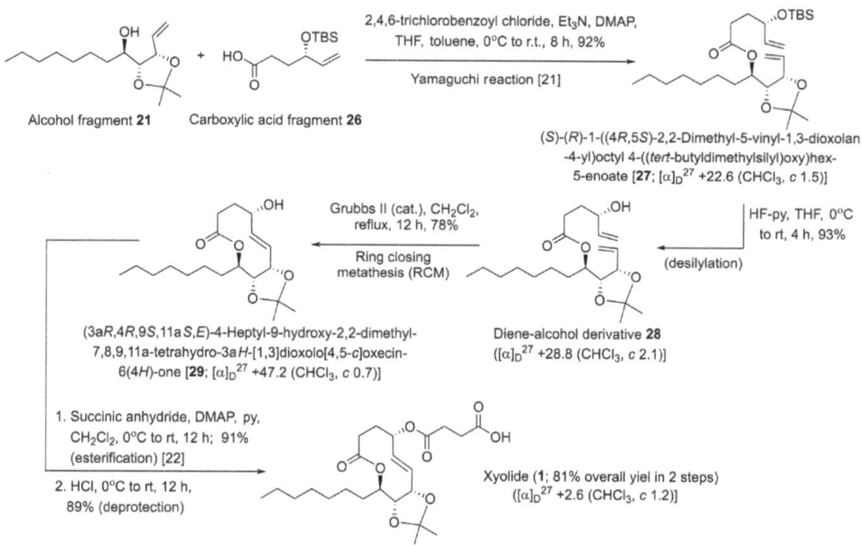

Scheme 6 Coupling of fragments **21** and **26**, and total synthesis of xyolide **1** [11].

They began their synthesis with the preparation of alcohol fragment **21** starting from D-(−)-ribose (**17**) (Scheme 4). The sugar was first protected with acetone in the presence of concentrated sulfuric acid as its lactol derivative **18** [12], which then underwent reaction with in situ generated methyl (triphenyl)phosphorane to produce olefin (*R*)-1-((4*R*,5*S*)-2,2-dimethyl-5-vinyl-1,3-dioxolan-4-yl)ethane-1,2-diol (**19**) [13]. Selective tosylation of the primary alcohol followed by treatment with sodium hydride afforded epoxide **20** [14], which was subjected to epoxide ring opening with hexylmagnesium bromide to obtain the target component **21** in 88% yield [15].

The investigators then synthesized another key component **26** from the known chiral epoxide **22** [16], which upon treatment with butyllithium and TMSI in THF at −78°C produced the allylic alcohol (*S*)-6-((4-methoxybenzyl)oxy)hex-1-en-3-ol (**23**) in 85% yield [17]. TBS-protected allylic alcohol **24** was treated with DDQ in CH_2Cl_2:water (9:1) to remove the PMB group to yield the corresponding primary alcohol **25** [18]. Compound **25** was then subjected to oxidation first with Dess-Martin periodinane in $CH_2Cl_2$to have the corresponding aldehyde [19], which was further oxidized under Pinnick conditions [20] to furnish the target component (*S*)-4-((*tert*-butyldimethylsilyl)oxy)hex-5-enoic acid (**26**) in 84% yield over the two steps (Scheme 5).

After preparing the two key components **21** and **26**, the investigators then coupled them through Yamaguchi esterification [21] to obtain diene **27** in 92% yield (Scheme 6). RCM was performed with diene-alcohol derivative **28** using second-generation Grubbs catalyst in refluxing CH_2Cl_2 to afford the 10-membered lactone **29** exclusively as the *E*-isomer. Finally, esterification of lactone **29** with succinic anhydride in the presence of pyridine and 4-(*N*,*N*-dimethylamino)pyridine [10] followed by removal of the isopropylidene group with hydrochloric acid furnished xyolide (**1**) as an amorphous solid in 81% yield over the two steps (Scheme 6). The physical and spectroscopic data for **1** were found to be in close agreement with those reported in the literature [1].

References

[1] E.G. Baraban, J.B. Morin, G.M. Phillips, A.J. Phillips, S.A. Strobel, J. Handelsman, Tetrahedron Lett. 54 (2013) 4058.
[2] B.V.S. Reddy, P. Sivaramakrishna Reddy, B. Phaneendra Reddy, J.S. Yadav, A.A.K. Al Ghamdi, Tetrahedron Lett. 54 (2013) 5758.
[3] D.A. Devalankar, P.V. Chouthaiwale, A. Sudalai, Tetrahedron: Asymmetry 23 (2012) 240.
[4] E.N. Jacobsen, I. Marko, W.S. Mungall, G. Schroeder, K.B. Sharpless, J. Am. Chem. Soc. 110 (1988) 1968 H.C. Kolb, M.S. Van Nieuwenhze, K.B. Sharpless, Chem. Rev. 94 (1994) 2483.
[5] P. Srihari, B. Kumaraswamy, P. Shankar, V. Ravishashidhar, J.S. Yadav, Tetrahedron Lett. 51 (2010) 6174.
[6] G. Zhong, Angew. Chem. Int. Ed. 42 (2003) 4247 B.V. Reddy, B.P. Reddy, T. Pandurangam, J.S. Yadav, Tetrahedron Lett. 52 (2011) 2306.
[7] B.P. Reddy, B.V.S. Reddy, T. Pandurangam, J.S. Yadav, Tetrahedron Lett. 53 (2012) 5749.
[8] T. Shimizu, T. Masuda, K. Hiramoto, T. Nakata, Org. Lett. 2 (2000) 2153.
[9] H.R. Grubbs, S. Chang, Tetrahedron 54 (1998) 4413.
[10] Y. Kobayashi, H. Okui, J. Org. Chem. 65 (2000) 612.
[11] D.K. Mohapatra, D.P. Reddy, D. Karhale, J.S. Yadav, Synlett 24 (2013) 2679.
[12] H.Y. Moon, W.J. Choi, H.O. Kim, L.S. Jeong, Tetrahedron: Asymmetry 13 (2002) 1189.

[13] W.J. Choi, H.R. Moon, H.L. Kim, B.N. Yoo, J.A. Lee, D.H. Shink, L.S. Jeong, J. Org. Chem. 69 (2004) 2634.
[14] P. Srihari, B. Kumaraswamy, J.S. Yadav, Tetrahedron 65 (2009) 6304.
[15] B.H. Lipshutz, S. Sengupta, Org. React. (N. Y.) 41 (1992) 135.
[16] P.C. Kumar, N.S. Sai, M. Ravinder, N.K. Singam, V.J. Rao, Synthesis (2011) 451.
[17] T.R. Pradhan, D.K. Mohapatra, Tetrahedron: Asymmetry 23 (2012) 709 P. Yu, C. Li, Z. Guozhu, Z. Liming, J. Am. Chem. Soc. 131 (2009) 5062; T.M. Trygstad, Y. Pang, C.J. Forsyth, J. Org. Chem. 74 (2009) 910.
[18] D.K. Mohapatra, E. Bhimireddy, P.S. Krishnarao, P.P. Das, J.S. Yadav, Org. Lett. 13 (2011) 744 Y. Oikawa, T. Yoshioka, O. Yonemitsu, Tetrahedron Lett. 23 (1982) 885.
[19] D.B. Dess, J.C. Martin, J. Org. Chem. 48 (1983) 415.
[20] E. Dalcanale, F. Montanari, J. Org. Chem. 51 (1986) 567 S.B. Balkrishna, W.E. Childers Jr, H.W. Pinnick, Tetrahedron 37 (1981) 2091.
[21] J. Inanaga, K. Hirata, H. Saeki, T. Katsuki, M. Yamaguchi, Bull. Chem. Soc. Jpn. 52 (1979) 1989.

Index

Note: Page numbers followed by *f* indicate figures and *t* indicate tables.

C

D

G

H

I

S

T

Kolibri 360 GmbH
Ferdinand-Jühlke-Straße 7
99095 Erfurt, Deutschland
produktsicherheit@kolibri360.de